2nd Edition

INTRODUCTION TO
Computational Mathematics

2nd Edition

INTRODUCTION TO Computational Mathematics

Xin-She Yang

Middlesex University London, UK

World Scientific

NEW JERSEY • LONDON • SINGAPORE • BEIJING • SHANGHAI • HONG KONG • TAIPEI • CHENNAI

Published by

World Scientific Publishing Co. Pte. Ltd.

5 Toh Tuck Link, Singapore 596224

USA office: 27 Warren Street, Suite 401-402, Hackensack, NJ 07601

UK office: 57 Shelton Street, Covent Garden, London WC2H 9HE

Library of Congress Cataloging-in-Publication Data

Yang, Xin-She.

Introduction to computational mathematics / by Xin-She Yang (Middlesex University London, UK). -- 2nd edition.

pages cm

Includes bibliographical references and index.

ISBN 978-9814635776 (hardcover : alk. paper) -- ISBN 978-9814635783 (pbk. : alk. paper)

1. Numerical analysis. 2. Algorithms. 3. Mathematical analysis--Foundations. 4. Programming (Mathematics) I. Title.

QA297.Y36 2015

518--dc23

2014038711

British Library Cataloguing-in-Publication Data

A catalogue record for this book is available from the British Library.

Printed in Singapore

Preface

Computational mathematics is essentially the foundation of modern scientific computing. Traditional ways of doing sciences consist of two major paradigms: by theory and by experiment. With the steady increase in computer power, there emerges a third paradigm of doing sciences: by computer simulation. Numerical algorithms are the very essence of any computer simulation, and computational mathematics is just the science of developing and analyzing numerical algorithms.

The science that studies numerical algorithms is numerical analysis or more broadly computational mathematics. Loosely speaking, numerical algorithms and analysis should include four categories of algorithms: numerical linear algebra, numerical optimization, numerical solutions of differential equations (ODEs and PDEs) and stochastic data modelling.

Many numerical algorithms were developed well before the computer was invented. For example, Newton's method for finding roots of nonlinear equations was developed in 1669, and Gauss quadrature for numerical integration was formulated in 1814. However, their true power and efficiency have been demonstrated again and again in modern scientific computing. Since the invention of the modern computer in the 1940s, many numerical algorithms have been developed since the 1950s. As the speed of computers increases, together with the increase in the efficiency of numerical algorithms, a diverse range of complex and challenging problems in mathematics, science and engineering can nowadays be solved numerically to very high accuracy. Numerical algorithms have become more important than ever.

The topics of computational mathematics are broad and the related literature is vast. It is often a daunting task for beginners to find the right book(s) and to learn the right algorithms that are widely used in computational mathematics. Even for lecturers and educators, it is no trivial task to decide what algorithms to teach and to provide balanced coverage of a wide range of topics, because there are so many algorithms to choose from.

The first edition of this book was published by World Scientific Publishing in 2008 and it was well received. Many universities courses used it as a main reference. Constructive feedbacks and helpful comments have also been received from the readers. This second edition has incorporated all these comments and consequently includes more algorithms and new algorithms to reflect the state-of-the-art developments such as computational intelligence and swarm intelligence.

Therefore, this new edition strives to provide extensive coverage of efficient algorithms commonly used in computational mathematics and modern scientific computing. It covers all the major topics including root-finding algorithms, numerical integration, interpolation, linear algebra, eigenvalues, numerical methods of ordinary differential equations (ODEs) and partial differential equations (PDEs), finite difference methods, finite element methods, finite volume methods, algorithm complexity, optimization, mathematical programming, stochastic models such as least squares and regression, machine learning such as neural networks and support vector machine, computational intelligence and swarm intelligence such as cuckoo search, bat algorithm, firefly algorithm as well as particle swarm optimization.

The book covers both traditional methods and new algorithms with dozens of worked examples to demonstrate how these algorithms work. Thus, this book can be used as a textbook and/or reference book, especially suitable for undergraduates and graduates in computational mathematics, engineering, computer science, computational intelligence, data science and scientific computing.

Xin-She Yang

London, 2014

Contents

V Stochastic Methods and Data Modelling 205

Part I

Mathematical Foundations

Chapter 1

Mathematical Foundations

Computational mathematics concerns a wide range of topics, from basic root-finding algorithms and linear algebra to advanced numerical methods for partial differential equations and nonlinear mathematical programming. In order to introduce various algorithms, we first review some mathematical foundations briefly.

1.1 The Essence of an Algorithm

Let us start by asking: what is an algorithm? In essence, an algorithm is a step-by-step procedure of providing calculations or instructions. Many algorithms are iterative. The actual steps and procedures will depend on the algorithm used and the context of interest. However, in this book, we place more emphasis on iterative procedures and ways for constructing algorithms.

For example, a simple algorithm of finding the square root of any positive number $k > 0$, or $x = \sqrt{k}$, can be written as

$$x_{n+1} = \frac{1}{2}(x_n + \frac{k}{x_n}), \tag{1.1}$$

starting from a guess solution $x_0 \neq 0$, say, $x_0 = 1$. Here, n is the iteration counter or index, also called the pseudo-time or generation counter. The above iterative equation comes from the re-arrangement of $x^2 = k$ in the following form

$$\frac{x}{2} = \frac{k}{2x}, \tag{1.2}$$

which can be rewritten as

$$x = \frac{1}{2}(x + \frac{k}{x}). \tag{1.3}$$

For example, for $k = 7$ with $x_0 = 1$, we have

$$x_1 = \frac{1}{2}(x_0 + \frac{7}{x_0}) = \frac{1}{2}(1 + \frac{7}{1}) = 4. \tag{1.4}$$

$$x_2 = \frac{1}{2}(x_1 + \frac{7}{x_1}) = 2.875, \quad x_3 \approx 2.654891304, \tag{1.5}$$

$$x_4 \approx 2.645767044, \quad x_5 \approx 2.6457513111. \tag{1.6}$$

We can see that x_5 after just 5 iterations (or generations) is very close to the true value of $\sqrt{7} = 2.64575131106459...$, which shows that this iteration method is very efficient.

The reason that this iterative process works is that the series $x_1, x_2, ..., x_n$ converges to the true value $\sqrt{k}$ due to the fact that

$$\frac{x_{n+1}}{x_n} = \frac{1}{2}(1 + \frac{k}{x_n^2}) \to 1, \quad x_n \to \sqrt{k}, \tag{1.7}$$

as $n \to \infty$. However, a good choice of the initial value x_0 will speed up the convergence. A wrong choice of x_0 could make the iteration fail; for example, we cannot use $x_0 = 0$ as the initial guess, and we cannot use $x_0 < 0$ either as $\sqrt{k} > 0$ (in this case, the iterations will approach another root $-\sqrt{k}$). So a sensible choice should be an educated guess. At the initial step, if $x_0^2 < k$, x_0 is the lower bound and k/x_0 is upper bound. If $x_0^2 > k$, then x_0 is the upper bound and k/x_0 is the lower bound. For other iterations, the new bounds will be x_n and k/x_n. In fact, the value x_{n+1} is always between these two bounds x_n and k/x_n, and the new estimate x_{n+1} is thus the mean or average of the two bounds. This guarantees that the series converges to the true value of $\sqrt{k}$. This method is similar to the well-known bisection method.

You may have already wondered why $x^2 = k$ was converted to Eq. (1.1)? Why do not we write it as the following iterative formula:

$$x_n = \frac{k}{x_n}, \tag{1.8}$$

starting from $x_0 = 1$? With this and $k = 7$, we have

$$x_1 = \frac{7}{x_0} = 7, \quad x_2 = \frac{7}{x_1} = 1, \quad x_3 = 7, \quad x_4 = 1, \quad x_5 = 7, \quad ..., \tag{1.9}$$

which leads to an oscillating feature at two distinct stages 1 and 7. You may wonder that it may be the problem of initial value x_0. In fact, for any initial value $x_0 \neq 0$, this above formula will lead to the oscillations between two values: x_0 and k. This clearly demonstrates that the way to design a good iterative formula is very important.

Mathematically speaking, an algorithm A is a procedure to generate a new and better solution x_{n+1} to a given problem from the current solution x_n at iteration or time t. That is,

$$x_{n+1} = A(x_n), \tag{1.10}$$

where A is a mathematical function of x_n. In fact, A can be a set of mathematical equations in general. In some literature, especially those in numerical analysis, n is often used for the iteration index. In many textbooks, the upper index form $x^{(n+1)}$ or x^{n+1} is commonly used. Here, x^{n+1} does not mean x to the power of $n+1$. Such notations will become useful and no confusion will occur when used appropriately. We will use such notations when appropriate in this book.

1.2 Big-O Notations

In analyzing the complexity of an algorithm, we usually estimate the order of computational efforts in terms of its problem size. This often requires the order notations, often in terms of big O and small o.

Loosely speaking, for two functions $f(x)$ and $g(x)$, if

$$\lim_{x \to x_0} \frac{f(x)}{g(x)} \to K, \tag{1.11}$$

where K is a finite, non-zero limit, we write

$$f = O(g). \tag{1.12}$$

The big O notation means that f is asymptotically equivalent to the order of $g(x)$. If the limit is unity or $K = 1$, we say $f(x)$ is order of $g(x)$. In this special case, we write

$$f \sim g, \tag{1.13}$$

which is equivalent to $f/g \to 1$ and $g/f \to 1$ as $x \to x_0$. Obviously, x_0 can be any value, including 0 and ∞. The notation $\sim$ does not necessarily mean $\approx$ in general, though it may give the same results, especially in the case when $x \to 0$. For example, $\sin x \sim x$ and $\sin x \approx x$ if $x \to 0$.

When we say f is order of 100 (or $f \sim 100$), this does not mean $f \approx 100$, but it can mean that f could be between about 50 and 150. The small o notation is often used if the limit tends to 0. That is

$$\lim_{x \to x_0} \frac{f}{g} \to 0, \tag{1.14}$$

or

$$f = o(g). \tag{1.15}$$

If $g > 0$, $f = o(g)$ is equivalent to $f \ll g$. For example, for $\forall x \in \mathcal{R}$, we have

$$e^x \approx 1 + x + O(x^2) \approx 1 + x + \frac{x^2}{2} + o(x).$$

Example 1.1: *A classic example is Stirling's asymptotic series for factorials*

$$n! \sim \sqrt{2\pi n}\,(\frac{n}{e})^n(1 + \frac{1}{12n} + \frac{1}{288n^2} - \frac{139}{51480n^3} - ...),$$

which can demonstrate the fundamental difference between an asymptotic series and the standard approximate expansions. For the standard power expansions, the error $R_k(h^k) \to 0$, but for an asymptotic series, the error of the truncated series R_k decreases compared with the leading term [here $\sqrt{2\pi n}(n/e)^n$]. However, R_n does not necessarily tend to zero. In fact,

$$R_2 = \frac{1}{12n} \cdot \sqrt{2\pi n}(n/e)^n,$$

is still very large as $R_2 \to \infty$ if $n \gg 1$. For example, for $n = 100$, we have $n! = 9.3326 \times 10^{157}$, while the leading approximation is $\sqrt{2\pi n}(n/e)^n = 9.3248 \times 10^{157}$. The difference between these two values is 7.7740×10^{154}, which is still very large, though three orders smaller than the leading approximation.

1.3 Differentiation and Integration

Differentiation is essentially to find the gradient of a function. For any curve $y = f(x)$, we define the gradient as

$$f'(x) \equiv \frac{dy}{dx} \equiv \frac{df(x)}{dx} = \lim_{h \to 0} \frac{f(x+h) - f(x)}{h}. \tag{1.16}$$

The gradient is also called the first derivative. The three notations $f'(x)$, dy/dx and $df(x)/dx$ are interchangeable. Conventionally, the notation dy/dx is called Leibnitz's notation, while the prime notation $'$ is called Lagrange's notation. Newton's dot notation $\dot{y} = dy/dt$ is now exclusively used for time derivatives. The choice of such notations is purely for clarity, convention and/or personal preference.

From the basic definition of the derivative, we can verify that differentiation is a linear operator. That is to say that for any two functions $f(x)$, $g(x)$ and two constants α and β, the derivative or gradient of a linear combination of the two functions can be obtained by differentiating the combination term by term. We have

$$[\alpha f(x) + \beta g(x)]' = \alpha f'(x) + \beta g'(x), \tag{1.17}$$

which can easily be extended to multiple terms.

If $y = f(u)$ is a function of u, and u is in turn a function of x, we want to calculate dy/dx. We then have

$$\frac{dy}{dx} = \frac{dy}{du} \cdot \frac{du}{dx}, \tag{1.18}$$

or

$$\frac{df[u(x)]}{dx} = \frac{df(u)}{du} \cdot \frac{du(x)}{dx}. \tag{1.19}$$

This is the well-known chain rule.

It is straightforward to verify the product rule

$$(uv)' = uv' + vu'. \tag{1.20}$$

If we replace v by $1/v = v^{-1}$ and apply the chain rule

$$\frac{d(v^{-1})}{dx} = -1 \times v^{-1-1} \times \frac{dv}{dx} = -\frac{1}{v^2}\frac{dv}{dx}, \tag{1.21}$$

we have the formula for quotients or the quotient rule

$$\frac{d(\frac{u}{v})}{dx} = \frac{d(uv^{-1})}{dx} = u(\frac{-1}{v^2})\frac{dv}{dx} + v^{-1}\frac{du}{dx} = \frac{v\frac{du}{dx} - u\frac{dv}{dx}}{v^2}. \tag{1.22}$$

For a smooth curve, it is relatively straightforward to draw a tangent line at any point; however, for a smooth surface, we have to use a tangent plane. For example, we now want to take the derivative of a function of two independent variables x and y, that is $z = f(x, y) = x^2+y^2/2$. The question is probably 'with respect to' what? x or y? If we take the derivative with respect to x, then will it be affected by y? The answer is we can take the derivative with respect to either x or y while taking the other variable as constant. That is, we can calculate the derivative with respect to x in the usual sense by assuming that $y =$ constant. Since there is more than one variable, we have more than one derivative and the derivatives can be associated with either the x-axis or y-axis. We call such derivatives partial derivatives, and use the following notation

$$\frac{\partial z}{\partial x} \equiv \frac{\partial f(x, y)}{\partial x} \equiv f_x \equiv \frac{\partial f}{\partial x}\Big|_y = \lim_{h \to 0, y = \text{const}} \frac{f(x + h, y) - f(x, y)}{h}. \tag{1.23}$$

The notation $\frac{\partial f}{\partial x}|_y$ emphasises the fact that y = constant; however, we often omit $|_y$ and simply write $\frac{\partial f}{\partial x}$ as we know this fact is implied.

Similarly, the partial derivative with respect to y is defined by

$$\frac{\partial z}{\partial y} \equiv \frac{\partial f(x,y)}{\partial y} \equiv f_y \equiv \frac{\partial f}{\partial y}\Big|_x = \lim_{x=\text{const},k\to 0} \frac{f(x,y+k)-f(x,y)}{k}. \quad (1.24)$$

Then, the standard differentiation rules for univariate functions such as $f(x)$ apply. For example, for $z = f(x,y) = x^2 + y^2/2$, we have

$$\frac{\partial f}{\partial x} = \frac{dx^2}{dx} + 0 = 2x,$$

and

$$\frac{\partial f}{\partial y} = 0 + \frac{d(y^2/2)}{dy} = \frac{1}{2} \times 2y = y,$$

where the appearance of 0 highlights the fact that $dy/dx = dx/dy = 0$ as x and y are independent variables.

Differentiation is used to find the gradient for a given function. Now a natural question is how to find the original function for a given gradient. This is the integration process, which can be considered as the reverse of the differentiation process. Since we know that

$$\frac{d\sin x}{dx} = \cos x, \quad (1.25)$$

that is, the gradient of $\sin x$ is $\cos x$, we can easily say that the original function is $\sin x$ since its gradient is $\cos x$. We can write

$$\int \cos x \, dx = \sin x + C, \quad (1.26)$$

where C is the constant of integration. Here $\int dx$ is the standard notation showing the integration is with respect to x, and we usually call this the integral. The function $\cos x$ is called the integrand.

The integration constant comes from the fact that a family of curves shifted by a constant will have the same gradient at their corresponding points. This means that the integration can be determined up to an arbitrary constant. For this reason, we call it an indefinite integral.

Integration is more complicated than differentiation in general. Even when we know the derivative of a function, we have to be careful. For example, we know that $(x^{n+1})' = (n+1)x^n$ or $(\frac{1}{n+1}x^{n+1})' = x^n$ for any n integers, so we can write

$$\int x^n dx = \frac{1}{n+1} x^{n+1} + C. \quad (1.27)$$

However, there is a possible problem when $n = -1$ because $1/(n+1)$ will become $1/0$. In fact, the above integral is valid for any n except $n = -1$. When $n = -1$, we have

$$\int \frac{1}{x} dx = \ln x + C. \tag{1.28}$$

If we know that the gradient of a function $F(x)$ is $f(x)$ or $F'(x) = f(x)$, it is possible and sometimes useful to express where the integration starts and ends, and we often write

$$\int_a^b f(x)dx = \Big[F(x)\Big]_a^b = F(b) - F(a). \tag{1.29}$$

Here a is called the lower limit of the integral, while b is the upper limit of the integral. In this case, the constant of integration has dropped out because the integral can be determined accurately. The integral becomes a definite integral and it corresponds to the area under a curve $f(x)$ from a to $b \geq a$.

From the differentiation rule

$$\frac{d(uv)}{dx} = u\frac{dv}{dx} + v\frac{du}{dx}, \tag{1.30}$$

we integrate it with respect to x, we have

$$\int \frac{d(uv)}{dx} dx = uv = \int u\frac{dv}{dx}dx + \int v\frac{du}{dx}dx. \tag{1.31}$$

By rearranging, we have

$$\int u\frac{dv}{dx}dx = uv - \int v\frac{du}{dx}dx, \tag{1.32}$$

or

$$\int uv'dx = uv - \int vu'dx, \tag{1.33}$$

or simply

$$\int udv = uv - \int vdu. \tag{1.34}$$

This is the well-known formula for the technique known as integration by parts.

Example 1.2: *To calculate $\int xe^x dx$, we can set $u = x$ and $v' = e^x$, which gives $u' = 1$ and $v = e^x$. Now we have*

$$\int xe^x dx = xe^x - \int e^x \cdot 1dx = xe^x - e^x + C.$$

However, this will not work if we set $u = e^x$ and $v' = x$. Thus, care should be taken when using integration by parts.

Obviously, there are many other techniques for integration such as substitution and transformation. Integration can also be multivariate, which leads to multiple integrals. We will not discuss this further, and interested readers can refer to more advanced textbooks.

1.4 Vector and Vector Calculus

Vector analysis is an important part of computational mathematics. Many quantities such as force, velocity, and deformation in sciences are vectors which have both a magnitude and a direction. Vectors are a special class of matrices. Here, we will briefly review the basic concepts in linear algebra.

A vector $\boldsymbol{u}$ is a set of ordered numbers $\boldsymbol{u} = (u_1, u_2, ..., u_n)$, where its components $u_i (i = 1, ..., n) \in \Re$ are real numbers. All these vectors form an n-dimensional vector space $\mathcal{V}^n$. A simple example is the position vector $\boldsymbol{p} = (x, y, z)$ where x, y, z are the 3-D Cartesian coordinates.

To add any two vectors $\boldsymbol{u} = (u_1, ..., u_n)$ and $\boldsymbol{v} = (v_1, ..., v_n)$, we simply add their corresponding components,

$$\boldsymbol{u} + \boldsymbol{v} = (u_1 + v_1, u_2 + v_2, ..., u_n + v_n), \tag{1.35}$$

and the sum is also a vector. The addition of vectors has commutability ($\boldsymbol{u} + \boldsymbol{v} = \boldsymbol{v} + \boldsymbol{u}$) and associativity $[(\boldsymbol{u} + \boldsymbol{v}) + \boldsymbol{w} = \boldsymbol{u} + (\boldsymbol{v} + \boldsymbol{w})]$. This is because each of the components is obtained by simple addition, which means it has the same properties.

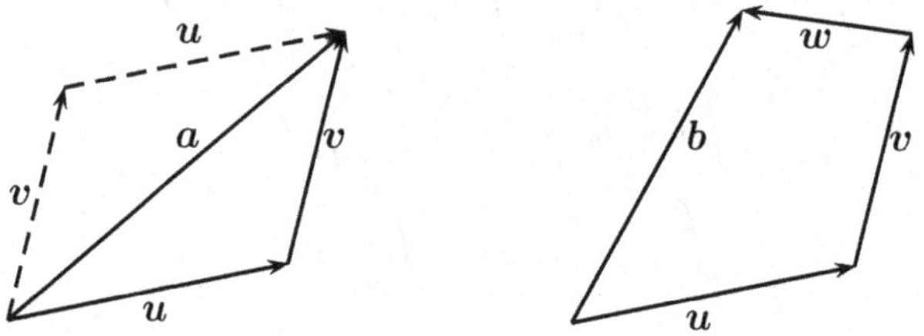

Fig. 1.1 Addition of vectors: (a) parallelogram $\boldsymbol{a} = \boldsymbol{u} + \boldsymbol{v}$; (b) vector polygon $\boldsymbol{b} = \boldsymbol{u} + \boldsymbol{v} + \boldsymbol{w}$.

The zero vector $\boldsymbol{0}$ is a special case where all its components are zeros. The multiplication of a vector $\boldsymbol{u}$ with a scalar or constant $\alpha \in \Re$ is carried out by the multiplication of each component,

$$\alpha \boldsymbol{u} = (\alpha u_1, \alpha u_2, ..., \alpha u_n). \tag{1.36}$$

Thus, we have

$$-\boldsymbol{u} = (-u_1, -u_2, ..., -u_n). \tag{1.37}$$

The dot product or inner product of two vectors $\boldsymbol{x}$ and $\boldsymbol{y}$ is defined as

$$\boldsymbol{x} \cdot \boldsymbol{y} = x_1 y_1 + x_2 y_2 + ... + x_n y_n = \sum_{i=1}^{n} x_i y_i, \tag{1.38}$$

which is a real number. The length or norm of a vector $\boldsymbol{x}$ is the root of the dot product of the vector itself,

$$|\boldsymbol{x}| = ||\boldsymbol{x}|| = \sqrt{\boldsymbol{x} \cdot \boldsymbol{x}} = \sqrt{\sum_{i=1}^{n} x_i^2}. \tag{1.39}$$

When $||\boldsymbol{x}|| = 1$, then it is a unit vector. It is straightforward to check that the dot product has the following properties:

$$\boldsymbol{u} \cdot \boldsymbol{v} = \boldsymbol{v} \cdot \boldsymbol{u}, \qquad \boldsymbol{u} \cdot (\boldsymbol{v} + \boldsymbol{w}) = \boldsymbol{u} \cdot \boldsymbol{v} + \boldsymbol{u} \cdot \boldsymbol{w}, \tag{1.40}$$

and

$$(\alpha \boldsymbol{u}) \cdot (\beta \boldsymbol{v}) = (\alpha\beta) \boldsymbol{u} \cdot \boldsymbol{v}, \tag{1.41}$$

where $\alpha, \beta \in \Re$ are constants.

The angle θ between two vectors $\boldsymbol{u}$ and $\boldsymbol{v}$ can be calculated using their dot product

$$\boldsymbol{u} \cdot \boldsymbol{v} = ||\boldsymbol{u}|| \, ||\boldsymbol{v}|| \cos(\theta), \qquad 0 \leq \theta \leq \pi, \tag{1.42}$$

which leads to

$$\cos(\theta) = \frac{\boldsymbol{u} \cdot \boldsymbol{v}}{||\boldsymbol{u}|| \, ||\boldsymbol{v}||}. \tag{1.43}$$

If the dot product of these two vectors is zero or $\cos(\theta) = 0$ (i.e., $\theta = \pi/2$), then we say that these two vectors are orthogonal.

Since $\cos(\theta) \leq 1$, then we get the useful Cauchy-Schwartz inequality:

$$||\boldsymbol{u} \cdot \boldsymbol{v}|| \leq ||\boldsymbol{u}|| \, ||\boldsymbol{v}||. \tag{1.44}$$

The dot product of two vectors is a scalar or a number. On the other hand, the cross product or outer product of two vectors is a new vector

$$\boldsymbol{u} = \boldsymbol{x} \times \boldsymbol{y} = \begin{vmatrix} \boldsymbol{i} & \boldsymbol{j} & \boldsymbol{k} \\ x_1 & x_2 & x_3 \\ y_1 & y_2 & y_3 \end{vmatrix}$$

$$= \begin{vmatrix} x_2 & x_3 \\ y_2 & y_3 \end{vmatrix} \boldsymbol{i} + \begin{vmatrix} x_3 & x_1 \\ y_3 & y_1 \end{vmatrix} \boldsymbol{j} + \begin{vmatrix} x_1 & x_2 \\ y_1 & y_2 \end{vmatrix} \boldsymbol{k}$$

$$= (x_2 y_3 - x_3 y_2)\boldsymbol{i} + (x_3 y_1 - x_1 y_3)\boldsymbol{j} + (x_1 y_2 - x_2 y_1)\boldsymbol{k}. \tag{1.45}$$

In fact, the norm of $||\boldsymbol{x} \times \boldsymbol{y}||$ is the area of the parallelogram formed by $\boldsymbol{x}$ and $\boldsymbol{y}$. We have

$$||\boldsymbol{x} \times \boldsymbol{y}|| = ||\boldsymbol{x}|| \, ||\boldsymbol{y}|| \sin\theta, \tag{1.46}$$

where θ is the angle between the two vectors. In addition, the vector $\boldsymbol{u} = \boldsymbol{x} \times \boldsymbol{y}$ is perpendicular to both $\boldsymbol{x}$ and $\boldsymbol{y}$, following a right-hand rule.

It is straightforward to check that the cross product has the following properties:

$$\boldsymbol{x} \times \boldsymbol{y} = -\boldsymbol{y} \times \boldsymbol{x}, \quad (\boldsymbol{x}+\boldsymbol{y}) \times \boldsymbol{z} = \boldsymbol{x} \times \boldsymbol{z} + \boldsymbol{y} \times \boldsymbol{z}, \tag{1.47}$$

and

$$(\alpha \boldsymbol{x}) \times (\beta \boldsymbol{y}) = (\alpha\beta) \boldsymbol{x} \times \boldsymbol{y}, \qquad \alpha, b \in \Re. \tag{1.48}$$

A very special case is $\boldsymbol{u} \times \boldsymbol{u} = \boldsymbol{0}$. For unit vectors, we have

$$\boldsymbol{i} \times \boldsymbol{j} = \boldsymbol{k}, \qquad \boldsymbol{j} \times \boldsymbol{k} = \boldsymbol{i}, \qquad \boldsymbol{k} \times \boldsymbol{i} = \boldsymbol{j}. \tag{1.49}$$

Example 1.3: *For two 3-D vectors* $\boldsymbol{u} = (4, 5, -6)$ *and* $\boldsymbol{v} = (2, -2, 1/2)$, *their dot product is*

$$\boldsymbol{u} \cdot \boldsymbol{v} = 4 \times 2 + 5 \times (-2) + (-6) \times 1/2 = -5.$$

As their moduli are

$$||\boldsymbol{u}|| = \sqrt{4^2 + 5^2 + (-6)^2} = \sqrt{77},$$

$$||\boldsymbol{v}|| = \sqrt{2^2 + (-2)^2 + (1/2)^2} = \sqrt{33}/2,$$

we can calculate the angle θ *between the two vectors. We have*

$$\cos\theta = \frac{\boldsymbol{u} \cdot \boldsymbol{v}}{||\boldsymbol{u}|| ||\boldsymbol{v}||} = \frac{-5}{\sqrt{77} \times \sqrt{33}/2} = -\frac{10}{11\sqrt{21}},$$

or

$$\theta = \cos^{-1}(-\frac{10}{11\sqrt{21}}) \approx 101.4°.$$

Their cross product is

$$\boldsymbol{w} = \boldsymbol{u} \times \boldsymbol{v}$$

$$= (5 \times 1/2 - (-2) \times (-6), (-6) \times 2 - 4 \times 1/2, 4 \times (-2) - 5 \times 2)$$

$$= (-19/2, -14, -18).$$

Similarly, we have

$$\boldsymbol{v} \times \boldsymbol{u} = (19/2, 14, 18) = -\boldsymbol{u} \times \boldsymbol{v}.$$

The norm of the cross product is

$$\|\boldsymbol{w}\| = \sqrt{(\frac{-19}{2})^2 + (-14)^2 + (-18)^2} \approx 24.70,$$

while

$$\|\boldsymbol{u}\|\|\boldsymbol{v}\|\sin\theta = \sqrt{77} \times \frac{\sqrt{33}}{2} \times \sin(101.4^\circ) \approx 24.70 = \|\boldsymbol{w}\|.$$

It is easy to verify that

$$\boldsymbol{u} \cdot \boldsymbol{w} = 4 \times (-19/2) + 5 \times (-14) + (-6) \times (-18) = 0,$$

and

$$\boldsymbol{v} \cdot \boldsymbol{w} = 2 \times (-19/2) + (-2) \times (-14) + 1/2 \times (-18) = 0.$$

Indeed, the vector $\boldsymbol{w}$ is perpendicular to both $\boldsymbol{u}$ and $\boldsymbol{v}$.

Any vector $\boldsymbol{v}$ in an n-dimensional vector space $\mathcal{V}^n$ can be written as a combination of a set of n independent basis vectors or orthogonal spanning vectors $\boldsymbol{e_1}, \boldsymbol{e_2}, ..., \boldsymbol{e_n}$, so that

$$\boldsymbol{v} = v_1\boldsymbol{e}_1 + v_2\boldsymbol{e}_2 + ... + v_n\boldsymbol{e}_n = \sum_{i=1}^{n} v_i\boldsymbol{e}_i, \tag{1.50}$$

where the coefficients/scalars $v_1, v_2, ..., v_n$ are the components of $\boldsymbol{v}$ relative to the basis $\boldsymbol{e_1}, \boldsymbol{e_2}, ..., \boldsymbol{e_n}$. The most common basis vectors are the orthogonal unit vectors. In a three-dimensional case, they are $\boldsymbol{i} = (1,0,0)$, $\boldsymbol{j} = (0,1,0)$, $\boldsymbol{k} = (0,0,1)$ for x-, y-, z-axis, respectively. Thus, we have

$$\boldsymbol{x} = x_1\boldsymbol{i} + x_2\boldsymbol{j} + x_3\boldsymbol{k}. \tag{1.51}$$

The three unit vectors satisfy $\boldsymbol{i} \cdot \boldsymbol{j} = \boldsymbol{j} \cdot \boldsymbol{k} = \boldsymbol{k} \cdot \boldsymbol{i} = 0$.

Two non-zero vectors $\boldsymbol{u}$ and $\boldsymbol{v}$ are said to be linearly independent if $\alpha\boldsymbol{u} + \beta\boldsymbol{v} = 0$ implies that $\alpha = \beta = 0$. If α, β are not all zeros, then these two vectors are linearly dependent. Two linearly dependent vectors are parallel ($\boldsymbol{u}/\!/\boldsymbol{v}$) to each other. Similarly, any three linearly dependent vectors $\boldsymbol{u}, \boldsymbol{v}, \boldsymbol{w}$ are in the same plane.

The differentiation of a vector is carried out over each component and treating each component as the usual differentiation of a scalar. Thus, from a position vector

$$\boldsymbol{P}(t) = x(t)\boldsymbol{i} + y(t)\boldsymbol{j} + z(t)\boldsymbol{k}, \tag{1.52}$$

we can write its velocity as

$$\boldsymbol{v} = \frac{d\boldsymbol{P}}{dt} = \dot{x}(t)\boldsymbol{i} + \dot{y}(t)\boldsymbol{j} + \dot{z}(t)\boldsymbol{k}, \tag{1.53}$$

and acceleration as

$$\boldsymbol{a} = \frac{d^2\boldsymbol{P}}{dt^2} = \ddot{x}(t)\boldsymbol{i} + \ddot{y}(t)\boldsymbol{j} + \ddot{z}(t)\boldsymbol{k}, \tag{1.54}$$

where $\dot{} = d/dt$. Conversely, the integral of $\boldsymbol{v}$ is

$$\boldsymbol{P} = \int_0^t \boldsymbol{v} dt + \boldsymbol{p}_0, \tag{1.55}$$

where $\boldsymbol{p}_0$ is a vector constant or the initial position at $t = 0$.

From the basic definition of differentiation, it is easy to check that the differentiation of vectors has the following properties:

$$\frac{d(\alpha \boldsymbol{a})}{dt} = \alpha \frac{d\boldsymbol{a}}{dt}, \quad \frac{d(\boldsymbol{a} \cdot \boldsymbol{b})}{dt} = \frac{d\boldsymbol{a}}{dt} \cdot \boldsymbol{b} + \boldsymbol{a} \cdot \frac{d\boldsymbol{b}}{dt}, \tag{1.56}$$

and

$$\frac{d(\boldsymbol{a} \times \boldsymbol{b})}{dt} = \frac{d\boldsymbol{a}}{dt} \times \boldsymbol{b} + \boldsymbol{a} \times \frac{d\boldsymbol{b}}{dt}. \tag{1.57}$$

Three important operators commonly used in vector analysis, especially in the formulation of mathematical models, are the gradient operator (grad or ∇), the divergence operator (div or $\nabla\cdot$) and the curl operator (curl or $\nabla\times$).

Sometimes, it is useful to calculate the directional derivative of ϕ at a point (x, y, z) in the direction of $\boldsymbol{n}$

$$\frac{\partial \phi}{\partial \boldsymbol{n}} = \boldsymbol{n} \cdot \nabla \phi = \frac{\partial \phi}{\partial x} \cos(\alpha) + \frac{\partial \phi}{\partial y} \cos(\beta) + \frac{\partial \phi}{\partial z} \cos(\gamma), \tag{1.58}$$

where $\boldsymbol{n} = (\cos\alpha, \cos\beta, \cos\gamma)$ is a unit vector and α, β, γ are the directional angles. Generally speaking, the gradient of any scalar function ϕ of x, y, z can be written in a similar way,

$$\text{grad}\phi = \nabla \phi = \frac{\partial \phi}{\partial x} \boldsymbol{i} + \frac{\partial \phi}{\partial y} \boldsymbol{j} + \frac{\partial \phi}{\partial z} \boldsymbol{k}. \tag{1.59}$$

This is the same as the application of the del operator ∇ to the scalar function ϕ

$$\nabla = \frac{\partial}{\partial x} \boldsymbol{i} + \frac{\partial}{\partial y} \boldsymbol{j} + \frac{\partial}{\partial z} \boldsymbol{k}. \tag{1.60}$$

The direction of the gradient operator on a scalar field gives a vector field.

As the gradient operator is a linear operator, it is straightforward to check that it has the following properties:

$$\nabla(\alpha\psi + \beta\phi) = \alpha\nabla\psi + \beta\nabla\phi, \qquad \nabla(\psi\phi) = \psi\nabla\phi + \phi\nabla\psi, \tag{1.61}$$

where α, β are constants and ψ, ϕ are scalar functions.

For a vector field

$$\boldsymbol{u}(x, y, z) = u(x, y, z)\boldsymbol{i} + v(x, y, z)\boldsymbol{j} + w(x, y, z)\boldsymbol{k}, \tag{1.62}$$

the application of the operator ∇ can lead to either a scalar field or vector field depending on how the del operator is applied to the vector field. The divergence of a vector field is the dot product of the del operator ∇ and $\boldsymbol{u}$

$$\text{div}\, \boldsymbol{u} \equiv \nabla \cdot \boldsymbol{u} = \frac{\partial u_1}{\partial x} + \frac{\partial u_2}{\partial y} + \frac{\partial u_3}{\partial z}, \tag{1.63}$$

and the curl of $\boldsymbol{u}$ is the cross product of the del operator and the vector field $\boldsymbol{u}$

$$\text{curl}\, \boldsymbol{u} \equiv \nabla \times \boldsymbol{u} = \begin{vmatrix} \boldsymbol{i} & \boldsymbol{j} & \boldsymbol{k} \\ \frac{\partial}{\partial x} & \frac{\partial}{\partial y} & \frac{\partial}{\partial z} \\ u_1 & u_2 & u_3 \end{vmatrix}. \tag{1.64}$$

One of the most commonly used operators in engineering and science is the Laplacian operator

$$\nabla^2 \phi = \nabla \cdot (\nabla \phi) = \frac{\partial^2 \phi}{\partial x^2} + \frac{\partial^2 \phi}{\partial y^2} + \frac{\partial^2 \phi}{\partial z^2}, \tag{1.65}$$

for Laplace's equation

$$\Delta \phi \equiv \nabla^2 \phi = 0. \tag{1.66}$$

Some important theorems are often rewritten in terms of the above three operators, especially in fluid dynamics and finite element analysis. For example, Gauss's theorem connects the integral of divergence with the related surface integral

$$\iiint_\Omega (\nabla \cdot \boldsymbol{Q})\, d\Omega = \iint_S \boldsymbol{Q} \cdot \boldsymbol{n}\, dS. \tag{1.67}$$

1.5 Matrices and Matrix Decomposition

Matrices are widely used in scientific computing, engineering and sciences, especially in the implementation of many algorithms. A matrix is a table or array of numbers or functions arranged in rows and columns. The elements or entries of a matrix $\boldsymbol{A}$ are often denoted as a_{ij}. For a matrix $\boldsymbol{A}$ with m rows and n columns,

$$\boldsymbol{A} \equiv [a_{ij}] = \begin{pmatrix} a_{11} & a_{12} & \dots & a_{1j} & \dots & a_{1n} \\ a_{21} & a_{22} & \dots & a_{2j} & \dots & a_{2n} \\ \vdots & \vdots & & & \ddots & \vdots \\ a_{m1} & a_{m2} & \dots & a_{mj} & \dots & a_{mn} \end{pmatrix}, \tag{1.68}$$

we say the size of $\boldsymbol{A}$ is m by n, or $m \times n$. $\boldsymbol{A}$ is a square matrix if $m = n$. For example,

$$\boldsymbol{A} = \begin{pmatrix} 1 & 2 & 3 \\ 4 & 5 & 6 \end{pmatrix}, \qquad \boldsymbol{B} = \begin{pmatrix} e^x & \sin x \\ -i\cos x & e^{i\theta} \end{pmatrix}, \tag{1.69}$$

and

$$\boldsymbol{u} = \begin{pmatrix} u \\ v \\ w \end{pmatrix}, \tag{1.70}$$

where $\boldsymbol{A}$ is a 2×3 matrix, $\boldsymbol{B}$ is a 2×2 square matrix, and $\boldsymbol{u}$ is a 3×1 column matrix or column vector.

The sum of two matrices $\boldsymbol{A}$ and $\boldsymbol{B}$ is possible only if they have the same size $m \times n$, and their sum, which is also $m \times n$, is obtained by adding their corresponding entries

$$\boldsymbol{C} = \boldsymbol{A} + \boldsymbol{B}, \qquad c_{ij} = a_{ij} + b_{ij}, \tag{1.71}$$

where $(i = 1, 2, ..., m; j = 1, 2, ..., n)$. The product of a matrix $\boldsymbol{A}$ with a scalar $\alpha \in \Re$ is obtained by multiplying each entry by α. The product of two matrices is possible only if the number of columns of $\boldsymbol{A}$ is the same as the number of rows of $\boldsymbol{B}$. That is to say, if $\boldsymbol{A}$ is $m \times n$ and $\boldsymbol{B}$ is $n \times r$, then the product $\boldsymbol{C}$ is $m \times r$,

$$c_{ij} = (AB)_{ij} = \sum_{k=1}^{n} a_{ik} b_{kj}. \tag{1.72}$$

If $\boldsymbol{A}$ is a square matrix, then we have $\boldsymbol{A}^n = \overbrace{\boldsymbol{AA...A}}^{n}$. The multiplications of matrices are generally not commutive, i.e., $\boldsymbol{AB} \neq \boldsymbol{BA}$. However, the multiplication has associativity $\boldsymbol{A(uv)} = \boldsymbol{(Au)v}$ and $\boldsymbol{A}(\boldsymbol{u}+\boldsymbol{v}) = \boldsymbol{Au}+\boldsymbol{Av}$.

The transpose $\boldsymbol{A}^T$ of $\boldsymbol{A}$ is obtained by switching the position of rows and columns, and thus $\boldsymbol{A}^T$ will be $n \times m$ if $\boldsymbol{A}$ is $m \times n$, $(a^T)_{ij} = a_{ji}, (i = 1, 2, ..., m; j = 1, 2, ..., n)$. Generally,

$$(\boldsymbol{A}^T)^T = \boldsymbol{A}, \qquad (\boldsymbol{AB})^T = \boldsymbol{B}^T \boldsymbol{A}^T. \tag{1.73}$$

The differentiation and integral of a matrix are carried out over each of its members or elements. For example, for a 2×2 matrix

$$\frac{d\boldsymbol{A}}{dt} = \dot{\boldsymbol{A}} = \begin{pmatrix} \frac{da_{11}}{dt} & \frac{da_{12}}{dt} \\ \frac{da_{21}}{dt} & \frac{da_{22}}{dt} \end{pmatrix}, \tag{1.74}$$

and

$$\int \boldsymbol{A} dt = \begin{pmatrix} \int a_{11} dt & \int a_{12} dt \\ \\ \int a_{21} dt & \int a_{22} dt \end{pmatrix}. \tag{1.75}$$

A diagonal matrix $\boldsymbol{A}$ is a square matrix whose every entry off the main diagonal is zero ($a_{ij} = 0$ if $i \neq j$). Its diagonal elements or entries may or may not have zeros. In general, it can be written as

$$\boldsymbol{D} = \begin{pmatrix} d_1 & 0 & \dots & 0 \\ 0 & d_2 & \dots & 0 \\ & & \ddots & \\ 0 & 0 & \dots & d_n \end{pmatrix}. \tag{1.76}$$

For example, the matrix

$$\boldsymbol{I} = \begin{pmatrix} 1 & 0 & 0 \\ 0 & 1 & 0 \\ 0 & 0 & 1 \end{pmatrix} \tag{1.77}$$

is a 3×3 identity or unitary matrix. In general, we have

$$\boldsymbol{A}\boldsymbol{I} = \boldsymbol{I}\boldsymbol{A} = \boldsymbol{A}. \tag{1.78}$$

A zero or null matrix $\mathbf{0}$ is a matrix with all of its elements being zero.

There are three important matrices: lower (upper) triangular matrix, tridiagonal matrix, and augmented matrix, and they are important in the solution of linear equations. A tridiagonal matrix often arises naturally from the finite difference and finite volume discretization of partial differential equations, and it can in general be written as

$$\boldsymbol{Q} = \begin{pmatrix} b_1 & c_1 & 0 & 0 & \dots & 0 & 0 \\ a_2 & b_2 & c_2 & 0 & \dots & 0 & 0 \\ 0 & a_3 & b_3 & c_3 & \dots & 0 & 0 \\ \vdots & & & & \ddots & & \vdots \\ 0 & 0 & 0 & 0 & \dots & a_n & b_n \end{pmatrix}. \tag{1.79}$$

An augmented matrix is formed by two matrices with the same number of rows. For example, the following system of linear equations

$$a_{11}u_1 + a_{12}u_2 + a_{13}u_3 = b_1,$$

$$a_{21}u_1 + a_{22}u_2 + a_{23}u_3 = b_2,$$

$$a_{31}u_1 + a_{32}u_2 + a_{33}u_3 = b_3, \tag{1.80}$$

can be written in a compact form in terms of matrices

$$\begin{pmatrix} a_{11} & a_{12} & a_{13} \\ a_{21} & a_{22} & a_{23} \\ a_{31} & a_{32} & a_{33} \end{pmatrix} \begin{pmatrix} u_1 \\ u_2 \\ u_3 \end{pmatrix} = \begin{pmatrix} b_1 \\ b_2 \\ b_3 \end{pmatrix}, \tag{1.81}$$

or

$$\boldsymbol{Au} = \boldsymbol{b}. \tag{1.82}$$

This can in turn be written as the following augmented form

$$[\boldsymbol{A}|\boldsymbol{b}] = \left(\begin{array}{ccc|c} a_{11} & a_{12} & a_{13} & b_1 \\ a_{21} & a_{22} & a_{23} & b_2 \\ a_{31} & a_{32} & a_{33} & b_3 \end{array}\right). \tag{1.83}$$

The augmented form is widely used in Gauss-Jordan elimination and linear programming.

A lower (upper) triangular matrix is a square matrix with all the elements above (below) the diagonal entries being zeros. In general, a lower triangular matrix can be written as

$$\boldsymbol{L} = \begin{pmatrix} l_{11} & 0 & ... & 0 \\ l_{21} & l_{22} & ... & 0 \\ & & \ddots & \\ l_{n1} & l_{n2} & ... & l_{nn} \end{pmatrix}, \tag{1.84}$$

while the upper triangular matrix can be written as

$$\boldsymbol{U} = \begin{pmatrix} u_{11} & u_{12} & ... & u_{1n} \\ 0 & u_{22} & ... & u_{2n} \\ & & \ddots & \\ 0 & 0 & ... & u_{nn} \end{pmatrix}. \tag{1.85}$$

Any $n \times n$ square matrix $\boldsymbol{A} = [a_{ij}]$ can be decomposed or factorized as a product of an $\boldsymbol{L}$ and a $\boldsymbol{U}$, that is

$$\boldsymbol{A} = \boldsymbol{LU}, \tag{1.86}$$

though some decomposition is not unique because we have n^2+n unknowns: $n(n+1)/2$ coefficients l_{ij} and $n(n+1)/2$ coefficients u_{ij}, but we can only provide n^2 equations from the coefficients a_{ij}. Thus, there are n free parameters. The uniqueness of decomposition is often achieved by imposing either $l_{ii} = 1$ or $u_{ii} = 1$ where $i = 1, 2, ..., n$.

Other LU variants include the LDU and LUP decompositions. An LDU decomposition can be written as

$$\boldsymbol{A} = \boldsymbol{LDU}, \tag{1.87}$$

where $\boldsymbol{L}$ and $\boldsymbol{U}$ are lower and upper matrices with all the diagonal entries being unity, and $\boldsymbol{D}$ is a diagonal matrix. On the other hand, the LUP decomposition can be expressed as

$$\boldsymbol{A} = \boldsymbol{LUP}, \qquad \text{or} \qquad \boldsymbol{A} = \boldsymbol{PLU}, \tag{1.88}$$

where $\boldsymbol{P}$ is a permutation matrix which is a square matrix and has exactly one entry 1 in each column and each row with 0's elsewhere. However, most numerical libraries and software packages use the following LUP decomposition

$$\boldsymbol{PA} = \boldsymbol{LU}, \tag{1.89}$$

which makes it easier to decompose some matrices. However, the requirement for LU decompositions is relatively strict. An invertible matrix $\boldsymbol{A}$ has an LU decomposition provided that the determinants of all its diagonal minors or leading submatrices are not zeros.

A simpler way of decomposing a square matrix $\boldsymbol{A}$ for solving a system of linear equations is to write

$$\boldsymbol{A} = \boldsymbol{D} + \boldsymbol{L} + \boldsymbol{U}, \tag{1.90}$$

where $\boldsymbol{D}$ is a diagonal matrix. $\boldsymbol{L}$ and $\boldsymbol{U}$ are the strictly lower and upper triangular matrices without diagonal elements, respectively. This decomposition is much simpler to implement than the LU decomposition because there is no multiplication involved here.

Example 1.4: *The following* 3×3 *matrix*

$$\boldsymbol{A} = \begin{pmatrix} 2 & 1 & 5 \\ 4 & -4 & 5 \\ 5 & 2 & -5 \end{pmatrix},$$

can be decomposed as $\boldsymbol{A} = \boldsymbol{LU}$. *That is*

$$\boldsymbol{A} = \begin{pmatrix} 1 & 0 & 0 \\ l_{21} & 1 & 0 \\ l_{31} & l_{32} & 0 \end{pmatrix} \begin{pmatrix} u_{11} & u_{12} & u_{13} \\ 0 & u_{22} & u_{23} \\ 0 & 0 & u_{33} \end{pmatrix},$$

which becomes

$$\boldsymbol{A} = \begin{pmatrix} u_{11} & u_{12} & u_{13} \\ l_{21}u_{11} & l_{21}u_{12} + u_{22} & l_{21}u_{13} + u_{23} \\ l_{31}u_{11} & l_{31}u_{12} + l_{32}u_{22} & l_{31}u_{13} + l_{32}u_{23} + u_{33} \end{pmatrix}$$

$$= \begin{pmatrix} 2 & 1 & 5 \\ 4 & -4 & 5 \\ 5 & 2 & -5 \end{pmatrix}.$$

This leads to $u_{11} = 2$, $u_{12} = 1$ *and* $u_{13} = 5$. *As* $l_{21}u_{11} = 4$, *so* $l_{21} = 4/u_{11} = 2$. *Similarly,* $l_{31} = 2.5$. *From* $l_{21}u_{12} + u_{22} = -4$, *we have* $u_{22} = -4 - 2 \times 1 = -6$. *From* $l_{21}u_{13} + u_{23} = 5$, *we have* $u_{23} = 5 - 2 \times 5 = -5$.

Using $l_{31}u_{12} + l_{32}u_{22} = 2$, *or* $2.5 \times 1 + l_{32} \times (-6) = 2$, *we get* $l_{32} = 1/12$. *Finally,* $l_{31}u_{13} + l_{32}u_{23} + u_{33} = -5$ *gives* $u_{33} = -5 - 2.5 \times 5 - 1/12 \times (-5) = -205/12$. *Therefore, we now have*

$$\begin{pmatrix} 2 & 1 & 5 \\ 4 & -4 & 5 \\ 5 & 2 & -5 \end{pmatrix} = \begin{pmatrix} 1 & 0 & 0 \\ 2 & 1 & 0 \\ 5/2 & 1/12 & 1 \end{pmatrix} \begin{pmatrix} 2 & 1 & 5 \\ 0 & -6 & -5 \\ 0 & 0 & -205/12 \end{pmatrix}.$$

The L+D+U decomposition can be written as

$$\boldsymbol{A} = \boldsymbol{D} + \boldsymbol{L} + \boldsymbol{U} = \begin{pmatrix} 2 & 0 & 0 \\ 0 & -4 & 0 \\ 0 & 0 & -5 \end{pmatrix} + \begin{pmatrix} 0 & 0 & 0 \\ 4 & 0 & 0 \\ 5 & 2 & 0 \end{pmatrix} + \begin{pmatrix} 0 & 1 & 5 \\ 0 & 0 & 5 \\ 0 & 0 & 0 \end{pmatrix}.$$

1.6 Determinant and Inverse

The determinant of a square matrix $\boldsymbol{A}$ is a number or scalar obtained by the following recursive formula, or the cofactors, or the Laplace expansion by column or row. For example, expanding by row k, we have

$$\det(\boldsymbol{A}) = |\boldsymbol{A}| = \sum_{j=1}^{n} (-1)^{k+j} a_{kj} M_{kj}, \tag{1.91}$$

where M_{ij} is the determinant of a minor matrix of $\boldsymbol{A}$ by deleting row i and column j. For a simple 2×2 matrix, its determinant simply becomes

$$\begin{vmatrix} a_{11} & a_{12} \\ a_{21} & a_{22} \end{vmatrix} = a_{11}a_{22} - a_{12}a_{21}. \tag{1.92}$$

The determinants of matrices have the following properties:

$$|\alpha \boldsymbol{A}| = \alpha |\boldsymbol{A}|, \qquad |\boldsymbol{A}^T| = |\boldsymbol{A}|, \qquad |\boldsymbol{A}\boldsymbol{B}| = |\boldsymbol{A}||\boldsymbol{B}|, \tag{1.93}$$

where $\boldsymbol{A}$ and $\boldsymbol{B}$ have the same size ($n \times n$).

An $n \times n$ square matrix is singular if $|\boldsymbol{A}| = 0$, and is nonsingular if and only if $|\boldsymbol{A}| \neq 0$. The trace of a square matrix $\mathrm{tr}(\boldsymbol{A})$ is defined as the sum of the diagonal elements,

$$\mathrm{tr}(\boldsymbol{A}) = \sum_{i=1}^{n} a_{ii} = a_{11} + a_{22} + \ldots + a_{nn}. \tag{1.94}$$

The rank of a matrix $\boldsymbol{A}$ is the number of linearly independent vectors forming the matrix. Generally speaking, the rank of $\boldsymbol{A}$ satisfies

$$\text{rank}(\boldsymbol{A}) \leq \min(m, n). \tag{1.95}$$

For an $n \times n$ square matrix $\boldsymbol{A}$, it is nonsingular if $\text{rank}(\boldsymbol{A}) = n$.

The inverse matrix $\boldsymbol{A}^{-1}$ of a square matrix $\boldsymbol{A}$ is defined as

$$\boldsymbol{A}^{-1}\boldsymbol{A} = \boldsymbol{A}\boldsymbol{A}^{-1} = \boldsymbol{I}. \tag{1.96}$$

More generally,

$$\boldsymbol{A}_l^{-1}\boldsymbol{A} = \boldsymbol{A}\boldsymbol{A}_r^{-1} = \boldsymbol{I}, \tag{1.97}$$

where $\boldsymbol{A}_l^{-1}$ is the left inverse while $\boldsymbol{A}_r^{-1}$ is the right inverse. If $\boldsymbol{A}_l^{-1} = \boldsymbol{A}_r^{-1}$, we say that the matrix $\boldsymbol{A}$ is invertible and its inverse is simply denoted by $\boldsymbol{A}^{-1}$. It is worth noting that the unit matrix $\boldsymbol{I}$ has the same size as $\boldsymbol{A}$.

The inverse of a square matrix exists if and only if $\boldsymbol{A}$ is nonsingular or $\det(\boldsymbol{A}) \neq 0$. From the basic definitions, it is straightforward to prove that the inverse of a matrix has the following properties

$$(\boldsymbol{A}^{-1})^{-1} = \boldsymbol{A}, \qquad (\boldsymbol{A}^T)^{-1} = (\boldsymbol{A}^{-1})^T, \tag{1.98}$$

and

$$(\boldsymbol{A}\boldsymbol{B})^{-1} = \boldsymbol{B}^{-1}\boldsymbol{A}^{-1}. \tag{1.99}$$

The inverse of a lower (upper) triangular matrix is also a lower (upper) triangular matrix. The inverse of a diagonal matrix

$$\boldsymbol{D} = \begin{pmatrix} d_1 & 0 & \dots & 0 \\ 0 & d_2 & \dots & 0 \\ & & \ddots & \\ 0 & 0 & \dots & d_n \end{pmatrix}, \tag{1.100}$$

can simply be written as

$$\boldsymbol{D}^{-1} = \begin{pmatrix} 1/d_1 & 0 & \dots & 0 \\ 0 & 1/d_2 & \dots & 0 \\ & & \ddots & \\ 0 & 0 & \dots & 1/d_n \end{pmatrix}, \tag{1.101}$$

where $d_i \neq 0$. If any of these elements d_i is zero, then the diagonal matrix is not invertible as it becomes singular. For a 2×2 matrix, its inverse is simply

$$\boldsymbol{A} = \begin{pmatrix} a & b \\ c & d \end{pmatrix}, \qquad \boldsymbol{A}^{-1} = \frac{1}{ad - bc}\begin{pmatrix} d & -b \\ -c & a \end{pmatrix}. \tag{1.102}$$

Example 1.5: *For two matrices,*

$$A = \begin{pmatrix} 4 & 5 & 0 \\ -2 & 2 & 5 \\ 2 & -3 & 1 \end{pmatrix}, \quad B = \begin{pmatrix} 2 & 3 \\ 0 & -2 \\ 5 & 2 \end{pmatrix},$$

their transpose matrices are

$$A^T = \begin{pmatrix} 4 & -2 & 2 \\ 5 & 2 & -3 \\ 0 & 5 & 1 \end{pmatrix}, \quad B^T = \begin{pmatrix} 2 & 0 & 5 \\ 3 & -2 & 2 \end{pmatrix}.$$

Let $D = AB$ *be their product; we have*

$$AB = D = \begin{pmatrix} D_{11} & D_{12} \\ D_{21} & D_{22} \\ D_{31} & D_{32} \end{pmatrix}.$$

The first two entries are

$$D_{11} = \sum_{j=1}^{3} A_{1j}B_{j1} = 2 \times 4 + 5 \times 0 + 0 \times 5 = 8,$$

and

$$D_{12} = \sum_{j=1}^{3} A_{1j}B_{j2} = 4 \times 3 + 5 \times (-2) + 0 \times 2 = 2.$$

Similarly, the other entries are:

$$D_{21} = 21, \quad D_{22} = 0, \quad D_{31} = 9, \quad D_{33} = 14.$$

Therefore, we get

$$AB = \begin{pmatrix} 4 & 5 & 0 \\ -2 & 2 & 5 \\ 2 & -3 & 1 \end{pmatrix} \begin{pmatrix} 2 & 3 \\ 0 & -2 \\ 5 & 2 \end{pmatrix} = D = \begin{pmatrix} 8 & 2 \\ 21 & 0 \\ 9 & 14 \end{pmatrix}.$$

However, the product BA *does not exist, though*

$$B^T A^T = \begin{pmatrix} 8 & 21 & 9 \\ 2 & 0 & 14 \end{pmatrix} = D^T = (AB)^T.$$

The inverse of A *is*

$$A^{-1} = \frac{1}{128} \begin{pmatrix} 17 & -5 & 25 \\ 12 & 4 & -20 \\ 2 & 22 & 18 \end{pmatrix},$$

and the determinant of $\boldsymbol{A}$ *is*

$$\det(\boldsymbol{A}) = 128.$$

It is straightforward to verify that

$$\boldsymbol{A}\boldsymbol{A}^{-1} = \boldsymbol{A}^{-1}\boldsymbol{A} = \boldsymbol{I} = \begin{pmatrix} 1 & 0 & 0 \\ 0 & 1 & 0 \\ 0 & 0 & 1 \end{pmatrix}.$$

For example, the first entry is obtained by

$$\sum_{j=1}^{3} A_{1j} A_{j1}^{-1} = 4 \times \frac{17}{128} + 5 \times \frac{12}{128} + 0 \times \frac{2}{128} = 1.$$

Other entries can be verified similarly. Finally, the trace of $\boldsymbol{A}$ *is*

$$\mathrm{tr}(\boldsymbol{A}) = A_{11} + A_{22} + A_{33} = 4 + 2 + 1 = 7.$$

The algorithmic complexity of most algorithms for obtaining the inverse of a general square matrix is $O(n^3)$. That is why most modern algorithms try to avoid the direct inverse of a large matrix. Solution of a large matrix system is instead carried out either by partial inverse via decomposition or by iteration (or a combination of these two methods). If the matrix can be decomposed into triangular matrices either by LU factorization or direction decomposition, the aim is then to invert a triangular matrix, which is simpler and more efficient.

For a triangular matrix, the inverse can be obtained using algorithms of $O(n^2)$ complexity. Similarly, the solution of a linear system with a lower (upper) triangular matrix $\boldsymbol{A}$ can be obtained by forward (back) substitutions. In general, for a lower triangular matrix

$$\boldsymbol{A} = \begin{pmatrix} \alpha_{11} & 0 & \dots & 0 \\ \alpha_{12} & \alpha_{22} & \dots & 0 \\ & & \ddots & \\ \alpha_{n1} & \alpha_{n2} & \dots & \alpha_{nn} \end{pmatrix}, \qquad (1.103)$$

the forward substitutions for the system $\boldsymbol{A}\boldsymbol{u} = \boldsymbol{b}$ can be carried out as follows:

$$u_1 = \frac{b_1}{\alpha_{11}},$$

$$u_2 = \frac{1}{\alpha_{22}}(b_2 - \alpha_{21} u_1),$$

$$u_i = \frac{1}{\alpha_{ii}}(b_i - \sum_{j=1}^{i-1} \alpha_{ij} u_j), \tag{1.104}$$

where $i = 2, ..., n$. We see that it takes 1 division to get u_1, 3 floating point calculations to get u_2, and $(2i - 1)$ to get u_i. So the total algorithmic complexity is $O(1 + 3 + ... + (2n - 1)) = O(n^2)$. Similar arguments apply to the upper triangular systems.

The inverse $\boldsymbol{A}^{-1}$ of a lower triangular matrix can in general be written as

$$\boldsymbol{A}^{-1} = \begin{pmatrix} \beta_{11} & 0 & ... & 0 \\ \beta_{21} & \beta_{22} & ... & 0 \\ & & \ddots & \\ \beta_{n1} & \beta_{n2} & ... & \beta_{nn} \end{pmatrix} = \boldsymbol{B} = \begin{pmatrix} \boldsymbol{B_1} & \boldsymbol{B_2} & ... & \boldsymbol{B_n} \end{pmatrix}, \tag{1.105}$$

where $\boldsymbol{B}_j$ are the j-th column vector of $\boldsymbol{B}$. The inverse must satisfy $\boldsymbol{A}\boldsymbol{A}^{-1} = \boldsymbol{I}$ or

$$\boldsymbol{A}\begin{pmatrix} \boldsymbol{B}_1 & \boldsymbol{B_2} & ... & \boldsymbol{B_n} \end{pmatrix} = \boldsymbol{I} = \begin{pmatrix} \boldsymbol{e}_1 & \boldsymbol{e}_2 & ... & \boldsymbol{e}_n \end{pmatrix}, \tag{1.106}$$

where $\boldsymbol{e}_j$ is the j-th unit vector of size n with the j-th element being 1 and all other elements being zero. That is $\boldsymbol{e}_j^T = \begin{pmatrix} 0 & 0 & ... & 1 & 0 & ... & 0 \end{pmatrix}$. In order to obtain $\boldsymbol{B}$, we have to solve n linear systems

$$\boldsymbol{A}\boldsymbol{B}_1 = \boldsymbol{e}_1, \quad \boldsymbol{A}\boldsymbol{B}_2 = \boldsymbol{e}_2, \quad ..., \quad \boldsymbol{A}\boldsymbol{B}_n = \boldsymbol{e}_n. \tag{1.107}$$

As $\boldsymbol{A}$ is a lower triangular matrix, the solution of $\boldsymbol{A}\boldsymbol{B}_j = \boldsymbol{e}_j$ can easily be obtained by direct forward substitutions discussed earlier in this section.

1.7 Matrix Exponential

Sometimes, we need to calculate exp[$\boldsymbol{A}$], where $\boldsymbol{A}$ is a square matrix. In this case, we have to deal with matrix exponentials. The exponential of a square matrix $\boldsymbol{A}$ is defined as

$$e^{\boldsymbol{A}} \equiv \sum_{n=0}^{\infty} \frac{1}{n!}\boldsymbol{A}^n$$

$$= \boldsymbol{I} + \boldsymbol{A} + \frac{1}{2!}\boldsymbol{A}^2 + ..., \tag{1.108}$$

where $\boldsymbol{I}$ is an identity matrix with the same size as $\boldsymbol{A}$, and $\boldsymbol{A}^2 = \boldsymbol{A}\boldsymbol{A}$ and so on. This (rather odd) definition in fact provides a method of calculating the matrix exponential. The matrix exponentials are very useful in solving systems of differential equations.

Example 1.6: *For a simple matrix*

$$\boldsymbol{A} = \begin{pmatrix} t & 0 \\ 0 & t \end{pmatrix},$$

its exponential is simply

$$e^{\boldsymbol{A}} = \begin{pmatrix} e^t & 0 \\ 0 & e^t \end{pmatrix}.$$

For a more complicated matrix

$$\boldsymbol{B} = \begin{pmatrix} t & a \\ a & t \end{pmatrix},$$

we have

$$e^{\boldsymbol{B}} = \begin{pmatrix} \frac{1}{2}(e^{t+a} + e^{t-a}) & \frac{1}{2}(e^{t+a} - e^{t-a}) \\ \frac{1}{2}(e^{t+a} - e^{t-a}) & \frac{1}{2}(e^{t+a} + e^{t-a}) \end{pmatrix}.$$

As you can see, it is quite complicated but still straightforward to calculate the matrix exponentials. Fortunately, it can easily be done using most computer software packages. By using the power expansions and the basic definition, we can prove the following useful identities

$$e^{t\boldsymbol{A}} \equiv \sum_{n=0}^{\infty} \frac{1}{n!}(t\boldsymbol{A})^n = \boldsymbol{I} + t\boldsymbol{A} + \frac{t^2}{2!}\boldsymbol{A}^2 + ..., \tag{1.109}$$

$$\ln(\boldsymbol{I} + \boldsymbol{A}) \equiv \sum_{n=1}^{\infty} \frac{(-1)^{n-1}}{n!}\boldsymbol{A}^n = \boldsymbol{A} - \frac{1}{2}\boldsymbol{A}^2 + \frac{1}{3}\boldsymbol{A}^3 + ..., \tag{1.110}$$

$$e^{\boldsymbol{A}}e^{\boldsymbol{B}} = e^{\boldsymbol{A}+\boldsymbol{B}} \qquad (\text{if } \boldsymbol{AB} = \boldsymbol{BA}), \tag{1.111}$$

$$\frac{d}{dt}(e^{t\boldsymbol{A}}) = \boldsymbol{A}e^{t\boldsymbol{A}} = e^{t\boldsymbol{A}}\boldsymbol{A}, \tag{1.112}$$

$$(e^{\boldsymbol{A}})^{-1} = e^{-\boldsymbol{A}}, \qquad \det(e^{\boldsymbol{A}}) = e^{\text{tr}\boldsymbol{A}}. \tag{1.113}$$

1.8 Hermitian and Quadratic Forms

The matrices we have discussed so far are real matrices because all their elements are real. In general, the entries or elements of a matrix can be complex numbers, and the matrix becomes a complex matrix. For a matrix $\boldsymbol{A}$, its complex conjugate $\boldsymbol{A}^*$ is obtained by taking the complex conjugate of each of its elements. The Hermitian conjugate $\boldsymbol{A}^\dagger$ is obtained by taking the transpose of its complex conjugate matrix. That is to say, for

$$\boldsymbol{A} = \begin{pmatrix} a_{11} \; a_{12} \; ... \\ a_{21} \; a_{21} \; ... \\ ... \;\; ... \;\; ... \end{pmatrix}, \tag{1.114}$$

we have

$$\boldsymbol{A}^* = \begin{pmatrix} a_{11}^* \; a_{12}^* \; ... \\ a_{21}^* \; a_{22}^* \; ... \\ ... \;\; ... \;\; ... \end{pmatrix}, \tag{1.115}$$

and

$$\boldsymbol{A}^\dagger = (\boldsymbol{A}^*)^T = (\boldsymbol{A}^T)^* = \begin{pmatrix} a_{11}^* \; a_{21}^* \; ... \\ a_{12}^* \; a_{22} \; ... \\ ... \;\; ... \;\; ... \end{pmatrix}. \tag{1.116}$$

A square matrix $\boldsymbol{A}$ is called orthogonal if and only if $\boldsymbol{A}^{-1} = \boldsymbol{A}^T$. If a square matrix $\boldsymbol{A}$ satisfies $\boldsymbol{A}^* = \boldsymbol{A}$, it is called an Hermitian matrix. It is an anti-Hermitian matrix if $\boldsymbol{A}^* = -\boldsymbol{A}$. If the Hermitian matrix of a square matrix $\boldsymbol{A}$ is equal to the inverse of the matrix (or $\boldsymbol{A}^\dagger = \boldsymbol{A}^{-1}$), it is called a unitary matrix.

Example 1.7: *For a complex matrix*

$$\boldsymbol{A} = \begin{pmatrix} 2+3i\pi & 1+9i & 0 \\ e^{i\pi} & -2i & i\sin\theta \end{pmatrix},$$

its complex conjugate $\boldsymbol{A}^*$ *is*

$$\boldsymbol{A}^* = \begin{pmatrix} 2-3i\pi & 1-9i & 0 \\ e^{-i\pi} & 2i & -i\sin\theta \end{pmatrix}.$$

The Hermitian conjugate of $\boldsymbol{A}$ *is*

$$\boldsymbol{A}^\dagger = \begin{pmatrix} 2-3i\pi & e^{-i\pi} \\ 1-9i & 2i \\ 0 & -i\sin\theta \end{pmatrix} = (\boldsymbol{A}^*)^T.$$

For the rotation matrix

$$A = \begin{pmatrix} \cos\theta & \sin\theta \\ -\sin\theta & \cos\theta \end{pmatrix},$$

its inverse and transpose are

$$A^{-1} = \frac{1}{\cos^2\theta + \sin^2\theta} \begin{pmatrix} \cos\theta & -\sin\theta \\ \sin\theta & \cos\theta \end{pmatrix},$$

and

$$A^T = \begin{pmatrix} \cos\theta & -\sin\theta \\ \sin\theta & \cos\theta \end{pmatrix}.$$

Since $\cos^2\theta + \sin^2\theta = 1$, *we have* $A^T = A^{-1}$. *Therefore, the original rotation matrix* A *is orthogonal.*

A very useful concept in computational mathematics and computing is quadratic forms. For a real vector $q^T = (q_1, q_2, q_3, ..., q_n)$ and a real symmetric square matrix A, a quadratic form $\psi(q)$ is a scalar function defined by

$$\psi(q) = q^T A q = \begin{pmatrix} q_1 & q_2 & \cdots & q_n \end{pmatrix} \begin{pmatrix} A_{11} & A_{12} & \ldots & A_{1n} \\ A_{21} & A_{22} & \ldots & A_{2n} \\ \ldots & \ldots & \ldots & \ldots \\ A_{n1} & A_{n2} & \ldots & A_{nn} \end{pmatrix} \begin{pmatrix} q_1 \\ q_2 \\ \vdots \\ q_n \end{pmatrix}, \qquad (1.117)$$

which can be written as

$$\psi(q) = \sum_{i=1}^{n} \sum_{j=1}^{n} q_i A_{ij} q_j. \qquad (1.118)$$

Since ψ is a scalar, it should be independent of the coordinates.

In the case of a square matrix A, ψ might be more easily evaluated in certain intrinsic coordinates $Q_1, Q_2, ..., Q_n$. An important result concerning the quadratic form is that it can always be written through appropriate transformations as

$$\psi(q) = \sum_{i=1}^{n} \lambda_i Q_i^2 = \lambda_1 Q_1^2 + \lambda_2 Q_2^2 + ... + \lambda_n Q_n^2, \qquad (1.119)$$

where λ_i are the eigenvalues of the matrix A determined by

$$\det|A - \lambda I| = 0, \qquad (1.120)$$

and Q_i are the intrinsic components along directions of the eigenvectors in this case.

The natural extension of quadratic forms is the Hermitian form which is the quadratic form for a complex Hermitian matrix $\boldsymbol{A}$. Furthermore, the matrix $\boldsymbol{A}$ can consist of linear operators and functionals in addition to numbers.

Example 1.8: *For a vector $\boldsymbol{q} = (q_1, q_2)$ and the square matrix*

$$\boldsymbol{A} = \begin{pmatrix} 2 & -5 \\ -5 & 2 \end{pmatrix},$$

we have a quadratic form

$$\psi(\boldsymbol{q}) = \begin{pmatrix} q_1 & q_2 \end{pmatrix} \begin{pmatrix} 2 & -5 \\ -5 & 2 \end{pmatrix} \begin{pmatrix} q_1 \\ q_2 \end{pmatrix} = 2q_1^2 - 10q_1q_2 + 2q_2^2.$$

The eigenvalues of the matrix $\boldsymbol{A}$ is determined by

$$\begin{vmatrix} 2-\lambda & -5 \\ -5 & 2-\lambda \end{vmatrix} = 0,$$

whose solutions are $\lambda_1 = 7$ and $\lambda_2 = -3$ (see the next section for further details). Their corresponding eigenvectors are

$$\boldsymbol{v}_1 = \begin{pmatrix} -\sqrt{2}/2 \\ \sqrt{2}/2 \end{pmatrix}, \qquad \boldsymbol{v}_2 = \begin{pmatrix} \sqrt{2}/2 \\ \sqrt{2}/2 \end{pmatrix}.$$

We can see that $\boldsymbol{v}_1 \cdot \boldsymbol{v}_2 = 0$, which means that these two eigenvectors are orthogonal. Writing the quadratic form in terms of the intrinsic coordinates, we have

$$\psi(\boldsymbol{q}) = 7Q_1^2 - 3Q_2^2.$$

Furthermore, if we assume $\psi(\boldsymbol{q}) = 1$ as a simple constraint, then the equation $7Q_1^2 - 3Q_2^2 = 1$ corresponds to a hyperbola.

1.9 Eigenvalues and Eigenvectors

The eigenvalue λ of any $n \times n$ square matrix $\boldsymbol{A}$ is determined by

$$\boldsymbol{A}\boldsymbol{u} = \lambda \boldsymbol{u}, \tag{1.121}$$

or

$$(\boldsymbol{A} - \lambda \boldsymbol{I})\boldsymbol{u} = 0, \tag{1.122}$$

where $\boldsymbol{I}$ is a unitary matrix with the same size as $\boldsymbol{A}$. Any non-trivial solution requires that

$$\det|\boldsymbol{A} - \lambda \boldsymbol{I}| = 0, \tag{1.123}$$

or

$$\begin{vmatrix} a_{11}-\lambda & a_{12} & \dots & a_{1n} \\ a_{21} & a_{22}-\lambda & \dots & a_{2n} \\ \vdots & & \ddots & \\ a_{n1} & a_{n2} & \dots & a_{nn}-\lambda \end{vmatrix} = 0, \tag{1.124}$$

which again can be written as a polynomial

$$\lambda^n + \alpha_{n-1}\lambda^{n-1} + \dots + \alpha_0 = (\lambda - \lambda_1)\dots(\lambda - \lambda_n) = 0, \tag{1.125}$$

where λ_is are the eigenvalues which could be complex numbers. In general, the determinant is zero, which leads to a polynomial of order n in λ. For each eigenvalue λ, there is a corresponding eigenvector $\boldsymbol{u}$ whose direction can be uniquely determined. However, the length of the eigenvector is not unique because any non-zero multiple of $\boldsymbol{u}$ will also satisfy equation (1.121), and thus can be considered as an eigenvector. For this reason, it is usually necessary to apply additional conditions by setting the length as unity, and subsequently the eigenvector becomes a unit eigenvector.

Generally speaking, a real $n \times n$ matrix $\boldsymbol{A}$ has n eigenvalues $\lambda_i (i = 1, 2, ..., n)$, however, these eigenvalues are not necessarily distinct. If the real matrix is symmetric, that is to say $\boldsymbol{A}^T = \boldsymbol{A}$, then the matrix has n distinct eigenvectors, and all the eigenvalues are real numbers.

The eigenvalues λ_i are related to the trace and determinant of the matrix

$$\text{tr}(\boldsymbol{A}) = \sum_{i=1}^{n} a_{ii} = \lambda_1 + \lambda_2 + \dots + \lambda_n = \sum_{i=1}^{n} \lambda_i, \tag{1.126}$$

and

$$\det(\boldsymbol{A}) = |\boldsymbol{A}| = \prod_{i=1}^{n} \lambda_i. \tag{1.127}$$

Example 1.9: *The eigenvalues of the square matrix*

$$\boldsymbol{A} = \begin{pmatrix} 4 & 9 \\ 2 & -3 \end{pmatrix},$$

can be obtained by solving

$$\begin{vmatrix} 4-\lambda & 9 \\ 2 & -3-\lambda \end{vmatrix} = 0.$$

We have

$$(4-\lambda)(-3-\lambda) - 18 = (\lambda - 6)(\lambda + 5) = 0.$$

Thus, the eigenvalues are $\lambda = 6$ and $\lambda = -5$. Let $\boldsymbol{v} = (v_1\ v_2)^T$ be the eigenvector; we have for $\lambda = 6$

$$|\boldsymbol{A} - \lambda \boldsymbol{I}| = \begin{pmatrix} -2 & 9 \\ 2 & -9 \end{pmatrix} \begin{pmatrix} v_1 \\ v_2 \end{pmatrix} = 0,$$

which means that

$$-2v_1 + 9v_2 = 0, \qquad 2v_1 - 9v_2 = 0.$$

These two equations are virtually the same (not linearly independent), so the solution is

$$v_1 = \frac{9}{2} v_2.$$

Any vector parallel to $\boldsymbol{v}$ is also an eigenvector. In order to get a unique eigenvector, we have to impose an extra requirement, that is, the length of the vector is unity. We now have

$$v_1^2 + v_2^2 = 1,$$

or

$$(\frac{9v_2}{2})^2 + v_2^2 = 1,$$

which gives $v_2 = \pm 2/\sqrt{85}$, and $v_1 = \pm 9/\sqrt{85}$. As these two vectors are in opposite directions, we can choose any of the two directions. So the eigenvector for the eigenvalue $\lambda = 6$ is

$$\boldsymbol{v} = \begin{pmatrix} 9/\sqrt{85} \\ \\ 2/\sqrt{85} \end{pmatrix}.$$

Similarly, the corresponding eigenvector for the eigenvalue $\lambda = -5$ is $\boldsymbol{v} = (-\sqrt{2}/2\ \ \sqrt{2}/2)^T$.

Furthermore, the trace and determinant of $\boldsymbol{A}$ are

$$\mathrm{tr}(\boldsymbol{A}) = 4 + (-3) = 1, \qquad \det(\boldsymbol{A}) = 4 \times (-3) - 2 \times 9 = -30.$$

The sum of the eigenvalues is

$$\sum_{i=1}^{2} \lambda_i = 6 + (-5) = 1 = \mathrm{tr}(\boldsymbol{A}),$$

while the product of the eigenvalues is

$$\prod_{i=1}^{2} \lambda_i = 6 \times (-5) = -30 = \det(\boldsymbol{A}).$$

For any real square matrix $\boldsymbol{A}$ with the eigenvalues $\lambda_i = \mathrm{eig}(\boldsymbol{A})$, the eigenvalues of $\alpha \boldsymbol{A}$ are $\alpha \lambda_i$ where $\alpha \neq 0 \in \Re$. This property becomes handy when rescaling the matrices in some iteration formulae so that the rescaled scheme becomes more stable. This is also the major reason why the pivoting and removing/rescaling of exceptionally large elements works.

1.10 Definiteness of Matrices

A square symmetric matrix $\boldsymbol{A}$ is said to be positive definite if all its eigenvalues are strictly positive ($\lambda_i > 0$ where $i = 1, 2, ..., n$). By multiplying (1.121) by $\boldsymbol{u}^T$, we have

$$\boldsymbol{u}^T \boldsymbol{A}\boldsymbol{u} = \boldsymbol{u}^T \lambda \boldsymbol{u} = \lambda \boldsymbol{u}^T \boldsymbol{u}, \tag{1.128}$$

which leads to

$$\lambda = \frac{\boldsymbol{u}^T \boldsymbol{A}\boldsymbol{u}}{\boldsymbol{u}^T \boldsymbol{u}}. \tag{1.129}$$

This means that

$$\boldsymbol{u}^T \boldsymbol{A}\boldsymbol{u} > 0, \qquad \text{if } \lambda > 0. \tag{1.130}$$

In fact, for any vector $\boldsymbol{v}$, the following relationship holds

$$\boldsymbol{v}^T \boldsymbol{A}\boldsymbol{v} > 0. \tag{1.131}$$

Since $\boldsymbol{v}$ can be a unit vector, thus all the diagonal elements of $\boldsymbol{A}$ should be strictly positive as well. If all the eigenvalues are non-negative or $\lambda_i \geq 0$, then the matrix is called positive semi-definite. In general, an indefinite matrix can have both positive and negative eigenvalues.

The inverse of a positive definite matrix is also positive definite. For a linear system $\boldsymbol{A}\boldsymbol{u} = \boldsymbol{f}$ where $\boldsymbol{f}$ is a known column vector, if $\boldsymbol{A}$ is positive definite, then the system can be solved more efficiently by matrix decomposition methods.

Example 1.10: *In general, a 2×2 symmetric matrix $\boldsymbol{A}$*

$$\boldsymbol{A} = \begin{pmatrix} \alpha & \beta \\ \beta & \gamma \end{pmatrix},$$

is positive definite if

$$\alpha u_1^2 + 2\beta u_1 u_2 + \gamma u_2^2 > 0,$$

for all $\boldsymbol{u} = (u_1, u_2)^T \neq 0$. The inverse of $\boldsymbol{A}$ is

$$\boldsymbol{A}^{-1} = \frac{1}{\alpha\gamma - \beta^2} \begin{pmatrix} \gamma & -\beta \\ -\beta & \alpha \end{pmatrix},$$

which is also positive definite.

As the eigenvalues of

$$\boldsymbol{A} = \begin{pmatrix} 1 & 2 \\ 2 & 1 \end{pmatrix},$$

are $\lambda = 3, -1$, *the matrix is indefinite. For another matrix*

$$B = \begin{pmatrix} 4 & 6 \\ 6 & 20 \end{pmatrix},$$

we can find its eigenvalues using a similar method as discussed earlier, and the eigenvalues are $\lambda = 2, 22$. *So matrix* $\boldsymbol{B}$ *is positive definite. The inverse of* $\boldsymbol{B}$

$$B^{-1} = \frac{1}{44}\begin{pmatrix} 20 & -6 \\ -6 & 4 \end{pmatrix},$$

is also positive definite because $\boldsymbol{B}^{-1}$ *has two eigenvalues:* $\lambda = 1/2, 1/22$.

We have briefly reviewed the basic algebra of vectors and matrices as well as some basic calculus, now can move onto more algorithm-related topics such as algorithm complexity.

Chapter 2

Algorithmic Complexity, Norms and Convexity

When analyzing an algorithm, we often discuss its computational complexity. This also allows us to compare one algorithm with other algorithms in terms of various performance measures.

2.1 Computational Complexity

The efficiency of an algorithm is often measured by its algorithmic complexity or computational complexity. In the literature, this complexity is also called Kolmogorov complexity. For a given problem size n, the complexity is denoted using Big-O notations such as $O(n^2)$ or $O(n \log n)$.

For the sorting algorithm for a given number of n data entries, sorting these numbers into either ascending or descending order will take the computational time as a function of the problem size n. $O(n)$ means a linear complexity, while $O(n^2)$ has a quadratic complexity. That is, if n is doubled, then the computational time or computational efforts will double for linear complexity, but it will quadruple for quadratic complexity.

For example, the bubble sorting algorithm starts at the beginning of the data set by comparing the first two elements. If the first is smaller than the second, then swap them. This comparison and swap process continues for each possible pair of adjacent elements. There are $n \times n$ pairs as we need two loops over the whole data set, then the algorithm complexity is $O(n^2)$.

On the other hand, the quicksort algorithm uses a divide-and-conquer approach via partition. By first choosing a pivot element, we then put all the elements into two sublists with all the smaller elements before the pivot and all the greater elements after it. Then, the sublists are recursively sorted in a similar manner. This algorithm will result in a complexity of $O(n \log n)$. The quicksort is much more efficient than the bubble algorithm.

For $n = 1000$, then the bubble algorithm will need about $O(n^2) \approx O(10^6)$ calculations, while the quicksort only requires $O(n \log n) \approx O(3 \times 10^3)$ calculations (at least two orders less in this simple case).

2.2 NP-Complete Problems

In mathematical programming, an easy or tractable problem is a problem whose solution can be obtained by computer algorithms with a solution time (or number of steps) as a polynomial function of problem size n. Algorithms with polynomial-time are considered efficient.

A problem is called a P-problem or polynomial-time problem if the number of steps needed to find the solution is bounded by a polynomial in n and it has at least one algorithm to solve it.

On the other hand, a hard or intractable problem requires a solution time that is an exponential function of n, and thus exponential-time algorithms are considered inefficient. A problem is called nondeterministic polynomial (NP) if its solution can only be guessed and evaluated in polynomial time, and there is no known rule to make such a guess (hence, nondeterministic). Consequently, guessed solutions cannot be guaranteed to be optimal or even near optimal.

In fact, no known algorithms exist to solve NP-hard problems, and only approximate solutions or heuristic solutions are possible. Thus, heuristic and metaheuristic methods are very promising in obtaining approximate solutions or nearly optimal/suboptimal solutions. We will introduce some popular nature-inspired metaheuristic algorithms, especially those based on swarm intelligence, in the last two chapters of the book in Part VI.

A problem is called NP-complete if it is an NP-hard problem and all other problems in NP are reducible to it via certain reduction algorithms. The reduction algorithm has a polynomial time. An example of an NP-hard problem is the Travelling Salesman Problem (TSP), and its objective is to find the shortest route or minimum travelling cost to visit all n given cities exactly once and then return to the starting city.

The solvability of NP-complete problems (whether by polynomial time or not) is still an unsolved problem, which is why Clay Mathematical Institute is offering a million dollar reward for a formal proof. Most real-world problems are NP-hard, and thus any advance in dealing with NP problems will have a profound impact on many applications.

The analysis of an algorithm often involves the calculations of norms and other quantities. In addition, the Hessian matrices are often used in optimization while the spectral radius of a matrix is widely used in the stability analysis of an iteration procedure. We will now review these fundamental concepts.

2.3 Vector and Matrix Norms

For a vector $\boldsymbol{v}$, its p-norm is denoted by $\|\boldsymbol{v}\|_p$ and defined as

$$\|\boldsymbol{v}\|_p = (\sum_{i=1}^{n} |v_i|^p)^{1/p}, \tag{2.1}$$

where p is a positive integer. From this definition, it is straightforward to show that the p-norm satisfies the following conditions: $\|\boldsymbol{v}\| \geq 0$ for all $\boldsymbol{v}$, and $\|\boldsymbol{v}\| = 0$ if and only if $\boldsymbol{v} = \boldsymbol{0}$. This is the non-negativeness condition. In addition, for any real number α, we have the scaling condition: $\|\alpha \boldsymbol{v}\| = \alpha \|\boldsymbol{v}\|$.

The three most common norms are one-, two- and infinity-norms when $p = 1, 2$, and ∞, respectively. For $p = 1$, the one-norm is just the simple sum of the absolute value of each component $|v_i|$, while the 2-norm (or two-norm) $\|\boldsymbol{v}\|_2$ for $p = 2$ is the standard Euclidean norm because $\|\boldsymbol{v}\|_2$ is the length of the vector $\boldsymbol{v}$

$$\|\boldsymbol{v}\|_2 = \sqrt{\boldsymbol{v} \cdot \boldsymbol{v}} = \sqrt{v_1^2 + v_2^2 + ... + v_n^2}, \tag{2.2}$$

where the notation $\boldsymbol{u} \cdot \boldsymbol{v}$ is the inner product of two vectors $\boldsymbol{u}$ and $\boldsymbol{v}$.

For the special case $p = \infty$, we denote $v_{\max}$ as the maximum absolute value of all the components v_i, or $v_{\max} \equiv \max |v_i| = \max(|v_1|, |v_2|, ..., |v_n|)$.

$$\|\boldsymbol{v}\|_\infty = \lim_{p\to\infty} (\sum_{i=1}^{n} |v_i|^p)^{1/p} = \lim_{p\to\infty} \Big(v_{\max}^p \sum_{i=1}^{n} |\frac{v_i}{v_{\max}}|^p\Big)^{1/p}$$

$$= \lim_{p\to\infty} (v_{\max}^p)^{\frac{1}{p}} (\sum |\frac{v_i}{v_{\max}}|^p)^{\frac{1}{p}} = v_{\max} \lim_{p\to\infty} (\sum_{i=1}^{n} |\frac{v_i}{v_{\max}}|^p)^{\frac{1}{p}}. \tag{2.3}$$

Since $|v_i/v_{\max}| \leq 1$ and for all terms $|v_i/v_{\max}| < 1$, we have

$$|v_i/v_{\max}|^p \to 0, \text{ for } p \to \infty.$$

Thus, the only non-zero term in the sum is the one with $|v_i/v_{\max}| = 1$, which means that

$$\lim_{p\to\infty} \sum_{i=1}^{n} |v_i/v_{\max}|^p = 1. \tag{2.4}$$

Therefore, we finally have

$$\|\boldsymbol{v}\|_\infty = v_{\max} = \max|v_i|. \tag{2.5}$$

For the uniqueness of norms, it is necessary for the norms to satisfy the triangle inequality

$$\|\boldsymbol{u}+\boldsymbol{v}\| \le \|\boldsymbol{u}\| + \|\boldsymbol{v}\|. \tag{2.6}$$

It is straightforward to check that for $p = 1, 2$, and ∞ from their definitions, they indeed satisfy the triangle inequality. The equality occurs when $\boldsymbol{u} = \boldsymbol{v}$. It remains as an exercise to check this inequality is true for any $p > 0$.

Example 2.1: *For two 4-dimensional vectors* $\boldsymbol{u} = \begin{pmatrix} 5 & 2 & 3 & -2 \end{pmatrix}^T$ *and* $\boldsymbol{v} = \begin{pmatrix} -2 & 0 & 1 & 2 \end{pmatrix}^T$, *then the p-norms of* $\boldsymbol{u}$ *are*

$$\|\boldsymbol{u}\|_1 = |5| + |2| + |3| + |-2| = 12,$$

$$\|\boldsymbol{u}\|_2 = \sqrt{5^2 + 2^2 + 3^2 + (-2)^2} = \sqrt{42},$$

and

$$\|\boldsymbol{u}\|_\infty = \max(5, 2, 3, -2) = 5.$$

Similarly, $\|\boldsymbol{v}\|_1 = 5$, $\|\boldsymbol{v}\|_2 = 3$ *and* $\|\boldsymbol{v}\|_\infty = 2$. *We know that*

$$\boldsymbol{u}+\boldsymbol{v} = \begin{pmatrix} 5+(-2) \\ 2+0 \\ 3+1 \\ -2+2 \end{pmatrix} = \begin{pmatrix} 3 \\ 2 \\ 4 \\ 0 \end{pmatrix},$$

and its corresponding norms are $\|\boldsymbol{u}+\boldsymbol{v}\|_1 = 9$, $\|\boldsymbol{u}+\boldsymbol{v}\|_2 = \sqrt{29}$ *and* $\|\boldsymbol{u}+\boldsymbol{v}\|_\infty = 4$. *It is straightforward to check that*

$$\|\boldsymbol{u}+\boldsymbol{v}\|_1 = 9 < 12 + 5 = \|\boldsymbol{u}\|_1 + \|\boldsymbol{v}\|_1,$$

$$\|\boldsymbol{u}+\boldsymbol{v}\|_2 = \sqrt{29} < \sqrt{42} + 3 = \|\boldsymbol{u}\|_2 + \|\boldsymbol{v}\|_2,$$

and

$$\|\boldsymbol{u}+\boldsymbol{v}\|_\infty = 4 < 5 + 4 = \|\boldsymbol{u}\|_\infty + \|\boldsymbol{v}\|_\infty.$$

Matrices are the extension of vectors, so we can define the corresponding norms. For an $m \times n$ matrix $\boldsymbol{A} = [a_{ij}]$, a simple way to extend the norms is to use the fact that $\boldsymbol{Au}$ is a vector for any vector $\|\boldsymbol{u}\| = 1$. So the p-norm is defined as

$$\|\boldsymbol{A}\|_p = (\sum_{i=1}^{m}\sum_{j=1}^{n} |a_{ij}|^p)^{1/p}. \tag{2.7}$$

Alternatively, we can consider that all the elements or entries a_{ij} form a vector. A popular norm, called Frobenius form (also called the Hilbert-Schmidt norm), is defined as

$$\|\boldsymbol{A}\|_F = \Big(\sum_{i=1}^{m}\sum_{j=1}^{n} a_{ij}^2\Big)^{1/2}. \tag{2.8}$$

In fact, Frobenius norm is a 2-norm.

Other popular norms are based on the absolute column sum or row sum. For example,

$$\|\boldsymbol{A}\|_1 = \max_{1 \le j \le n} \Big(\sum_{i=1}^{m} |a_{ij}|\Big), \tag{2.9}$$

which is the maximum of the absolute column sum, while

$$\|\boldsymbol{A}\|_\infty = \max_{1 \le i \le m} \Big(\sum_{j=1}^{n} |a_{ij}|\Big), \tag{2.10}$$

is the maximum of the absolute row sum. The max norm is defined as

$$\|\boldsymbol{A}\|_{\max} = \max\{|a_{ij}|\}. \tag{2.11}$$

From the definitions of these norms, we know that they satisfy the non-negativeness condition $\|\boldsymbol{A}\| \ge 0$, the scaling condition $\|\alpha \boldsymbol{A}\| = |\alpha|\|\boldsymbol{A}\|$, and the triangle inequality $\|\boldsymbol{A} + \boldsymbol{B}\| \le \|\boldsymbol{A}\| + \|\boldsymbol{B}\|$.

Example 2.2: *For the matrix* $\boldsymbol{A} = \begin{pmatrix} 2 & 3 \\ 4 & -5 \end{pmatrix}$, *it is easy to calculate that*

$$\|\boldsymbol{A}\|_F = \|\boldsymbol{A}\|_2 = \sqrt{2^2 + 3^2 + 4^2 + (-5)^2} = \sqrt{54},$$

$$\|\boldsymbol{A}\|_\infty = \max \begin{bmatrix} |2| + |3| \\ \\ |4| + |-5| \end{bmatrix} = 9,$$

and $\|\boldsymbol{A}\|_{\max} = 5$.

2.4 Distribution of Eigenvalues

For any $n \times n$ matrix $\boldsymbol{A} = [a_{ij}]$, there is an important theorem, called Gerschgorin theorem, concerning the locations of all the eigenvalues λ_i of $\boldsymbol{A}$.

Let us first define a number (or radius) r_i by

$$r_i \equiv \sum_{j=1, j\neq i}^{n} |a_{ij}| = \sum_{j=1}^{n} |a_{ij}| - |a_{ii}|, \qquad (i = 1, 2, ..., n), \tag{2.12}$$

and then denote Ω_i as the circle, $|z - a_{ii}| \leq r_i$, centred at a_{ii} with a radius r_i in the complex plane $z \in \mathcal{C}$. Such circles are often called Gerschgorin's circles or discs.

Since the eigenvalues λ_i (counting the multiplicity of roots) and their corresponding eigenvectors $\boldsymbol{u}^{(i)}$ are determined by

$$\boldsymbol{A}\boldsymbol{u}^{(i)} = \lambda_i \boldsymbol{u}^{(i)}, \tag{2.13}$$

for all $i = 1, 2, ..., n$, we have each component $u_k^{(i)} (k = 1, 2, ..., n)$ satisfying

$$\sum_{j=1}^{n} a_{kj} u_j^{(i)} = \lambda_k u_k^{(i)}, \tag{2.14}$$

where $\boldsymbol{u}^{(i)} = (u_1^{(i)}, u_2^{(i)}, ..., u_n^{(i)})^T$ and $u_j^{(i)}$ is the j-th component of the vector $\boldsymbol{u}^{(i)}$. Furthermore, we also define the largest absolute component of $\boldsymbol{u}^{(i)}$ (or its infinity norm) as

$$|\boldsymbol{u}^{(i)}| = \|\boldsymbol{u}^{(i)}\|_\infty = \max_{1 \leq j \leq n} \{|u_j^{(i)}|\}. \tag{2.15}$$

As the length of an eigenvector is not zero, we get $|\boldsymbol{u}^{(i)}| > 0$. We now have

$$a_{kk} u_k^{(i)} + \sum_{j \neq k} a_{kj} u_j^{(i)} = \lambda_k u_k^{(i)}, \tag{2.16}$$

whose norm leads to

$$|\lambda_k - a_{kk}| = \left| \frac{\sum_{j\neq k} a_{kj} u_j^{(i)}}{u_k^{(i)}} \right| \leq \frac{\sum_{j\neq k} |a_{kj}| |\boldsymbol{u}^{(i)}|}{|\boldsymbol{u}^{(i)}|} = \sum_{j \neq k} |a_{kj}| = r_k.$$

This is equivalent to the following simple inequality (for any eigenvalue λ)

$$|\lambda - a_{ii}| \leq r_i, \tag{2.17}$$

which is essentially the Gerschgorin circle theorem. Geometrically speaking, this important theorem states that the eigenvalues λ_i of $\boldsymbol{A}$ must be inside one of these circles Ω_i. In addition, if p of these circles form a connected set $\mathcal{S}$ which is disjointed from the remaining $n - p$ circles, it can be proven that there are exactly p of the eigenvalues are inside the set $\mathcal{S}$, here we have to count the multiplicity of roots. Furthermore, if $\boldsymbol{A}$ is symmetric and real (or $\boldsymbol{A} = \boldsymbol{A}^T$), all the eigenvalues are real, and thus they all fall on the real axis.

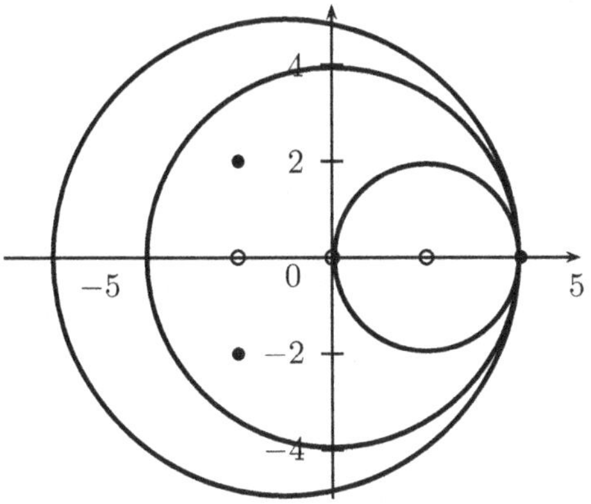

Fig. 2.1 Gerschgorin circles of eigenvalues.

First, let us look at a simple example.

Example 2.3: *For example, the matrix* $\boldsymbol{A}$

$$\boldsymbol{A} = \begin{pmatrix} 2 & 2 & 0 \\ 2 & -2 & 4 \\ 4 & 0 & 0 \end{pmatrix},$$

has three eigenvalues λ. *These eigenvalues should satisfy*

$$|\lambda - 2| \leq r_1 = |2| + |0| = 2,$$

$$|\lambda - (-2)| \leq r_2 = |2| + |4| = 6,$$

$$|\lambda - 0| \leq |4| + |0| = 4.$$

These circles are shown in Fig. 2.1.

Following the method of finding the eigenvalues discussed earlier in this book, we have the eigenvalues of $\boldsymbol{A}$

$$\lambda_i = 4, \; -2 + 2i, \; -2 - 2i,$$

which are marked as solid dots in the same figure. We can see that all these eigenvalues λ_i *are within the union of all the Greschgorin discs.*

Tridiagonal matrices are important in many applications. For example, the finite difference scheme (to be discussed in later chapters) in an equally-spaced grid often results in a simple tridiagonal matrix

$$A = \begin{pmatrix} b & a & & & 0 \\ a & b & a & & \\ & a & b & \ddots & \\ & & \ddots & \ddots & a \\ 0 & & & a & b \end{pmatrix}, \tag{2.18}$$

In this case, the radius r_i becomes

$$r_1 = r_n - |a|,$$

$$r_i = |a| + |a| = 2|a|, \qquad (i = 2, 3, ..., n-1). \tag{2.19}$$

All the eigenvalues of $\boldsymbol{A}$ will satisfy

$$|\lambda - b| \leq r_i, \tag{2.20}$$

or

$$-2|a| + b \leq \lambda \leq b + 2|a|, \tag{2.21}$$

where we have used that $r_1 = r_n = |a| \leq 2|a|$.

In many applications, we also have to use the inverse $\boldsymbol{A}^{-1}$ of $\boldsymbol{A}$. The eigenvalues Λ of the inverse $\boldsymbol{A}^{-1}$ are simply the reciprocals of the eigenvalues λ of $\boldsymbol{A}$. That is, $\Lambda = 1/\lambda$. Now we have

$$-2|a| + b \leq \frac{1}{\Lambda} \leq b + 2|a|, \tag{2.22}$$

or

$$\frac{1}{b + 2|a|} \leq \Lambda \leq \frac{1}{b - 2|a|}. \tag{2.23}$$

For example, if $a = 1$ and $b = 4$, we have $1/6 \leq \Lambda \leq 1/2$. However, if $a = -1$ and $b = -4$, we have $-1/2 \leq \Lambda \leq -1/6$.

Eigenvalues and eigenvectors have important applications. For example, they can be used to carry out eigendecomposition of the original square matrix $\boldsymbol{A}$. If $\boldsymbol{A}$ has n eigenvalues $\lambda_i (i = 1, 2, ..., n)$ with n linearly independent eigenvectors $\boldsymbol{u}_i$. Then, $\boldsymbol{A}$ can be decomposed in terms of its eigenvalues and eigenvectors as

$$\boldsymbol{A} = \boldsymbol{P \Lambda P}^{-1}, \tag{2.24}$$

where $\boldsymbol{P}$ is an $n \times n$ matrix formed by using the eigenvector $\boldsymbol{u}_i$ for eigenvalue λ_i to form its i-th column. That is

$$\boldsymbol{P} = \big(\boldsymbol{u}_1 \ \boldsymbol{u}_2 \ ... \ \boldsymbol{u}_n \big), \tag{2.25}$$

which is sometimes called the modal matrix.

Furthermore, the diagonal eigenvalue matrix is given by

$$\boldsymbol{\Lambda} = \begin{pmatrix} \lambda_1 & & & 0 \\ & \lambda_2 & & \\ & & \ddots & \\ 0 & & & \lambda_n \end{pmatrix}. \tag{2.26}$$

This kind of matrix decomposition is often referred to as eigendecomposition. It is also often called spectral decomposition because the eigenvalues form the spectrum of $\boldsymbol{A}$.

In the case of a symmetric matrix, all its eigenvalues are real, and the matrix $\boldsymbol{P}$ becomes orthogonal. An orthogonal matrix $\boldsymbol{Q}$ is a real square matrix satisfying

$$\boldsymbol{Q}\boldsymbol{Q}^T = \boldsymbol{Q}^T\boldsymbol{Q} = \boldsymbol{I}, \tag{2.27}$$

where $\boldsymbol{I}$ is an identity matrix with the same size $n \times n$ as $\boldsymbol{Q}$. For any invertible matrix, we have $\boldsymbol{Q}\boldsymbol{Q}^{-1} = \boldsymbol{Q}^{-1}\boldsymbol{Q} = \boldsymbol{I}$, which means that

$$\boldsymbol{Q}^{-1} = \boldsymbol{Q}^T. \tag{2.28}$$

A distinct advantage of an orthogonal matrix is that its inverse can easily be obtained by simple transposition. Therefore, for a symmetric matrix $\boldsymbol{A}$, we have $\boldsymbol{Q} = \boldsymbol{P}$ and

$$\boldsymbol{A} = \boldsymbol{Q}\boldsymbol{\Lambda}\boldsymbol{Q}^T. \tag{2.29}$$

Eigendecomposition can be used to invert the matrix, especially for a symmetric matrix. If a matrix $\boldsymbol{A}$ is invertable and all its eigenvalues are non-zero, we have

$$\boldsymbol{A}^{-1} = \boldsymbol{Q}\boldsymbol{\Lambda}^{-1}\boldsymbol{Q}^{-1}, \tag{2.30}$$

where

$$\boldsymbol{\Lambda}^{-1} = \begin{pmatrix} 1/\lambda_1 & & & 0 \\ & 1/\lambda_2 & & \\ & & \ddots & \\ 0 & & & 1/\lambda_n \end{pmatrix}. \tag{2.31}$$

Let us use a simple example to demonstrate this.

Example 2.4: *Since the following matrix*

$$\boldsymbol{A} = \begin{pmatrix} 4 & 12 & 0 \\ 12 & 11 & 0 \\ 0 & 0 & 30 \end{pmatrix},$$

is symmetric, all its eigenvalues are real. Using the standard method of computing the eigenvalues and eigenvectors, we have its eigenvalues

$$\lambda_1 = 20, \ \ \lambda_2 = -5, \ \ \lambda_3 = 30,$$

and their corresponding eigenvectors

$$\boldsymbol{u}_1 = \begin{pmatrix} 3/5 \\ 4/5 \\ 0 \end{pmatrix}, \ \ \boldsymbol{u}_2 = \begin{pmatrix} 4/5 \\ -3/5 \\ 0 \end{pmatrix}, \ \ \boldsymbol{u}_3 = \begin{pmatrix} 0 \\ 0 \\ 1 \end{pmatrix}.$$

It is straightforward to verify that these three eigenvectors are mutually orthogonal. For example,

$$\boldsymbol{u}_1 \cdot \boldsymbol{u}_2 = \boldsymbol{u}_1^T \boldsymbol{u}_2 = \frac{3}{5} \times \frac{4}{5} + \frac{4}{5} \times \frac{-3}{5} + 0 \times 0 = 0.$$

As the length of the eigenvectors are normalized to unity, these three eigenvectors are thus orthonormal, too. The modal matrix $\boldsymbol{Q}$ *can be formed by these three eigenvectors*

$$\boldsymbol{Q} = \begin{pmatrix} \boldsymbol{u}_1 & \boldsymbol{u}_2 & \boldsymbol{u}_3 \end{pmatrix} = \begin{pmatrix} 3/4 & 4/5 & 0 \\ 4/5 & -3/5 & 0 \\ 0 & 0 & 1 \end{pmatrix},$$

which is orthogonal. In fact, we have

$$\boldsymbol{Q}^{-1} = \begin{pmatrix} 3/5 & 4/5 & 0 \\ 4/5 & -3/5 & 0 \\ 0 & 0 & 1 \end{pmatrix} = \boldsymbol{Q}^T.$$

By defining $\boldsymbol{\Lambda}$ *as*

$$\boldsymbol{\Lambda} = \begin{pmatrix} \lambda_1 & 0 & 0 \\ 0 & \lambda_2 & 0 \\ 0 & 0 & \lambda_3 \end{pmatrix},$$

we have

$$\boldsymbol{A} = \begin{pmatrix} 4 & 12 & 0 \\ 12 & 11 & 0 \\ 0 & 0 & 30 \end{pmatrix} = \boldsymbol{Q \Lambda Q}^T$$

$$= \begin{pmatrix} 3/5 & 4/5 & 0 \\ 4/5 & -3/5 & 0 \\ 0 & 0 & 1 \end{pmatrix} \begin{pmatrix} 20 & 0 & 0 \\ 0 & -5 & 0 \\ 0 & 0 & 30 \end{pmatrix} \begin{pmatrix} 3/5 & 4/5 & 0 \\ 4/5 & -3/5 & 0 \\ 0 & 0 & 1 \end{pmatrix}.$$

In addition, the inverse of $\boldsymbol{A}$ *can be obtained by*

$$\boldsymbol{A}^{-1} = \boldsymbol{Q}\boldsymbol{\Lambda}^{-1}\boldsymbol{Q}^T = \frac{1}{100} \begin{pmatrix} -11 & 12 & 0 \\ 12 & -4 & 0 \\ 0 & 0 & 10/3 \end{pmatrix}.$$

Since the determinant of $\boldsymbol{A}$ has the property

$$\det(\boldsymbol{A}) = \det(\boldsymbol{Q}) \det(\boldsymbol{\Lambda}) \det(\boldsymbol{Q}^{-1}) = \det(\boldsymbol{\Lambda}), \tag{2.32}$$

it is not necessary to normalize the eigenvectors. In fact, if we repeat the above example with eigenvectors

$$\boldsymbol{u}_1 = \begin{pmatrix} 3 \\ 4 \\ 0 \end{pmatrix}, \quad \boldsymbol{u}_2 = \begin{pmatrix} 4 \\ -3 \\ 0 \end{pmatrix}, \quad \boldsymbol{u}_3 = \begin{pmatrix} 0 \\ 0 \\ 2 \end{pmatrix}, \tag{2.33}$$

the eigendecomposition will lead to the same matrix $\boldsymbol{A}$.

Another important question is how to construct a matrix for a given set of eigenvalues and their corresponding eigenvectors (not necessarily mutually orthogonal). This is basically the reverse of the eigendecomposition process. Now let us look at another example.

Example 2.5: *For given eigenvalues* $\lambda_1 = 1/4$, $\lambda_2 = -1/5$ *and* $\lambda_3 = -2$, *and their corresponding eigenvectors*

$$\boldsymbol{u}_1 = \begin{pmatrix} 3 \\ 4 \\ -2 \end{pmatrix}, \quad \boldsymbol{u}_2 = \begin{pmatrix} 4 \\ -3 \\ 0 \end{pmatrix}, \quad \boldsymbol{u}_3 = \begin{pmatrix} 3 \\ 4 \\ 25 \end{pmatrix}.$$

We will now construct the original matrix $\boldsymbol{A}$. *First we have to use the eigenvectors to form a modal matrix* $\boldsymbol{P}$

$$\boldsymbol{P} = \begin{pmatrix} \boldsymbol{u}_1 & \boldsymbol{u}_2 & \boldsymbol{u}_3 \end{pmatrix} = \begin{pmatrix} 3 & 4 & 3 \\ 4 & -3 & 4 \\ -2 & 0 & 25 \end{pmatrix}.$$

Its inverse is

$$\boldsymbol{P}^{-1} = \begin{pmatrix} 1/9 & 4/27 & -1/27 \\ 4/25 & -3/25 & 0 \\ 2/225 & 8/675 & 1/27 \end{pmatrix}.$$

Therefore, the matrix $\boldsymbol{A}$ *becomes*

$$\boldsymbol{A} = \boldsymbol{P}\boldsymbol{\Lambda}\boldsymbol{P}^{-1},$$

where the eigenvalue matrix $\boldsymbol{\Lambda}$ *is given by*

$$\boldsymbol{\Lambda} = \begin{pmatrix} \lambda_1 & 0 & 0 \\ 0 & \lambda_2 & 0 \\ 0 & 0 & \lambda_3 \end{pmatrix} = \begin{pmatrix} 1/4 & 0 & 0 \\ 0 & -1/5 & 0 \\ 0 & 0 & -2 \end{pmatrix}.$$

Finally, we have

$$\boldsymbol{A} = \boldsymbol{P}\boldsymbol{\Lambda}\boldsymbol{P}^{-1} = \begin{pmatrix} -49/500 & 17/125 & -1/4 \\ 17/125 & -7/375 & -1/3 \\ -1/2 & -2/3 & -11/6 \end{pmatrix}.$$

It is straightforward to verify that the eigenvalues of $\boldsymbol{A}$ *are* $1/4, -1/5$ *and* -2*, which are indeed the same as* $\boldsymbol{\Lambda}$.

2.5 Spectral Radius of Matrices

Another important concept related to the eigenvalues of a matrix is the spectral radius of a square matrix. If $\lambda_i (i = 1, 2, ..., n)$ are the eigenvalues (either real or complex) of a matrix $\boldsymbol{A}$, then the spectral radius $\rho(\boldsymbol{A})$ is defined as

$$\rho(\boldsymbol{A}) \equiv \max_{1 \le i \le n} \{|\lambda_i|\}, \tag{2.34}$$

which is the maximum absolute value of all the eigenvalues. Geometrically speaking, if we plot all the eigenvalues of the matrix $\boldsymbol{A}$ on the complex plane, and draw a circle with its centre at the origin $(0, 0)$ on a complex plane so that it encloses all the eigenvalues inside, then the minimum radius of such a circle is the spectral radius.

For any $0 < p \in \Re$, the eigenvectors have non-zero norms $||\boldsymbol{u}|| \neq 0$ and $||\boldsymbol{u}^p|| \neq 0$. Using $\boldsymbol{A}\boldsymbol{u} = \lambda \boldsymbol{u}$ and taking the norms, we have

$$|\lambda|^p ||\boldsymbol{u}^p|| = ||(\lambda \boldsymbol{u})^p|| = ||(\boldsymbol{A}\boldsymbol{u})^p|| \le ||\boldsymbol{A}^p|| \, ||\boldsymbol{u}^p||. \tag{2.35}$$

By dividing both sides of the above equation by $||\boldsymbol{u}^p|| \neq 0$, we reach the following inequality

$$|\lambda| \le ||\boldsymbol{A}^p||^{1/p}, \tag{2.36}$$

which is valid for any eigenvalue. Therefore, it should also be valid for the maximum absolute value or $\rho(\boldsymbol{A})$. We finally have

$$\rho(\boldsymbol{A}) \le ||\boldsymbol{A}^p||^{1/p}, \tag{2.37}$$

which becomes an equality when $p \to \infty$.

Example 2.6: *Let us now calculate the spectral radius of the following matrix*

$$A = \begin{pmatrix} -5 & 1/2 & 1/2 \\ 0 & -1 & -2 \\ 1 & 0 & -3/2 \end{pmatrix}.$$

From Gerschgorin's theorem, we know that

$$|\lambda - (-5)| \le |1/2| + |1/2| = 1, \quad |\lambda - (-1)| \le 2, \quad |\lambda - (-3/2)| \le 1.$$

These Gerschgorin discs are shown in Figure 2.2. Since two discs (Ω_2 and Ω_3) form a connected region (S), which means that there are exactly two eigenvalues inside this connected region, and there is a single eigenvalue inside the isolated disc (Ω_1) centred at $(-5, 0)$.

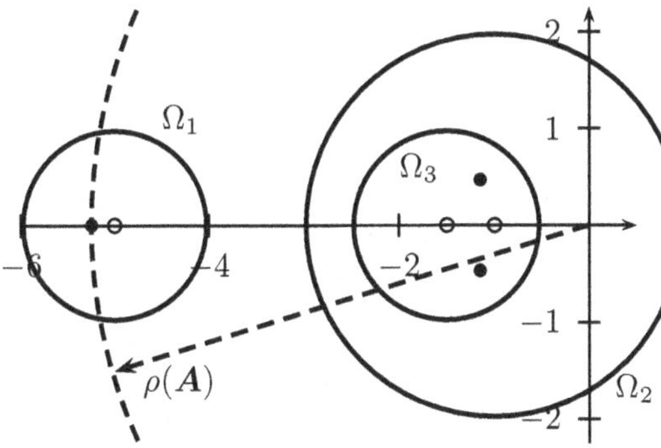

Fig. 2.2 Spectral radius of a square matrix.

We can calculate its eigenvalues by

$$\det(A - \lambda I) = 0,$$

and we have

$$\lambda_1 = -5.199, \quad \lambda_2 = -1.150 + 0.464i, \quad \lambda_3 = -1.150 - 0.464i.$$

These three eigenvalues are marked as solid dots in Figure 2.2 inside the three Gerschgorin discs. The spectral radius is

$$\rho(A) = \max\{|\lambda_i|\} = 5.199,$$

which is also shown in the same figure. Indeed, there are two eigenvalues (λ_2 and λ_3) inside the connected region (S) and there is a single (λ_1) inside the isolated disc (Ω_1).

The spectral radius is very useful in determining whether an iteration algorithm is stable or not. Most iteration schemes can be written as

$$\boldsymbol{u}^{(n+1)} = \boldsymbol{A}\boldsymbol{u}^{(n)} + \boldsymbol{b}, \tag{2.38}$$

where $\boldsymbol{b}$ is a known column vector and $\boldsymbol{A}$ is a square matrix with known coefficients. The iterations start from an initial guess $\boldsymbol{u}^{(0)}$ (often, set $\boldsymbol{u}^{(0)} = \boldsymbol{0}$), and proceed to the approximate solution $\boldsymbol{u}^{(n+1)}$. For the iteration procedure to be stable, it requires that

$$\rho(\boldsymbol{A}) \le 1.$$

If $\rho(\boldsymbol{A}) > 1$, then the algorithm will not be stable and any initial errors will be amplified in each iteration.

In the case of $\boldsymbol{A}$ is a lower (or upper) matrix

$$\boldsymbol{A} = \begin{pmatrix} a_{11} & 0 & ... & 0 \\ a_{21} & a_{22} & ... & 0 \\ \vdots & & \ddots & \\ a_{n1} & a_{n2} & ... & a_{nn} \end{pmatrix}, \tag{2.39}$$

then its eigenvalues are the diagonal entries: a_{11}, a_{22}, ..., a_{nn}. In addition, the determinant of the triangular matrix $\boldsymbol{A}$ is simply the product of its diagonal entries. That is

$$\det(\boldsymbol{A}) = |\boldsymbol{A}| = \prod_{i=1}^{n} a_{ii} = a_{11}a_{22}...a_{nn}. \tag{2.40}$$

Obviously, a diagonal matrix is just a special case of a triangular matrix. Thus, the properties for its inverse, eigenvalues and determinant are the same as the above.

These properties are convenient in determining the stability of an iteration scheme such as the Jacobi-type and Gauss-Seidel iteration methods where $\boldsymbol{A}$ may contain triangular matrices.

Example 2.7: *Determine if the following iteration is stable or not*

$$\begin{pmatrix} u_1 \\ u_2 \\ u_2 \end{pmatrix}_{n+1} = \begin{pmatrix} 2\ 2\ 3 \\ 7\ 6\ 5 \\ 0\ 4\ 5 \end{pmatrix} \begin{pmatrix} u_1 \\ u_2 \\ u_3 \end{pmatrix} + \begin{pmatrix} 2 \\ -2 \\ 1/2 \end{pmatrix}.$$

We know the eigenvalues of

$$A = \begin{pmatrix} 2 & 2 & 3 \\ 7 & 6 & 5 \\ 0 & 4 & 5 \end{pmatrix},$$

are $\lambda_1 = 0.6446 + 1.5773i, \lambda_2 = 0.6446 - 1.5773i, \lambda_3 = 11.7109$. *Then the spectral radius is*

$$\rho(A) = \max_{i \in \{1,2,3\}} (|\lambda_i|) \approx 11.7109 > 1,$$

therefore, the iteration process will not be convergent.

2.6 Hessian Matrix

The gradient vector of a multivariate function $f(\boldsymbol{x})$ is defined as a column vector

$$\boldsymbol{G}(\boldsymbol{x}) \equiv \nabla f(\boldsymbol{x}) \equiv \left(\frac{\partial f}{\partial x_1}, \frac{\partial f}{\partial x_2}, \ldots, \frac{\partial f}{\partial x_n} \right)^T, \tag{2.41}$$

where $\boldsymbol{x} = (x_1, x_2, ..., x_n)^T$ is a vector. As the gradient $\nabla f(\boldsymbol{x})$ of a linear function $f(\boldsymbol{x})$ is always a constant vector $\boldsymbol{k}$, then any linear function can be written as

$$f(\boldsymbol{x}) = \boldsymbol{k}^T \boldsymbol{x} + \boldsymbol{b}, \tag{2.42}$$

where $\boldsymbol{b}$ is a vector constant.

The second derivatives of a generic function $f(\boldsymbol{x})$ form an $n \times n$ matrix, called Hessian matrix, given by

$$\boldsymbol{H}(\boldsymbol{x}) \equiv \nabla^2 f(\boldsymbol{x}) \equiv \begin{pmatrix} \frac{\partial^2 f}{\partial x_1^2} & \cdots & \frac{\partial^2 f}{\partial x_1 \partial x_n} \\ \vdots & & \vdots \\ \frac{\partial^2 f}{\partial x_1 \partial x_n} & \cdots & \frac{\partial^2 f}{\partial {x_n}^2} \end{pmatrix}, \tag{2.43}$$

which is symmetric due to the fact that $\frac{\partial^2 f}{\partial x_i \partial x_j} = \frac{\partial^2 f}{\partial x_j \partial x_i}$. When the Hessian matrix $\boldsymbol{H}(\boldsymbol{x}) = \boldsymbol{A}$ is a constant matrix (the values of its entries are independent of $\boldsymbol{x}$), the function $f(\boldsymbol{x})$ is called a quadratic function, and can subsequently be written as the following generic form

$$f(\boldsymbol{x}) = \frac{1}{2} \boldsymbol{x}^T \boldsymbol{A} \boldsymbol{x} + \boldsymbol{k}^T \boldsymbol{x} + \boldsymbol{b}. \tag{2.44}$$

The use of the factor 1/2 in the expression is to avoid the appearance everywhere of a factor 2 in the derivatives, and this choice is purely for convenience.

Example 2.8: *The gradient of* $f(x,y,z) = xy - y\exp(-z) + z\cos(x)$ *is simply*

$$G = \begin{pmatrix} y - z\sin x & x - e^{-z} & ye^{-z} + \cos x \end{pmatrix}^T.$$

The Hessian matrix is

$$H = \begin{pmatrix} \frac{\partial^2 f}{\partial x^2} & \frac{\partial^2 f}{\partial x \partial y} & \frac{\partial^2 f}{\partial x \partial z} \\ \frac{\partial^2 f}{\partial y \partial x} & \frac{\partial^2 f}{\partial y^2} & \frac{\partial^2 f}{\partial y \partial z} \\ \frac{\partial^2 f}{\partial z \partial x} & \frac{\partial^2 f}{\partial z \partial y} & \frac{\partial^2 f}{\partial z^2} \end{pmatrix} = \begin{pmatrix} -z\cos x & 1 & -\sin x \\ 1 & 0 & e^{-z} \\ -\sin x & e^{-z} & -ye^{-z} \end{pmatrix}.$$

2.7 Convexity

Knowing the properties of a function can be useful for finding the maximum or minimum of the function. In fact, in mathematical optimization, nonlinear problems are often classified according to the convexity of the defining function(s). Geometrically speaking, an object is convex if for any two points within the object, every point on the straight line segment joining them is also within the object. Examples are a solid ball, a cube or a pyramid. Obviously, a hollow object is not convex. Three examples are given in Fig. 2.3.

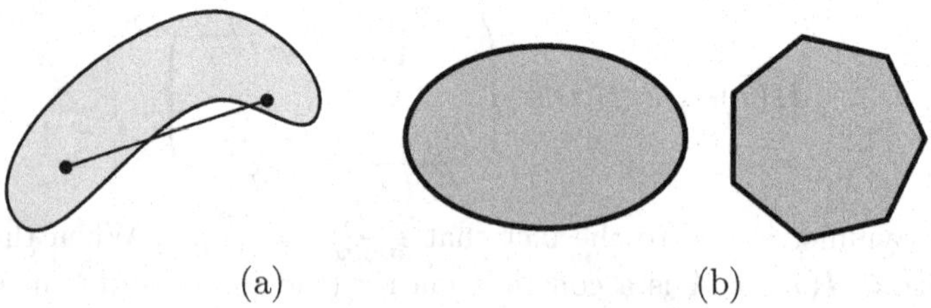

Fig. 2.3 Convexity: (a) non-convex, and (b) convex.

Mathematically speaking, a set $S \in \Re^n$ in a real vector space is called a convex set if

$$tx + (1-t)y \in S, \quad \forall (x,y) \in S, \; t \in [0,1]. \tag{2.45}$$

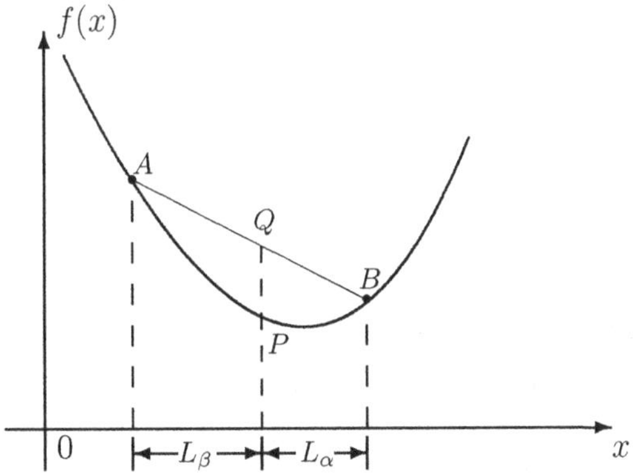

Fig. 2.4 Convexity of a function $f(x)$. Chord AB lies above the curve segment joining A and B. For any point P, we have $L_\alpha = \alpha L$, $L_\beta = \beta L$ and $L = |x_B - x_A|$.

A function $f(x)$ defined on a convex set Ω is called convex if and only if it satisfies

$$f(\alpha x + \beta y) \le \alpha f(x) + \beta f(y), \qquad \forall x, y \in \Omega, \tag{2.46}$$

and

$$\alpha \ge 0, \;\; \beta \ge 0, \;\; \alpha + \beta = 1. \tag{2.47}$$

Geometrically speaking, the chord AB lies above the curve segment APB joining A and B (see Fig. 2.4). For example, for any point P between A and B, we have $x_P = \alpha x_A + \beta x_B$ with

$$\alpha = \frac{L_\alpha}{L} = \frac{x_B - x_P}{x_B - x_A} \ge 0,$$

$$\beta = \frac{L_\beta}{L} = \frac{x_P - x_A}{x_B - x_A} \ge 0, \tag{2.48}$$

which indeed gives $\alpha + \beta = 1$. In addition, we know that

$$\alpha x_A + \beta x_B = \frac{x_A(x_B - x_P)}{x_B - x_A} + \frac{x_B(x_P - x_A)}{x_B - x_A} = x_P. \tag{2.49}$$

The value of the function $f(x_P)$ at P should be less than or equal to the weighted combination $\alpha f(x_A) + \beta f(x_B)$ (or the value at point Q). That is

$$f(x_P) = f(\alpha x_A + \beta x_B) \le \alpha f(x_A) + \beta f(x_B). \tag{2.50}$$

Example 2.9: *For example, the convexity of $f(x) = x^2 - 1$ requires*

$$(\alpha x + \beta y)^2 - 1 \le \alpha(x^2 - 1) + \beta(y^2 - 1), \;\; \forall x, y \in \Re, \tag{2.51}$$

where $\alpha, \beta \geq 0$ *and* $\alpha + \beta = 1$. *This is equivalent to*

$$\alpha x^2 + \beta y^2 - (\alpha x + \beta y)^2 \geq 0, \tag{2.52}$$

where we have used $\alpha + \beta = 1$. *We now have*

$$\alpha x^2 + \beta y^2 - \alpha^2 x^2 - 2\alpha\beta xy - \beta^2 y^2$$

$$= \alpha(1-\alpha)(x-y)^2 = \alpha\beta(x-y)^2 \geq 0, \tag{2.53}$$

which is always true because $\alpha, \beta \geq 0$ *and* $(x-y)^2 \geq 0$. *Therefore,* $f(x) = x^2 - 1$ *is convex for* $\forall x \in \Re$.

A function $f(x)$ on Ω is concave if and only if $g(x) = -f(x)$ is convex. An interesting property of a convex function f is that the vanishing of the gradient $df/dx|_{x_*} = 0$ guarantees that the point x_* is the global minimum or maximum of f, depending on the sign of $f''(x)$, or the positive definiteness of its Hessian in general. If a function is not convex or concave, then it is much more difficult to find its global minima or maxima.

Chapter 3

Ordinary Differential Equations

Most mathematical models in physics, engineering, and applied mathematics are formulated in terms of differential equations. If the variables or quantities (such as velocity, temperature, pressure) change with other independent variables such as spatial coordinates and time, their relationship can in general be written as a differential equation or even a set of differential equations.

3.1 Ordinary Differential Equations

An ordinary differential equation (ODE) is a relationship between a function $y(x)$ (of an independent variable x) and its derivatives y', y'', ..., $y^{(n)}$. It can be written in a generic form as

$$\Psi(x, y, y', y'', ..., y^{(n)}) = 0. \tag{3.1}$$

The solution of the equation is a function $y = f(x)$, satisfying the equation for all x in a given domain Ω.

The order of the differential equation is equal to the order n of the highest derivative in the equation. For example, the Riccati equation:

$$y' + a(x)y^2 + b(x)y = c(x), \tag{3.2}$$

is a first-order ODE, and the Euler Equation

$$x^2y'' + a_1xy' + a_0y = 0, \tag{3.3}$$

is a second-order ODE. The degree of the equation is defined as the power to which the highest derivative occurs. Therefore, both the Riccati equation and Euler equation are of the first degree.

An ordinary differential equation is called linear if it can be arranged into the following standard form

$$a_n(x)y^{(n)} + ... + a_1(x)y' + a_0(x)y = \phi(x), \tag{3.4}$$

where all the coefficients depend on x only, not on y or any derivatives. If any of the coefficients is a function of y or its derivatives, then the equation is nonlinear. If the right-hand side is zero or $\phi(x) \equiv 0$, the equation is homogeneous. It is called nonhomogeneous if $\phi(x) \neq 0$.

The solution of an ordinary differential equation is not always straightforward, and it is usually very complicated for nonlinear equations. Even for linear equations, the solutions are straightforward in only a very few simple cases. The solutions of differential equations generally fall into three types: the closed form, series form and integral form. A closed-form solution is the type of solution that can be expressed in terms of elementary functions and some arbitrary constants. Series solutions are the ones that can be expressed in terms of a series when a closed-form is not possible for certain types of equations. The integral form of solutions or quadrature is sometimes the only form of solution that is possible. If all these forms are not possible, the alternative is to use approximations and numerical solutions.

3.2 First-Order ODEs

The general form of a first-order linear differential equation can be written as

$$y' + a(x)y = b(x), \tag{3.5}$$

where $a(x)$ and $b(x)$ are known functions. Multiplying both sides of the equation by $\exp[\int a(x)dx]$, which is often called the integrating factor, we have

$$y' e^{\int a(x)dx} + a(x) y e^{\int a(x)dx} = b(x) e^{\int a(x)dx}, \tag{3.6}$$

which can be written as

$$[y e^{\int a(x)dx}]' = b(x) e^{\int a(x)dx}. \tag{3.7}$$

By simple integration, we have

$$y e^{\int a(x)dx} = \int b(x) e^{\int a(x)dx} dx + C. \tag{3.8}$$

So its solution becomes

$$y(x) = e^{-\int a(x)dx} \int b(x) e^{\int a(x)dx} dx + C e^{-\int a(x)dx}, \tag{3.9}$$

where C is an integration constant.

Nonlinear ODEs are much more complicated, though some nonlinear ordinary differential equations can be converted into a first-order linear equation (3.5) by a transform or change of variables. For example, Bernoulli's equation can be written as

$$y' + p(x)y = q(x)y^n, \qquad n \neq 1. \tag{3.10}$$

In the case of $n = 1$, it reduces to a standard first-order linear equation. Dividing both sides by y^n and using the change of variables

$$u(x) = \frac{1}{y^{n-1}}, \qquad u' = \frac{(1-n)y'}{y^n}, \tag{3.11}$$

we have

$$u' + (1-n)p(x)u = (1-n)q(x), \tag{3.12}$$

which is a standard first-order linear ODE whose general solution is given earlier in this section.

3.3 Higher-Order ODEs

Higher-order ODEs are more complicated to solve even for linear equations. For the special case of higher-order ODEs where all the coefficients $a_n, ..., a_1$ and a_0 are constants,

$$a_n y^{(n)} + ... + a_1 y' + a_0 y = f(x), \tag{3.13}$$

its general solution $y(x)$ consists of two parts: the complementary function $y_c(x)$ and the particular integral or particular solution $y_p^*(x)$. We have

$$y(x) = y_c(x) + y_p^*(x). \tag{3.14}$$

The complementary function is the solution of the linear homogeneous equation with constant coefficients which can be written in a generic form as

$$a_n y_c^{(n)} + a_{n-1} y_c^{(n-1)} + ... + a_1 y_c' + a_0 = 0. \tag{3.15}$$

Assuming $y = Ae^{\lambda x}$ where A is an arbitrary constant, we get the polynomial equation of characteristics

$$a_n \lambda^n + a_{n-1}\lambda^{(n-1)} + ... + a_1 \lambda + a_0 = 0, \tag{3.16}$$

which has n roots in the general case. Then, the solution can be expressed as the summation of various terms $y_c(x) = \sum_{k=1}^{n} c_k e^{\lambda_k x}$ if the polynomial has

n distinct zeros $\lambda_1, ..., \lambda_n$. Complex roots always occur in pairs $\lambda = r \pm i\omega$, and the corresponding linearly independent terms can then be replaced by $e^{rx}[A\cos(\omega x) + B\sin(\omega x)]$.

Example 3.1: *In order to solve the differential equation*

$$y'''(x) - 5y''(x) - y'(x) + 5y(x) = e^{2x},$$

we have to find its complementary function y_c *and its particular integral* $y^*(x)$. *The complementary function is given by*

$$y'''(x) - 5y''(x) - y'(x) + 5y(x) = 0.$$

By assuming $y_c = \alpha e^{\lambda x}$, *we have the characteristic equation*

$$\lambda^3 - 5\lambda^2 - \lambda + 5 = 0,$$

which is equivalent to

$$(\lambda + 1)(\lambda - 1)(\lambda - 5) = 0.$$

So the solutions are $\lambda = \pm 1, 5$, *which lead to three basic solutions:* $\exp(-x), \exp(x)$ *and* $\exp(5x)$. *So the general complementary function becomes*

$$y_c = Ae^{-x} + Be^{x} + Ce^{5x},$$

where A, B *and* C *are undetermined constants. As the function on the right-hand side is* e^{2x}, *we can assume the particular function takes the form*

$$y^*(x) = Ke^{2x}.$$

Substituting it into the original differential equation, we have

$$(8K - 20K - 2K + 5K)e^{2x} = e^{2x}.$$

Dividing both sides by $e^{2x} \neq 0$, *we have*

$$-9K = 1, \qquad or \qquad K = -\frac{1}{9}.$$

So the particular function becomes $y^* = -\frac{1}{9}e^{2x}$. *Finally, the general solution becomes*

$$y(x) = Ae^{-x} + Be^{x} + Ce^{5x} - \frac{1}{9}e^{2x}.$$

The constants A, B *and* C *will be determined by appropriate boundary or initial conditions.*

The particular solution $y_p^*(x)$ is any $y(x)$ that satisfies the original inhomogeneous equation (3.13). Depending on the form of the function $f(x)$, the particular solutions can take various forms. For most combinations of basic functions such as $\sin x, \cos x$, e^{kx}, and x^n, the method of the undetermined coefficients is widely used. For $f(x) = \sin(\alpha x)$ or $\cos(\alpha x)$, then we can try $y_p^* = A \sin \alpha x + B \sin \alpha x$. We then substitute it into the original equation (3.13) so that the coefficients A and B can be determined. For a polynomial $f(x) = x^n (n = 0, 1, 2, ..., N)$, we then try $y_p^* = A + Bx + ... + Qx^n$ (polynomial). For $f(x) = e^{kx} x^n$, $y_p^* = (A+Bx+...+Qx^n)e^{kx}$. Similarly, for $f(x) = e^{kx} \sin \alpha x$ or $f(x) = e^{kx} \cos \alpha x$, we can use $y_p^* = e^{kx}(A \sin \alpha x + B \cos \alpha x)$. More general cases and their particular solutions can be found in more specialized textbooks.

A very useful technique is to use the differential operator D. A differential operator D is defined as

$$D \equiv \frac{d}{dx}. \tag{3.17}$$

Since we know that $De^{\lambda x} = \lambda e^{\lambda x}$ and $D^n e^{\lambda x} = \lambda^n e^{\lambda x}$, so they are equivalent to the mapping: $D \mapsto \lambda$, and $D^n \mapsto \lambda^n$. Thus, any polynomial $P(D)$ will be mapped to its corresponding $P(\lambda)$. On the other hand, the integral operator $D^{-1} = \int dx$ is just the inverse of differentiation. The beauty of the differential operator method is that the expression of differential operators can be factorized in the same way as for polynomials, then each factor can be solved separately. The differential operator method is very useful in finding out both the complementary functions and particular integral.

Example 3.2: *For example, the particular integral of the differential equation in the previous example*

$$y'''(x) - 5y''(x) - y'(x) + 5y(x) = e^{2x},$$

can be found by

$$(D^3 - 5D^2 - D + 5)y_p^* = e^{2x},$$

or

$$y_p^* = \frac{1}{D^3 - 5D^2 - D + 5} e^{2x}.$$

Since $D^n \mapsto \lambda^n = 2^n$ *where* $n = 1, 2, ...,$ *we have*

$$y_p^* = \frac{e^{2x}}{2^3 - 5 \times 2^2 - 2 + 5} = -\frac{e^{2x}}{9},$$

which is exactly what we got in the previous example.

Higher-order differential equations can conveniently be written as a system of differential equations. In fact, an n-th-order linear equation can always be written as a linear system of n first-order differential equations. A linear system of ODEs is more suitable for mathematical analysis and numerical integration.

3.4 Linear System

For a linear equation of order n [see (3.15)], it can always be written as a linear system

$$\frac{dy}{dx} = y_1, \quad \frac{dy_1}{dx} = y_2, \ \ldots, \quad \frac{dy_{n-1}}{dx} = y_{n-1},$$

$$a_n(x)y'_{n-1} = -a_{n-1}(x)y_{n-1} + \ldots + a_0(x)y + \phi(x), \tag{3.18}$$

which is a system for $u = [y \ \ y_1 \ \ y_2 \ \ \ldots \ \ y_{n-1}]^T$.

For a second-order linear differential equation, we can always write it in the following form

$$\frac{du}{dx} = f(u, v, x), \qquad \frac{dv}{dx} = g(u, v, x). \tag{3.19}$$

If the independent variable x does not appear explicitly in f and g, then the system is said to be autonomous. Such a system has important properties. For simplicity and in keeping with the convention, we use $t = x$ and $\dot{u} = du/dt$ in the following discussion. In general, a homogeneous linear system of n-th order can be written as

$$\begin{pmatrix} \dot{u}_1 \\ \dot{u}_2 \\ \vdots \\ \dot{u}_n \end{pmatrix} = \begin{pmatrix} a_{11} & a_{12} & \ldots & a_{1n} \\ a_{21} & a_{22} & \ldots & a_{2n} \\ \vdots & & & \vdots \\ a_{n1} & a_{n2} & \ldots & a_{nn} \end{pmatrix} \begin{pmatrix} u_1 \\ u_2 \\ \vdots \\ u_n \end{pmatrix}, \tag{3.20}$$

or

$$\dot{\mathbf{u}} = \mathbf{A}\mathbf{u}. \tag{3.21}$$

If $\boldsymbol{u} = \boldsymbol{v}\exp(\lambda t)$, then this becomes an eigenvalue problem for matrix $\boldsymbol{A}$,

$$(\boldsymbol{A} - \lambda\boldsymbol{I})\boldsymbol{v} = 0, \tag{3.22}$$

which will have non-null solutions only if

$$\det(\boldsymbol{A} - \lambda\boldsymbol{I}) = 0. \tag{3.23}$$

Here $\boldsymbol{v}$ is the eigenvector for a given eigenvalue λ. Let us look at an example.

Example 3.3: *For a simple homogeneous system*

$$\frac{du}{dt} = \alpha u + \beta w, \qquad \frac{dw}{dt} = \gamma w,$$

we can write it as

$$\begin{pmatrix} \dot{u} \\ \dot{w} \end{pmatrix} = \begin{pmatrix} \alpha & \beta \\ 0 & \gamma \end{pmatrix} \begin{pmatrix} u \\ w \end{pmatrix}.$$

Here we assume the parameters α, β and γ are real constants. As the matrix

$$\boldsymbol{A} = \begin{pmatrix} \alpha & 0 \\ \beta & \gamma \end{pmatrix},$$

is a triangular matrix, its eigenvalues are the diagonal elements. So the eigenvalues of $\boldsymbol{A}$ are

$$\lambda_1 = \alpha, \qquad \lambda_2 = \gamma.$$

Following the procedure of finding the eigenvectors in Chapter 1, we have the eigenvector

$$\boldsymbol{v}_1 = \begin{pmatrix} 1 \\ 0 \end{pmatrix},$$

for $\lambda_1 = \alpha$. Similarly, the eigenvector for $\lambda_2 = \gamma$ is

$$\boldsymbol{v}_2 = \frac{1}{\sqrt{1 + \frac{(\alpha-\gamma)^2}{\beta^2}}} \begin{pmatrix} 1 \\ -(\alpha - \gamma)/\beta \end{pmatrix}.$$

The general solution of this linear system is

$$\begin{pmatrix} u \\ w \end{pmatrix} = K_1 \boldsymbol{v}_1 e^{\alpha t} + K_2 \boldsymbol{v}_2 e^{\gamma t},$$

where K_1 and K_2 are constants. The basic solutions corresponding to each eigenvalue and eigenvector are straight-line solutions because they correspond to straight lines in the phase diagram.

For the case of $\alpha = 2$, $\beta = -2$ and $\gamma = 3$, or

$$\boldsymbol{A} = \begin{pmatrix} 2 & -2 \\ 0 & 3 \end{pmatrix},$$

we have

$$\boldsymbol{v}_1 = \begin{pmatrix} 1 \\ 0 \end{pmatrix}, \qquad \boldsymbol{v}_2 = \begin{pmatrix} 2/\sqrt{5} \\ -1/\sqrt{5} \end{pmatrix}.$$

Therefore, the general solution becomes

$$\begin{pmatrix} u \\ w \end{pmatrix} = K_1 \begin{pmatrix} 1 \\ 0 \end{pmatrix} e^{2t} + K_2 \begin{pmatrix} 2/\sqrt{5} \\ -1/\sqrt{5} \end{pmatrix} e^{3t},$$

which leads to

$$u = K_1 e^{2t} + \frac{2K_2}{\sqrt{5}} e^{3t}, \qquad w = -\frac{K_2}{\sqrt{5}} e^{3t}.$$

As the constants K_1 and K_2 are undetermined, it is in fact not necessary to calculate eigenvectors in general. Here we just show how to construct the general solution.

Obviously, such systems can be analyzed by using dynamical system theories and numerical methods. We will discuss this in more detail later when appropriate.

3.5 Sturm-Liouville Equation

One of the commonly used second-order ordinary differential equations is the Sturm-Liouville equation in the interval $x \in [a, b]$

$$\frac{d}{dx}[p(x)\frac{dy}{dx}] + q(x)y + \lambda r(x)y = 0, \tag{3.24}$$

with the boundary conditions

$$y(a) + \alpha y'(a) = 0, \qquad y(b) + \beta y'(b) = 0, \tag{3.25}$$

where the known function $p(x)$ is differentiable, and the known functions $q(x), r(x)$ are continuous. The parameter λ to be determined can only take certain values λ_n, called the eigenvalues, if the problem has solutions. For obvious reasons, this problem is called the Sturm-Liouville eigenvalue problem.

For each eigenvalue λ_n, there is a corresponding solution ψ_{λ_n}, called eigenfunction. The Sturm-Liouville theory states that for two different eigenvalues $\lambda_m \neq \lambda_n$, their eigenfunctions are orthogonal. That is

$$\int_a^b \psi_{\lambda_m}(x)\psi_{\lambda_n}(x)r(x)dx = 0, \tag{3.26}$$

or more generally

$$\int_a^b \psi_{\lambda_m}(x)\psi_{\lambda_n}(x)r(x)dx = \delta_{mn}, \tag{3.27}$$

where $\delta_{mn} = 1$ is a delta function; that is, $\delta_{mn} = 1$ if $m = n$ and $\delta_{mn} = 0$ otherwise. It is possible to arrange the eigenvalues in an increasing order

$$\lambda_1 < \lambda_2 < .... < \lambda_n < ... \to \infty. \tag{3.28}$$

Some nonlinear equations can be transformed into a standard linear equation. For example, the Riccati equation can be written in general as

$$y' = p(x) + q(x)y + r(x)y^2, \qquad r(x) \neq 0. \tag{3.29}$$

If $r(x) = 0$, then it reduces to a first-order linear ODE. By using the transform

$$y(x) = -\frac{u'(x)}{r(x)u(x)}, \tag{3.30}$$

or

$$u(x) = e^{-\int r(x)y(x)dx}, \tag{3.31}$$

we have

$$u'' - P(x)u' + Q(x)u = 0, \tag{3.32}$$

where

$$P(x) = -\frac{r'(x) + r(x)q(x)}{r(x)}, \qquad Q(x) = r(x)p(x). \tag{3.33}$$

Chapter 4

Partial Differential Equations

Partial differential equations are much more complicated compared with ordinary differential equations. There is no universal solution technique for nonlinear equations; even their numerical solutions are usually not straightforward to obtain. Thus, we will mainly focus on some linear partial differential equations that are of special interest to computational sciences.

4.1 Partial Differential Equations

A partial differential equation (PDE) is a relationship containing one or more partial derivatives. Similar to the concept of ODEs, the highest n-th partial derivative is referred to as the order n of the partial differential equation. The general form of a partial differential equation can be written as

$$\psi(u, x, y, ..., \frac{\partial u}{\partial x}, \frac{\partial u}{\partial y}, \frac{\partial^2 u}{\partial x^2}, \frac{\partial^2 u}{\partial y^2}, \frac{\partial^2 u}{\partial x \partial y}, ...) = 0 \tag{4.1}$$

where u is the dependent variable and $x, y, ...$ are the independent variables. A simple example of partial differential equations is the linear first-order partial differential equation, which can be written as

$$a(x,y)\frac{\partial u}{\partial x} + b(x,y)\frac{\partial u}{\partial y} = f(x,y) \tag{4.2}$$

for two independent variables and one dependent variable u. If the right-hand side is zero or simply $f(x,y) = 0$, then the equation is said to be homogeneous. The equation is said to be linear if a, b and f are functions of x, y only, not u itself.

For simplicity in notations in the studies of partial differential equations, compact subscript forms are often used in the literature. They are

$$u_x \equiv \frac{\partial u}{\partial x}, \quad u_y \equiv \frac{\partial u}{\partial y}, \quad u_{xx} \equiv \frac{\partial^2 u}{\partial x^2},$$

$$u_{yy} \equiv \frac{\partial^2 u}{\partial y^2}, \quad u_{xy} \equiv \frac{\partial^2 u}{\partial x \partial y}, \quad \ldots \tag{4.3}$$

and thus we can write (4.2) as

$$au_x + bu_y = f. \tag{4.4}$$

In the rest of the chapters in this book, we will use this notation whenever no confusion occurs.

4.1.1 *First-Order Partial Differential Equation*

A first-order partial differential equation of linear type can be written as

$$a(x,y)u_x + b(x,y)u_y = f(x,y), \tag{4.5}$$

which can be solved using the method of characteristics in terms of a new parameter s

$$\frac{dx}{ds} = a, \quad \frac{dy}{ds} = b, \quad \frac{du}{ds} = f, \tag{4.6}$$

which essentially forms a system of first-order ordinary differential equations.

The simplest example of a first-order linear partial differential equation is the first-order hyperbolic equation

$$u_t + cu_x = 0, \tag{4.7}$$

where c is a constant. It has a general solution of

$$u = \psi(x - ct), \tag{4.8}$$

which is a travelling wave along x-axis with a constant speed c. If the initial shape is $u(x,0) = \psi(x)$, then $u(x,t) = \psi(x - ct)$ at time t; therefore the shape of the wave does not change, though its position is constantly changing.

4.1.2 *Classification of Second-Order Equations*

A linear second-order partial differential equation can be written in the generic form in terms of two independent variables x and y,

$$au_{xx} + bu_{xy} + cu_{yy} + gu_x + hu_y + ku = f, \tag{4.9}$$

where a, b, c, g, h, k and f are functions of x and y only. If $f(x, y, u)$ is also a function of u, then we say that this equation is quasi-linear.

If $\Delta = b^2 - 4ac < 0$, the equation is elliptic. One famous example is the Laplace equation $u_{xx} + u_{yy} = 0$.

If $\Delta > 0$, it is hyperbolic. One example is the wave equation $u_{tt} = v^2 u_{xx}$ where v is the wave speed.

If $\Delta = 0$, it is parabolic. The diffusion and heat conduction equations are of the parabolic type $u_t = \kappa u_{xx}$.

Three types of classic partial differential equations are widely used and they occur in a vast range of applications. In fact, almost all books or studies on partial differential equations deal with these three basic types.

4.2 Mathematical Models

In mathematical modelling, some PDEs occur more often than others. In this section, we outline some commonly used PDEs.

4.2.1 *Parabolic Equation*

Time-dependent problems, such as diffusion and transient heat conduction, are governed by parabolic equations. The heat conduction equation

$$u_t = ku_{xx}, \tag{4.10}$$

is a famous example. Written in the n-dimensional case $x_1 = x, x_2 = y, x_3 = z, ...,$ it can be extended to the general reaction-diffusion equation

$$u_t = k\nabla^2 u + f(u, x_1, .., x_n, t). \tag{4.11}$$

4.2.2 *Poisson's Equation*

In heat transfer problems, the steady state of heat conduction with a source is governed by the Poisson equation

$$k\nabla^2 u = f(x, y, t), \qquad (x, y) \in \Omega, \tag{4.12}$$

or

$$u_{xx} + u_{yy} = q(x, y, t), \tag{4.13}$$

for two independent variables x and y. Here k is thermal diffusivity and $q = f(x, y, t)/k$ is the heat source. If there is no heat source ($q = 0$), this becomes the Laplace equation. The solution, or a function, is said to be harmonic if it satisfies Laplace's equation.

In order to determine the temperature distribution (u) completely and uniquely, appropriate boundary conditions are needed. A simple boundary condition is to specify the temperature $u = u_0$ on the boundary $\partial\Omega$. This type of problem is the Dirichlet problem. On the other hand, if the temperature is not known, but the gradient $\partial u/\partial n$ is known on the boundary where n is the outward-pointing unit normal, then this forms the Neumann problem. Furthermore, some problems may have a mixed type of boundary condition in the combination of

$$\alpha u + \beta \frac{\partial u}{\partial n} = \gamma,$$

which may naturally occur as a radiation or cooling boundary condition.

4.2.3 *Wave Equation*

The vibrations of a string and travelling sound waves are governed by the hyperbolic wave equation. The 1-D wave equation in its simplest form is

$$u_{tt} = c^2 u_{xx}, \tag{4.14}$$

where c is the velocity of the wave. Using a transformation of the pair of independent variables

$$\xi = x + ct, \qquad \eta = x - ct, \tag{4.15}$$

for $t > 0$ and $-\infty < x < \infty$, the wave equation becomes

$$u_{\xi\eta} = 0. \tag{4.16}$$

Integrating twice and then substituting back in terms of x and t, we have

$$u(x, t) = f(x + ct) + g(x - ct), \tag{4.17}$$

where f and g are arbitrary functions of $x + ct$ and $x - ct$, respectively. We can see that there are two directions in which the wave can travel. One wave moves to the right and one travels to the left at a constant speed c.

Let us look a simple example of acoustic vibrations or standing waves.

Example 4.1: *As an example, let us look at the acoustics of a flute. The air pressure change $u(x,t)$ inside the pipe of a flute is governed by the wave equation*

$$\frac{\partial^2 u}{\partial t^2} = a^2 \frac{\partial^2 u}{\partial x^2},$$

where a is the speed of sound. The vibrations of the air form a series of standing waves with different frequencies, depending on the length of the flute and its boundary conditions. The boundary conditions can be either open or closed. If the end is open, we have $u(0,t) = 0$ or $u(L,t) = 0$ where L is the length of the flute. Here we have implicitly assumed that the pressure outside the pipe is constant, and thus its change is zero. If the end is closed, we have $u(0,t) = 0$ or $\frac{\partial u}{\partial x}(L,t) = 0$ which corresponds to the boundary conditions when we blow across the neck of an empty bottle. For simplicity, we will only discuss the case when both ends are open. This problem is exactly the same as the vibration of a string with both ends fixed.

We now use the separation of variables (see the next section for details) by letting

$$u(x,t) = \phi(x)T(t),$$

where $\phi(x)$ is a function of x only, while $T(t)$ is a function of time t. Substituting it into the wave equation, we have

$$\phi T'' = a^2 \phi'' T,$$

which can be written as

$$\frac{1}{a^2}\frac{T''}{T} = \frac{\phi''}{\phi}.$$

Since the left-hand side only depends on t while the right-hand side only depends on x, so both sides should be equal to the same constant, say, $-\lambda$. Here the negative sign is purely for convenience. That is

$$\frac{1}{a^2}\frac{T''}{T} = \frac{\phi''}{\phi} = -\lambda,$$

which is equivalent to equations

$$T'' = -a^2 \lambda T, \qquad \phi'' = -\lambda \phi.$$

The general solution for ϕ is

$$\phi(x) = A\sin(\sqrt{\lambda}x) + B\cos(\sqrt{\lambda}x),$$

where A and B are constants. As the boundary conditions on u lead to

$$\phi(0) = 0, \qquad \phi(L) = 0,$$

we now have

$$\phi(x) = \sin(\sqrt{\lambda_n} x) = \sin\frac{n\pi x}{L}, \qquad \lambda_n = \frac{n^2\pi^2}{L^2}.$$

where the eigenvalues are $\lambda_n (n = 1, 2, ...)$. Similarly, the equation for T becomes

$$T'' + \omega_n^2 T = 0, \qquad w_n = \frac{n\pi a}{L},$$

whose basic solution is

$$T(t) = C_n \cos(\omega_n t) + D_n \sin(\omega_n t),$$

where C_n and D_n are coefficients. Therefore, the basic frequencies of the harmonics are

$$f_n = \frac{\omega_n}{2\pi} = \frac{na}{2L}, \qquad (n = 1, 2, ...).$$

Here, $n = 1$ corresponds to the fundamental frequency, $n = 2$ corresponds to the second harmonics and so on and so forth. So changing the effective length L by cutting holes in the pipe can create differences in pitch. The distance between holes will affect the notes of the flute. By changing the force of the air flow, we effectively change n, also resulting in differences in pitch.

Finally, the general solution for u is given by superposing all the basic solutions, and we have

$$u(x, t) = \sum_{n=1}^{\infty} \sin\frac{n\pi a}{L}[C_n \cos\frac{n\pi at}{L} + D_n \sin\frac{n\pi at}{L}].$$

It is worth pointing out that for a pipe with one end open and the other other end closed, the frequencies are given by

$$f_n = \frac{na}{4L}, \qquad (n = 1, 3, 5, ...).$$

So the pitch can also be changed by closing one end.

4.3 Solution Techniques

Each type of PDEs usually requires a different solution technique. However, some methods work for most linear partial differential equations with appropriate boundary conditions on a regular domain. These methods include separation of variables, the method of series expansion, and transform methods such as the Laplace transform and the Fourier transform.

4.3.1 Separation of Variables

The separation of variables attempts a solution of the form

$$u = X(x)Y(y)T(t), \tag{4.18}$$

where $X(x), Y(y), T(t)$ are functions of x, y, t, respectively. By determining these functions that satisfy the partial differential equation and the required boundary conditions in terms of eigenvalue problems, the solution is then obtained.

As a classic example, we now try to solve the 1-D heat conduction equation in the domain $x \in [0, L]$ and $t > 0$

$$u_t = k u_{xx}, \tag{4.19}$$

with the initial value and boundary conditions

$$u(0,t) = 0, \qquad \frac{\partial u}{\partial x}\Big|_{x=L} = 0, \qquad u(x,0) = \psi. \tag{4.20}$$

Letting $u(x,t) = X(x)T(t)$, we have

$$\frac{X''(x)}{X} = \frac{T'(t)}{kT}. \tag{4.21}$$

As the left-hand side depends only on x and the right-hand side only depends on t, therefore, both sides must be equal to the same constant, and the constant can be taken as $-\lambda^2$. The negative sign is just for convenience because we will see below that the finiteness of the solution $T(t)$ requires that eigenvalues $\lambda^2 > 0$ or λ are real. Hence, we now get two ordinary differential equations

$$X''(x) + \lambda^2 X(x) = 0, \tag{4.22}$$

and

$$T'(t) + k\lambda^2 T(t) = 0, \tag{4.23}$$

where λ is the eigenvalue. The solution for $T(t)$ is

$$T = A_n e^{-\lambda^2 kt}. \tag{4.24}$$

The basic solution for $X(x)$ is simply

$$X(t) = \alpha \cos \lambda x + \beta \sin \lambda x. \tag{4.25}$$

So the fundamental solution for u is

$$u = (\alpha \cos \lambda x + \beta \sin \lambda x)e^{-\lambda^2 kt}, \tag{4.26}$$

where we have absorbed the coefficient A_n into α and β because they are the undetermined coefficients anyway. As the value of λ varies with the boundary conditions, in fact, it forms an eigenvalue problem. The general solution for u should be derived by superposing solutions of (4.26), and we have

$$u = \sum_{n=1}^{\infty} X_n T_n = \sum_{n=1}^{\infty} (\alpha_n \cos \lambda_n x + \beta_n \sin \lambda_n x) e^{-\lambda_n^2 kt}. \tag{4.27}$$

From the boundary condition $u(0,t) = 0$ at $x = 0$, we have

$$0 = \sum_{n=1}^{\infty} \alpha_n e^{-\lambda_n^2 kt}, \tag{4.28}$$

which leads to $\alpha_n = 0$ since $\exp(-\lambda^2 kt) > 0$.

From $\frac{\partial u}{\partial x}\Big|_{x=L} = 0$, we have

$$\lambda_n \cos \lambda_n L = 0, \tag{4.29}$$

which requires

$$\lambda_n L = \frac{(2n+1)\pi}{2}, \qquad (n = 0, 1, 2, ...). \tag{4.30}$$

Therefore, λ cannot be continuous, and it only takes an infinite number of discrete values, called eigenvalues. Each eigenvalue $\lambda = \lambda_n = \frac{(2n+1)\pi}{2L}$, $(n = 0, 1, 2, ...)$ has a corresponding eigenfunction $X_n = \sin(\lambda_n x)$. Substituting into the solution for $T(t)$, we have

$$T_n(t) = A_n e^{-\frac{[(2n+1)\pi]^2}{4L^2} kt}. \tag{4.31}$$

By expanding the initial condition into a Fourier series so as to determine the coefficients, we have

$$u(x,t) = \sum_{n=0}^{\infty} \beta_n \sin(\frac{(2n+1)\pi x}{2L}) e^{-[\frac{(2n+1)\pi}{2L}]^2 kt},$$

$$\beta_n = \frac{2}{L} \int_0^L \psi(x) \sin[\frac{(2n+1)\pi\xi}{2L}] d\xi. \tag{4.32}$$

In the special case when the initial condition $u(x, t = 0) = \psi = u_0$ is constant, the requirement for $u = u_0$ at $t = 0$ becomes

$$u_0 = \sum_{n=0}^{\infty} \beta_n \sin \frac{(2n+1)\pi x}{2L}. \tag{4.33}$$

Using the orthogonal relationship

$$\int_0^L \sin\frac{m\pi x}{L}\sin\frac{n\pi x}{L}dx = 0, \qquad m \neq n, \tag{4.34}$$

and

$$\int_0^L (\sin\frac{n\pi x}{L})^2 dx = \frac{L}{2}, \qquad (n = 1, 2, ...), \tag{4.35}$$

and multiplying both sides of equation (4.33) by $\sin\frac{(2n+1)\pi x}{2L}$, we have the integral

$$\beta_n\frac{L}{2} = \int_0^L \sin\frac{(2n+1)\pi x}{2L}u_0 dx = \frac{2L}{(2n+1)\pi}\ (n = 0, 1, 2, ...), \tag{4.36}$$

which leads to

$$\beta_n = \frac{4u_0}{(2n+1)\pi}, \qquad n = 0, 1, 2, ... \tag{4.37}$$

Therefore, the solution becomes

$$u = \frac{4u_0}{\pi}\sum_{n=0}^{\infty}\frac{1}{(2n+1)}e^{-\frac{(2n+1)^2\pi^2 kt}{4L^2}}\sin\frac{(2n+1)\pi x}{2L}. \tag{4.38}$$

This solution is essentially the same as the classic heat conduction problem discussed by Carslaw and Jaeger in 1959. This same solution can also be obtained using the Fourier series of u_0 in $0 < x < L$.

4.3.2 *Laplace Transform*

The integral transform can reduce the number of the independent variables. For the 1-D time-dependent case, it transforms a partial differential equation into an ordinary differential equation by solving the ordinary differential equation and inverting back to obtain the solution for the original partial differential equation. As an example, we now solve the heat conduction in semi-infinite interval $[0, \infty)$,

$$u_t = ku_{xx}, \qquad u(x, 0) = 0, \ \ u(0, t) = T_0. \tag{4.39}$$

Let $\bar{u}(x, s) = \int_0^\infty u(x, t)e^{-st}dt$ be the Laplace transform of $u(x, t)$, the equation then becomes

$$s\bar{u} = k\frac{d^2\bar{u}}{dx^2}, \qquad \bar{u}_{x=0} = \frac{T_0}{s},$$

which is an ordinary differential equation whose general solution can be written as

$$\bar{u} = Ae^{-\sqrt{\frac{s}{k}}x} + Be^{\sqrt{\frac{s}{k}}x}.$$

The finiteness of the solution as $x \to \infty$ requires that $B = 0$, and the boundary conditions lead to

$$\bar{u} = \frac{T_0}{s} e^{-\sqrt{\frac{s}{k}} x}.$$

By the inversion of the Laplace transform, we have

$$u = T_0 \operatorname{erfc}(\frac{x}{2\sqrt{kt}}),$$

where $\operatorname{erfc}(x)$ is the complementary error function which is given by

$$\operatorname{erfc}(x) = 1 - \operatorname{erf}(x) = \frac{2}{\sqrt{\pi}} \int_x^{\infty} e^{-\eta^2} d\eta. \tag{4.40}$$

Here the error function $\operatorname{erf}(x)$ is defined by

$$\operatorname{erf}(x) = \frac{2}{\sqrt{\pi}} \int_0^x e^{-\eta^2} d\eta. \tag{4.41}$$

4.3.3 *Similarity Solution*

Sometimes, the diffusion equation

$$u_t = \kappa u_{xx}, \tag{4.42}$$

can be solved by using the so-called similarity method by defining a similar variable

$$\eta = \frac{x}{\sqrt{\kappa t}} \tag{4.43}$$

or

$$\zeta = \frac{x^2}{\kappa t}. \tag{4.44}$$

We can assume that the solution to the equation has the form

$$u = (\kappa t)^{\alpha} f[\frac{x^2}{(\kappa t)^{\beta}}]. \tag{4.45}$$

By substituting it into the diffusion equation, the coefficients α and β can be determined. For most applications, we can assume $\alpha = 0$ so that $u = f(\zeta)$. In this case, we have

$$4\zeta u'' + 2u' + \zeta\beta(\kappa t)^{\beta-1} u' = 0, \tag{4.46}$$

where $f' = df/d\zeta$. In deriving this equation, we have used the chain rules of differentiations $\frac{\partial}{\partial x} = \frac{\partial}{\partial \zeta}\frac{\partial \zeta}{\partial x}$ and $\frac{\partial}{\partial t} = \frac{\partial}{\partial \zeta}\frac{\partial \zeta}{\partial t}$.

Since the original equation does not have time-dependent terms *explicitly*, this means that all the exponents for any t-terms must be zero. Therefore, we have

$$\beta = 1. \tag{4.47}$$

Now the diffusion equation becomes

$$\zeta f''(\zeta) = -(\frac{1}{2} + \frac{\zeta}{4})f'. \tag{4.48}$$

Using $(\ln f')' = f''/f'$ and integrating the above equation once, we get

$$f' = \frac{K e^{-\zeta/4}}{\sqrt{\zeta}}. \tag{4.49}$$

Integrating it again and using $\zeta = 4\xi^2$, we obtain

$$u = A \int_{\xi_0}^{\xi} e^{-\xi^2} d\xi = C\mathrm{erf}(\frac{x}{\sqrt{4\kappa t}}) + D, \tag{4.50}$$

where C and D are constants that can be determined from appropriate boundary conditions. For the same problem as (4.39), the boundary condition as $x \to \infty$ implies that $C + D = 0$, while $u(0, t) = T_0$ means that $D = -C = T_0$. Therefore, we finally have

$$u = T_0[1 - \mathrm{erf}(\frac{x}{\sqrt{4\kappa t}})] = T_0\mathrm{erfc}(\frac{x}{\sqrt{4\kappa t}}). \tag{4.51}$$

There are other important methods for solving partial differential equations. These include series expansion methods, asymptotic methods, approximate methods, Green's function, conformal mapping, Fourier transform, Z-transform, hybrid method, perturbation methods and naturally numerical methods.

We have shown examples of second-order PDEs. There are other equations such the reaction-diffusion equation and Navier-Stokes equations that occur frequently in mathematical physics, engineering and computational sciences. Readers can find further details in more advanced literature.

Part II

Numerical Algorithms

Chapter 5

Roots of Nonlinear Equations

Many problems such as finding an optimal solution to a particular problem are often related to finding the critical points and extreme points. In order to find these critical points, we have to solve the stationary conditions when the first derivatives are zero. On the other hand, we often have to solve a nonlinear equation to find its roots. Therefore, root-finding algorithms are important. Close-form solutions are rare, and in most cases, only approximate solutions are possible. In this chapter, we will introduce the fundamentals of root-finding algorithms.

5.1 Bisection Method

The bisection method is a classic method of finding roots of a nonlinear function $f(x)$ in the interval $[a, b]$. It works in the following way as shown in Fig. 5.1.

The iteration procedure starts with two initial bounds x_a (lower bound), and x_b (upper bound) so that the true root $x = x_*$ lies between these two bounds. This requires that $f(x_a)$ and $f(x_b)$ have different signs. As shown in Fig. 5.1, $f(x_a) > 0$ and $f(x_b) < 0$, but $f(x_a)f(x_b) < 0$. The obvious choice is $x_a = a$ and $x_b = b$. The next estimate is just the midpoint of A and B, and we have

$$x_n = \frac{1}{2}(x_a + x_b). \tag{5.1}$$

We then have to test the sign of $f(x_n)$. If $f(x_n) < 0$ (having the same sign as $f(x_b)$), we then update the new upper bound as $x_b = x_n$. If $f(x_n) > 0$ (having the same sign as $f(x_a)$), we update the new lower bound as $x_a = x_n$. In a special case when $f(x_n) = 0$, we have found the true root. The

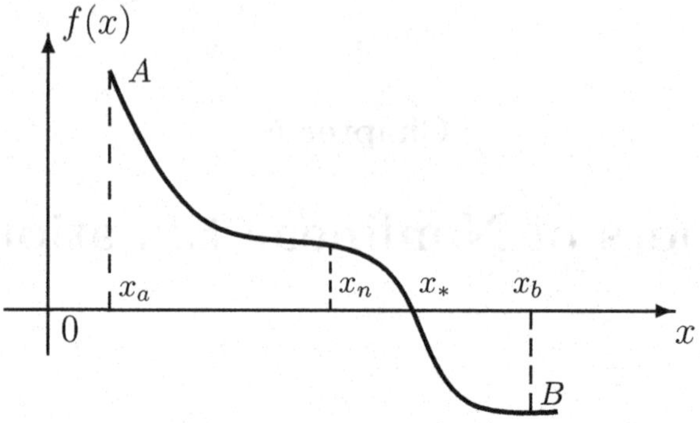

Fig. 5.1 Bisection method for finding the root x_* of $f(x_*) = 0$ between two bounds x_a and x_b in the domain $x \in [a, b]$.

iterations continue in the same manner until a given accuracy is achieved or the prescribed number of iterations is reached.

Example 5.1: *If we want to find the root of*

$$f(x) = x - \cos x = 0.$$

We can use $x_a = 0$ *and* $x_b = 1$ *because* $\cos x \leq 1$. *The first bisection point is*

$$x_1 = \frac{1}{2}(x_a + x_b) = \frac{1}{2}(0 + 1) = 0.5.$$

Since $f(x_a) < 0$, $f(x_b) > 0$ *and* $f(x_1) = -0.377582 < 0$, *so we update the new lower bound* $x_a = x_1 = 0.5$. *Now, the second bisection point is*

$$x_2 = \frac{1}{2}(0.5 + 1) = 0.75,$$

which gives $f(x_2) = 0.018311 > 0$, *so we now update the upper bound* $x_b = x_2 = 0.75$. *The third bisection point is*

$$x_3 = \frac{1}{2}(0.5 + 0.75) = 0.625.$$

Since $f(x_3) = -0.1859 < 0$, *so we now update the lower bound* $x_a = 0.625$. *The fourth bisection point is*

$$x_4 = \frac{1}{2}(0.625 + 0.75) = 0.6875.$$

Again $f(x_4) = -0.08533 < 0$, *we update the lower bound again* $x_a = 0.6875$. *Finally, the fifth bisection point is*

$$x_5 = \frac{1}{2}(0.6875 + 0.75) = 0.71875,$$

which is within about 3% *from the true value*

$$x_* \approx 0.739085133215.$$

If we continue in the same manner, we have

$$x_6 \approx 0.734375, \quad x_7 \approx 0.7421875.$$

The 7th iteration gives a value of 0.7421875 *which is within about* 0.5% *of the true root.*

In general, the convergence of the bisection method is very slow, other iteration methods such as Newton's method are much more efficient in most cases.

5.2 Simple Iterations

The essence of root-finding algorithms is to use an iterative procedure to obtain the approximate (though sometimes quite accurate) solutions, starting from some initial guess solution. For example, even ancient Babylonians knew how to find the square root of 2 using an iterative method. From the simple example we introduced in Section 1 of Chapter 1, we know that we can numerically compute the square root of any real number k (so that $x = \sqrt{k}$) using the equation

$$x_{n+1} = \frac{1}{2}(x_n + \frac{k}{x_n}). \tag{5.2}$$

If we start from an initial value, say, $x_0 = 1$ at $n = 0$, we can carry out the iterations to meet the accuracy we want.

Example 5.2: *In order to find* $\sqrt{100}$, *we have* $k = 100$ *with an initial guess* $x_0 = 1$. *The first five iterations are as follows:*

$$x_1 = \frac{1}{2}(x_0 + \frac{100}{x_0}) = 50.5, \quad x_2 = \frac{1}{2}(x_1 + \frac{100}{x_1}) \approx 26.2400990099,$$

$$x_3 \approx 15.025530119, \quad x_4 \approx 10.840434673,$$

$$x_5 \approx 10.032578511, \quad x_6 \approx 10.0000528956,$$

$$x_7 \approx 10.000000001.$$

We can see that x_5 *after 5 iterations is very close (*$\sim 0.3\%$*) to its true value* $\sqrt{100} = 10$, *which shows that the iteration method is very efficient.*

The above iteration procedure is equivalent to find the root of

$$x^2 - k = 0. \tag{5.3}$$

As pointed out in Chapter 1, the reason that this iterative process works is that the series $x_1, x_2, ..., x_n$ converges towards the true value $\sqrt{k}$ due to the fact that

$$\frac{x_{n+1}}{x_n} = \frac{1}{2}(1 + k/x_n^2) \to 1, \qquad \text{as } x_n \to \sqrt{k}.$$

The value x_{n+1} is always between these two bounds x_n and k/x_n, and the new estimate x_{n+1} is thus the mean or average of the two bounds. This guarantees that the series converges towards the true value of $\sqrt{k}$. This method is similar to the bisection method discussed in the previous section.

However, a good choice of the initial value x_0 will speed up the convergence. A wrong choice of x_0 could make the iteration fail; for example, we cannot use $x_0 = 0$ as the initial guess, and we cannot use $x_0 < 0$ either as $\sqrt{k} > 0$ (in this case, the iterations will approach another root $-\sqrt{k}$). This disadvantage means that a good initial starting point is very important. In fact, many local algorithms including Newton's method have such disadvantage, but they can usually converge very quickly in practice.

You may wonder if we can use the above iteration procedure to find the higher-order roots such as $\sqrt[9]{k}$ or $\sqrt[100]{k}$? Obviously, this depends on the clever construction of an iterative formula, and thus it leaves as an exercise to find the root of $\sqrt[m]{k}$ where $m \geq 2$ and $k > 0$.

5.3 Newton's Method

Newton's method is a widely used classic method for finding the zeros of a nonlinear univariate function of $f(x)$ on the interval $[a, b]$. It is also referred to as the Newton-Raphson method. At any given point x_n shown in Fig. 5.2, we can approximate the function by a Taylor series for $\Delta x = x_{n+1} - x_n$ about x_n,

$$f(x_{n+1}) = f(x_n + \Delta x) \approx f(x_n) + f'(x_n)\Delta x, \tag{5.4}$$

which leads to

$$x_{n+1} - x_n = \Delta x \approx \frac{f(x_{n+1}) - f(x_n)}{f'(x_n)}, \tag{5.5}$$

or

$$x_{n+1} \approx x_n + \frac{f(x_{n+1}) - f(x_n)}{f'(x_n)}. \tag{5.6}$$

Since we try to find an approximation to $f(x) = 0$ with $f(x_{n+1})$, we can use the approximation $f(x_{n+1}) \approx 0$ in the above expression. Thus we have the standard Newton iterative formula

$$x_{n+1} = x_n - \frac{f(x_n)}{f'(x_n)}. \tag{5.7}$$

The iteration procedure starts from an initial guess x_0 and continues until a certain criterion is met. A good initial guess will use fewer steps, however, if there is no obvious initial good starting point, you can start at any point on the interval $[a, b]$. But if the initial value is too far from the true zero, the iteration process may fail. So it is a good idea to limit the number of iterations.

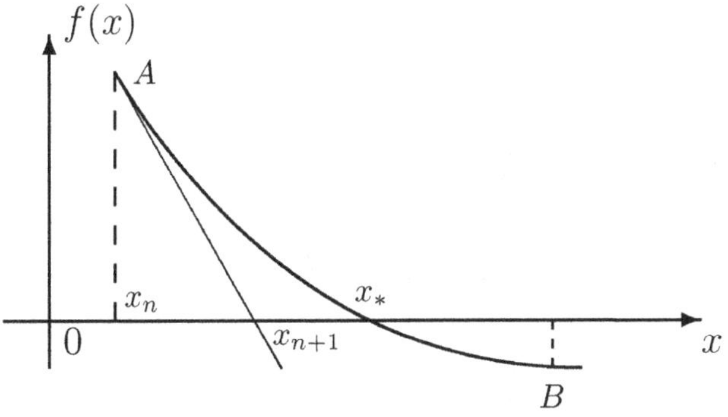

Fig. 5.2 Newton's method of approximating the root x_* by x_{n+1} from the previous value x_n.

Example 5.3: *To find the root of*

$$f(x) = x^2 - e^{-x^2} = 0,$$

we use Newton's method starting from $x_0 = 1$. We know that

$$f'(x) = 2x(1 + e^{-x^2}),$$

and thus the iteration formula becomes

$$x_{n+1} = x_n - \frac{x_n^2 - e^{-x_n^2}}{2x_n(1 + e^{-x_n^2})}.$$

Using $x_0 = 1$, we have

$$x_1 = 1 - \frac{1^2 - e^{-1^2}}{2 \times 1 \times (1 + e^{-1^2})} \approx 0.768941421.$$

Similarly, we have

$$x_2 \approx 0.7531845777, \qquad x_3 \approx 0.753089168.$$

We can see that x_3, after only three iterations, is very close to the true root $x_* \approx 0.7530891649$.

We have seen that Newton's method is very efficient and is thus so widely used. This method can be modified for solving unconstrained optimisation problems because it is equivalent to finding the root of the first derivative $f'(\boldsymbol{x}) = 0$ once the objective function $f(\boldsymbol{x})$ is given. The detail is given in Part IV.

5.4 Iteration Methods

In many applications, we may have to find the roots of a function of multiple variables, and Newton's method can be extended to serve this purpose. For nonlinear multivariate functions

$$\boldsymbol{F}(\boldsymbol{x}) = [F_1(\boldsymbol{x}), F_2(\boldsymbol{x}), ..., F_N(\boldsymbol{x})]^T, \tag{5.8}$$

where $\boldsymbol{x} = (x, y, ..., z)^T = (x_1, x_2, ..., x_p)^T$, an iteration method is usually needed to find the roots of

$$\boldsymbol{F}(\boldsymbol{x}) = 0. \tag{5.9}$$

In general, the Newton-Raphson iteration procedure is widely used. We first approximate $\boldsymbol{F}(\boldsymbol{x})$ by a linear residual function $\boldsymbol{R}(\boldsymbol{x}; \boldsymbol{x}^n)$ in the neighbourhood of an existing approximation $\boldsymbol{x}^n$ to $\boldsymbol{x}$, and we have

$$\boldsymbol{R}(\boldsymbol{x}, \boldsymbol{x}^n) = \boldsymbol{F}(\boldsymbol{x}^n) + \boldsymbol{J}(\boldsymbol{x}^n)(\boldsymbol{x} - \boldsymbol{x}^n), \tag{5.10}$$

and

$$\boldsymbol{J}(\boldsymbol{x}) = \nabla \boldsymbol{F}, \tag{5.11}$$

where $\boldsymbol{J}$ is the Jacobian of $\boldsymbol{F}$. That is

$$\boldsymbol{J}_{ij} = \frac{\partial F_i}{\partial x_j}. \tag{5.12}$$

Here we have used the notation $\boldsymbol{x}^n$ for the vector $\boldsymbol{x}$ at the n-th iteration, which should not be confused with the power $\boldsymbol{u}^n$ of a vector $\boldsymbol{u}$. This might be confusing, but such notations are widely used in the literature of numerical analysis. An alternative (and better) notation is to denote $\boldsymbol{x}^n$ as $\boldsymbol{x}^{(n)}$, which shows the vector value at n-th iteration using a bracket. However, we will use both notations if no confusion arises.

To find the next approximation $\boldsymbol{x}^{n+1}$ from the current estimate $\boldsymbol{x}^n$, we have to try to satisfy $\boldsymbol{R}(\boldsymbol{x}^{n+1}, \boldsymbol{u}^n) = 0$, which is equivalent to solving a linear system with $\boldsymbol{J}$ being the coefficient matrix

$$\boldsymbol{x}^{n+1} = \boldsymbol{x}^n - \boldsymbol{J}^{-1}\boldsymbol{F}(\boldsymbol{x}^n), \tag{5.13}$$

under a given termination criterion

$$\|\boldsymbol{x}^{n+1} - \boldsymbol{x}^n\| \leq \epsilon,$$

where $\epsilon > 0$ is a small tolerance. Iterations require an initial starting vector $\boldsymbol{x}^0$, which is often set to $\boldsymbol{x}^0 = 0$.

Example 5.4: *To find the roots of the system*

$$x - e^{-y} = 0, \qquad x^2 - y - z = 0, \qquad z - 2y^2 = 0,$$

we first write it as

$$\boldsymbol{F}(\boldsymbol{x}) = \begin{pmatrix} F_1 \\ F_2 \\ F_3 \end{pmatrix} = \begin{pmatrix} x_1 - e^{-x_2} \\ x_1^2 - x_2 - x_3 \\ x_3 - 2x_2^2 \end{pmatrix}, \qquad \boldsymbol{x} = \begin{pmatrix} x_1 \\ x_2 \\ x_3 \end{pmatrix} = \begin{pmatrix} x \\ y \\ z \end{pmatrix}.$$

The Newton-Raphson iteration formula becomes

$$\boldsymbol{x}^{n+1} = \boldsymbol{x}^n - \boldsymbol{J}^{-1}\boldsymbol{F}(\boldsymbol{x}^n),$$

where the Jacobian $\boldsymbol{J}$ *is*

$$\boldsymbol{J} = \begin{pmatrix} \frac{\partial F_1}{\partial x_1} & \frac{\partial F_1}{\partial x_2} & \frac{\partial F_1}{\partial x_3} \\ \\ \frac{\partial F_2}{\partial x_1} & \frac{\partial F_2}{\partial x_2} & \frac{\partial F_2}{\partial x_3} \\ \\ \frac{\partial F_3}{\partial x_1} & \frac{\partial F_3}{\partial x_2} & \frac{\partial F_3}{\partial x_3} \end{pmatrix} = \begin{pmatrix} 1 & e^{-x_2} & 0 \\ 2x_1 & -1 & -1 \\ 0 & -4x_2 & 1 \end{pmatrix},$$

whose inverse is

$$\boldsymbol{A} = \boldsymbol{J}^{-1} = \frac{1}{-1 - 4x_2 - 2x_1 e^{-x_2}} \begin{pmatrix} -1 - 4x_2 & -e^{-x_2} & -e^{-x_2} \\ -2x_1 & 1 & 1 \\ -8x_1x_2 & 4x_2 & -(1 + 2x_1 e^{-x_2}) \end{pmatrix}$$

$$= \frac{1}{1 + 4x_2 + 2x_1 e^{-x_2}} \begin{pmatrix} 1 + 4x_2 & e^{-x_2} & e^{-x_2} \\ 2x_1 & -1 & -1 \\ 8x_1x_2 & -4x_2 & 1 + 2x_1 e^{-x_2} \end{pmatrix}.$$

Therefore, the iteration equation becomes

$$\boldsymbol{x}^{n+1} = \boldsymbol{x}^n - \boldsymbol{u}^n$$

where

$$\boldsymbol{u}^n = \boldsymbol{J}^{-1}\boldsymbol{F}(\boldsymbol{x}^n)$$

$$= \frac{1}{\Delta}\begin{pmatrix} 1 & e^{-x_2} & e^{-x_2} \\ 2x_1 & -1 & -1 \\ 8x_1x_2 & -4x_2 & (1+2x_1e^{-x_2}) \end{pmatrix}\begin{pmatrix} x_1 - e^{-x_2} \\ x_1^2 - x_2 - x_3 \\ x_3 - 2x_2^2 \end{pmatrix}$$

$$= \frac{1}{\Delta}\begin{pmatrix} (1+4x_2)(x_1 - e^{-x_2}) + e^{-x_2}(x_1^2 - x_2 - x_3) + e^{-x_2}(x_3 - 2x_2^2) \\ 2x_1(x_1 - e^{-x_2}) - x_1^2 + x_2 + 2x_2^2 \\ 8x_1x_2(x_1 - e^{-x_2}) - 4x_2(x_1^2 - x_2 - x_3) + (1+2x_1e^{-x_2})(x_3 - 2x_2^2) \end{pmatrix},$$

where $\Delta = 1 + 4x_2 + 2x_1e^{-x_2}$. *If we start with the initial guess* $\boldsymbol{x}^0 = (0,0,0)^T$, *we have the first estimate* $\boldsymbol{x}^1$

$$\boldsymbol{x}^1 = \begin{pmatrix} 0 \\ 0 \\ 0 \end{pmatrix} - \begin{pmatrix} -1 \\ 0 \\ 0 \end{pmatrix} = \begin{pmatrix} 1 \\ 0 \\ 0 \end{pmatrix},$$

and the second iteration gives

$$\boldsymbol{x}^2 = \begin{pmatrix} 1 \\ 0 \\ 0 \end{pmatrix} - \begin{pmatrix} 0.3333333333 \\ -0.333333333 \\ 0 \end{pmatrix} = \begin{pmatrix} 0.6666666667 \\ 0.333333333 \\ 0 \end{pmatrix}.$$

If we continue this way, the third iteration gives

$$\boldsymbol{x}^3 = \boldsymbol{x}^2 - \begin{pmatrix} -0.0595873520 \\ 0.0135691322 \\ -0.2041300459 \end{pmatrix} = \begin{pmatrix} 0.7262540187 \\ 0.3197642011 \\ 0.2041300459 \end{pmatrix}.$$

Finally, the fourth iteration gives

$$\boldsymbol{x}^4 = \boldsymbol{x}^3 - \begin{pmatrix} 0.0006479911 \\ -0.0009833882 \\ -0.0016260521 \end{pmatrix} = \begin{pmatrix} 0.7256060276 \\ 0.3207475893 \\ 0.2057560980 \end{pmatrix}.$$

The true roots occur at $(0.7256065972, 0.3207472882, 0.2057576457)$, *and we can see that even after only four iterations, the estimates are very close to the true values. In fact, we can continue to the iterations, the fifth iteration gives*

$$\boldsymbol{x}^5 \approx (0.72560659719, 0.32074728816, 0.20575764573)^T,$$

which is accurate to the 9th decimal place.

In this example, we have used the inverse of the Jacobian. However, if the size of the Jacobian is very large, the computation of $\boldsymbol{J}^{-1}$ can be very expensive. In this case, some iterative procedure to replace the inverse should be used. We will come back to this point later in this book.

5.5 Numerical Oscillations and Chaos

Any iteration formulas must be designed carefully; otherwise, they may oscillate or even not converge at all. Let us look at an example.

Example 5.5: *To find the root of*

$$f(x) = x - e^{-x^2} = 0,$$

we now try to use the following iteration

$$x_{n+1} = e^{-x_n^2},$$

starting at $x_0 = 1$. *We have*

$$x_1 = e^{-x_0^2} \approx 0.36788, \quad x_2 \approx 0.8734, \quad x_3 \approx 0.46633,$$

$$x_4 = 0.80456, \quad x_5 \approx 0.52345, ...$$

The results oscillate but the variations or differences are getting smaller. If we continue the iterations, we have

$$x_{20} \approx 0.66417, \quad x_{30} \approx 0.65520, \quad x_{40} \approx 0.65338,$$

$$x_{50} \approx 0.65301, \quad x_{70} \approx 0.652922, \quad x_{100} \approx 0.652918.$$

The true root of $x = \exp(-x^2)$ *is* 0.6529186404192.

We can see that this process converges very slowly. The stability requirement $\|f(x)\| = e^{-x^2} \leq 1$ is satisfied, however, this does not guarantee its efficiency. A similar statement is true for many nonlinear iteration formulas. As another example, let us now try to solve

$$x - \cos^2(x) = 0,$$

we have

$$x_{n+1} = \cos^2(x_n), \qquad x_0 = 1.$$

Using this formula, we have

$$x_1 \approx 0.29193, \quad x_2 = 0.91717, \quad x_3 = 0.36975,$$

$$x_4 \approx 0.86941, \quad x_5 \approx 0.41639, ...,$$

$$x_{10} \approx 0.79175, \quad x_{20} \approx 0.73069, \quad x_{50} \approx 0.66567,$$

$$x_{100} \approx 0.64466, \quad x_{145} \approx 0.64127, \quad x_{150} \approx 0.642077,$$

which are still oscillating around the true value

$$x_* = 0.641714370872.$$

Such oscillations are quite universal in many nonlinear iterations. In some cases, chaos may appear even if the formula is seemingly simple. For example, the well-known iteration formula

$$x_{n+1} = \lambda x_n(1 - x_n), \quad \lambda > 0, \quad x \in [0, 1], \tag{5.14}$$

may show chaotic behaviour when λ increases from 1 to 4 for any initial value, say, $x^{(0)} = 0.1$ or 0.9, excluding $x^{(0)} = 0$ or 1.

For a given value, say $\lambda = 2$, we can use a computer or a pocket calculator to do these calculations. If the initial value $y_1 = 0$ or $y_1 = 1$, then the system seems to be trapped in the state $y_n = 0$ ($n = 2, 3, ...$). However, if we use a slight difference value (say) $y_1 = 0.01$, then we have $y_1 = 0.01$, $y_2 = 0.0198$, ..., $y_{10} = 0.5000$. Then, the values are attracted to a single value or state $y_\infty = 0.5000$.

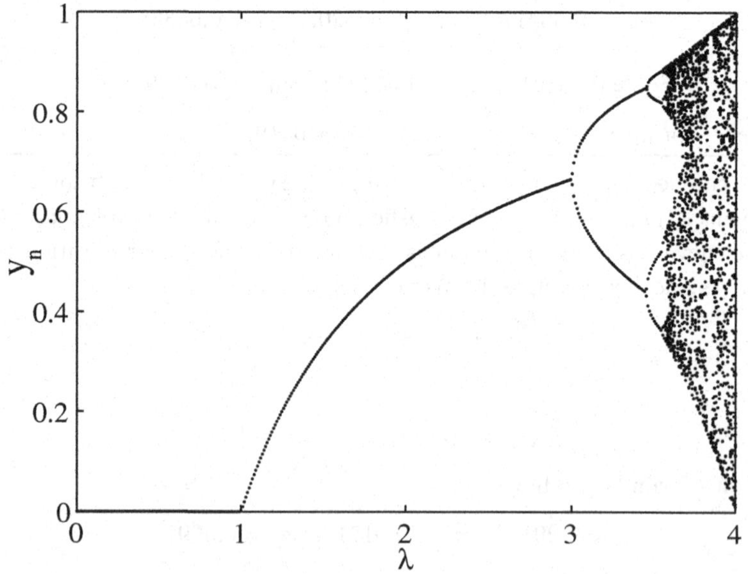

Fig. 5.3 Bifurcation diagram and chaos.

Following exactly the same process but using different values of λ ranging from $\lambda = 0.001$ to $\lambda = 4.000$, we can plot out the number of states (after

$N = 500$ iterations) and then we get a bifurcation map as shown in Figure 5.3. It gives a detailed map of how the system behaves. For $1 < \lambda < 3$, the system settles (or attracts) to a single state. For $3 < \lambda < 3.45$, the system bifurcates into two states. It seems that the system is attracted by these discrete states. For this reason, the map is also called the attractors of the dynamical system. The system becomes chaotic for $\lambda \geq \lambda_*$ where $\lambda_* \approx 3.57$ is the critical value.

It is a bit surprising for a seemingly simple and deterministic system

$$\lambda y_n(1 - y_n) \to y_{n+1}$$

because you may try many times to simulate the same system using the same initial value (say) $y_0 = 0.01$ and parameter λ. Then, we should get the same set of values $(y_1, y_2, ...)$. So where is the chaos, anyway? The problem is that this system is very sensitive to the small variation in the initial value y_0. If there is any tiny difference, say, $y_0 = 0.01 \pm 0.000000001$ or even 10^{-1000} difference, then the set of values you get will be completely different, resulting in the so-called 'butterfly effect'. Since there is always uncertainty in the real world, even computer simulations can only use a finite number of digits, so such chaos is intrinsic for some nonlinear dynamical systems.

Therefore, a nonlinear iteration procedure should carefully be designed in numerical algorithms. The widely used Newton-Raphson iteration is a very good example how a robust formula can improve the convergence so significantly.

Chapter 6

Numerical Integration

An interesting feature of differentiations and integrations is that you can get the explicit expressions of derivatives of most functions and complicated expressions if they exist, while it is very difficult and sometimes impossible to express an integral in an explicit form, even for seemingly simple integrands. For example, the error function, widely used in engineering and sciences, is defined by

$$\mathrm{erf}(x) = \frac{2}{\sqrt{\pi}} \int_0^x e^{-t^2} dt. \tag{6.1}$$

The integration of this simple integrand $\exp(-t^2)$ does not lead to any simple explicit expression, which is why it is often written as erf(), referred to as the error function. If we pick up a mathematical handbook, we can find that

$$\mathrm{erf}(0) = 0, \qquad \mathrm{erf}(\infty) = 1, \tag{6.2}$$

while

$$\mathrm{erf}(0.5) \approx 0.52049, \qquad \mathrm{erf}(1) \approx 0.84270. \tag{6.3}$$

If we want to calculate such integrals, numerical integration is the best alternative.

The beauty of a closed-form expression or a simple formula is that it will give tremendous insight into the problem. However, such closed-form formulas are rarely possible, especially for real-world processes. In most cases, only approximate solutions are possible. Such approximations can be obtained by using simplified models or solving the mathematical models using approximation techniques. Numerical integration is essential to many of these approximations.

There are many techniques for numerical integration, and we will introduce the Trapezium rule, Simpson's rule and Gaussian quadrature in the rest of this chapter.

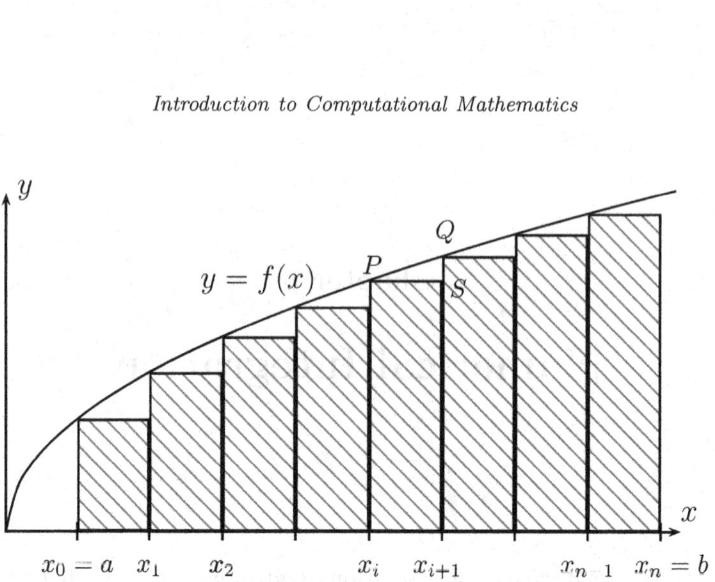

Fig. 6.1 Numerical integration: n thin strips to approximate the integral of $f(x)$.

6.1 Trapezium Rule

Now we want to numerically evaluate the following integral

$$\mathcal{I} = \int_a^b f(x)dx, \tag{6.4}$$

where a and b are fixed and finite. We know that the value of the integral is exactly the total area under the curve $y = f(x)$ between a and b. As both the integral and the area can be considered as the sum of the values over many small intervals, the simplest way of evaluating such numerical integration is to divide up the integral interval into n equal small sections and split the area into n thin strips so that $h \equiv \Delta x = (b - a)/n$, $x_0 = a$ and $x_i = ih + a (i = 1, 2, ..., n)$. The values of the functions at the dividing points x_i are denoted as $y_i = f(x_i)$, and the value at the midpoint between x_i and x_{i+1} is conventionally labeled as

$$y_{i+1/2} = f(x_{i+1/2}) = f_{i+1/2}, \qquad x_{i+1/2} = \frac{x_i + x_{i+1}}{2}. \tag{6.5}$$

The accuracy of such approximations depends on the number n and the way to approximate the curve in each interval. Figure 6.1 shows such an interval $[x_i, x_{i+1}]$ which is exaggerated in the figure for clarity. The curve segment between P and Q is approximated by a straight line with a slope

$$\frac{\Delta y}{\Delta x} = \frac{f(x_{i+1}) - f(x_i)}{h}, \tag{6.6}$$

which approaches $f'(x_{i+1/2})$ at the midpoint point when $h \to 0$.

The trapezium (formed by P, Q, x_{i+1}, and x_i) is a better approximation than the rectangle (P, S, x_{i+1} and x_i) because the former has an area

$$A_i = \frac{f(x_i) + f(x_{i+1})}{2} h, \tag{6.7}$$

which is close to the area

$$\mathcal{I}_i = \int_{x_i}^{x_{i+1}} f(x) dx, \tag{6.8}$$

under the curve in the small interval x_i and x_{i+1}. If we use the area A_i to approximate $\mathcal{I}_i$, we have the trapezium rule of numerical integration. Thus, the integral is simply the sum of all these small trapeziums, and we have

$$\mathcal{I} \approx \frac{h}{2}[f_0 + 2(f_1 + f_2 + ... + f_{n-1}) + f_n]$$

$$= h[f_1 + f_2 + ... + f_{n-1} + \frac{(f_0 + f_n)}{2}]. \tag{6.9}$$

From the Taylor series, we know that

$$\frac{f(x_i) + f(x_{i+1})}{2} \approx \frac{1}{2}\Big\{[f(x_{i+1/2}) - \frac{h}{2} f'(x_{i+1/2}) + \frac{1}{2!}(\frac{h}{2})^2 f''(x_{i+1/2})]$$

$$+ [f(x_{i+1/2}) + \frac{h}{2} f'(x_{i+1/2}) + \frac{1}{2!}(\frac{h}{2})^2] f''(x_{i+1/2})\Big\}$$

$$= f(x_{i+1/2}) + \frac{h^2}{8} f''(x_{i+1/2}) \tag{6.10}$$

where $O(h^2 f'')$ means that the value is the order of $h^2 f''$, or $O(h^2) = K h^2 f''$ where K is a constant. Therefore, the error of the estimate of $\mathcal{I}$ is $h \times O(h^2 f'') = O(h^3 f'')$.

6.2 Simpson's Rule

The trapezium rule is a relatively simple scheme for numerical integration with the error of $O(h^3 f'')$. If we want higher accuracy, we can either reduce h or use a better approximation for $f(x)$. A small h means a large n, which implies that we have to do the sum of many small sections, and it may increase the computational time.

On the other hand, we can use higher-order approximations for the curve. Instead of using straight lines or linear approximations for curve segments, we can use parabolas or quadratic approximations. For any consecutive three points x_{i-1}, x_i and x_{i+1}, we can construct a parabola in the form

$$f(x_i + t) = f_i + \alpha t + \beta t^2, \qquad t \in [-h, h]. \tag{6.11}$$

As this parabola must go through the three known points (x_{i-1}, f_{i-1}) at $t = -h$, (x_i, f_i) at $t = 0$ and (x_{i+1}, f_{i+1}) at $t = h$, we have the following equations for α and β

$$f_{i-1} = f_i - \alpha h + \beta h^2, \tag{6.12}$$

and

$$f_{i+1} = f_i + \alpha h + \beta h^2, \tag{6.13}$$

which lead to

$$\alpha = \frac{f_{i+1} - f_{i-1}}{2h}, \tag{6.14}$$

and

$$\beta = \frac{f_{i-1} - 2f_i + f_{i+1}}{h^2}. \tag{6.15}$$

We will see in later chapters that α is the centred approximation for the first derivative f_i' and β is the central difference scheme for the second derivative f_i''. Therefore, the integral from x_{i-1} to x_{i+1} can be approximated by

$$\mathcal{I}_i = \int_{x_{i-1}}^{x_{i+1}} f(x)dx \approx \int_{-h}^{h} [f_i + \alpha t + \beta t^2]dt$$

$$= \frac{h}{3}[f_{i-1} + 4f_i + f_{i+1}], \tag{6.16}$$

where we have substituted the expressions for α and β. To ensure the whole interval $[a, b]$ can be divided up to form three-point approximations without any point left out, n must be even. Therefore, the estimate of the integral becomes

$$\mathcal{I} \approx \frac{h}{3}[f_0 + 4(f_1 + f_3 + ... + f_{n-1})$$

$$+ 2(f_2 + f_4 + ... + f_{n-2}) + f_n], \tag{6.17}$$

which is the standard Simpson's rule.

As the approximation for the function $f(x)$ is quadratic, an order higher than the linear form, the error estimate of Simpson's rule is thus $O(h^4)$ or

$O(h^4 f'''')$ to be more specific. There are many variations of Simpson's rule with higher order accuracies such as $O(h^5 f^{(4)})$ and $O(h^7 f^{(6)})$.

Now let us look at an example.

Example 6.1: *We know the exact value of the integral*

$$I = \int_0^{\pi/2} \sin^2(x) dx = \frac{\pi}{4}.$$

Let us now estimate it using the Simpson rule with $n = 8$ and $h = (\pi/2 - 0)/8 = \pi/16$. We have

$$I \approx \frac{h}{3}[f_0 + 4(f_1 + f_3 + f_5 + f_7) + 2(f_2 + f_4 + f_6) + f_8].$$

Since $f_i = \sin^2(x_i) = \sin^2(i * h)$, we have $f_0 = 0$, $f_1 = 0.03806$, $f_2 = 0.14644$, $f_3 = 0.308658$, $f_4 = 0.5$, $f_5 = 0.69134$, $f_6 = 0.85355$, $f_7 = 0.86193$, and $f_8 = 1$. Now the integral estimate is

$$I \approx \frac{\pi}{48}[0 + 4 \times 2.00 + 2 \times 1.50 + 1] \approx 0.71994.$$

The true value is $\pi/4 = 0.78539$, so the error is about 9%. The order of the estimate is $O(h^4 f'''')$. Since $f'''' = -8\cos^2(x) + 8\sin^2(x)$ or $|f''''(0)| = 8$, so the error is $O(\pi^4/16^4 * 8) = O(0.01189)$. Thus, we can expect the numerical estimate to be accurate only to the first decimal place.

From the example, we have seen that the accuracy of Simpson's rule is only $O(h^4 f'''')$, and such estimate of integral usually requires very small h (or large n). This means the evaluations of the integrand at many points. Is there any way to go around this tedious slow process and evaluate the integral more accurately using fewer points of evaluation? The answer is yes, and the numerical technique is called the Gaussian integration or Gaussian quadrature.

6.3 Gaussian Integration

To get higher-order accuracy, we can use polynomials to construct various integration schemes. However, there is an easier way to do this. That is to use the Gauss-Legendre integration or simply Gaussian integration. Since any integral $\mathcal{I}$ with integration limits a and b can be transformed to an integral with limits -1 and $+1$ by using

$$\zeta = \frac{2(x-a)}{(b-a)} - 1, \tag{6.18}$$

so that

$$\mathcal{I} = \int_a^b g(x)dx = \frac{(b-a)}{2} \int_{-1}^{1} f(\zeta)d\zeta, \tag{6.19}$$

where we have used $dx = (b-a)d\zeta/2$. Therefore, we only have to study the integral

$$J = \int_{-1}^{1} f(\zeta)d\zeta. \tag{6.20}$$

The n values of the function or n integration points are given by a polynomial of $n-1$ degrees. For equal spacing h, this numerical integration technique is often referred to as the Newton-Cotes quadrature

$$J = \int_{-1}^{1} f(d\zeta)d\zeta = \sum_{i=1}^{n} w_i f(\zeta_i), \tag{6.21}$$

where w_i is the weighting coefficient attached to $f(\zeta_i)$. Such integral will have an error of $O(h^n)$. For example, $n=2$ with equal weighting corresponds to the trapezium rule because

$$J = f_{-1} + f_1. \tag{6.22}$$

For the case of $n=3$, we have

$$J = \frac{1}{3}[f_{-1} + 4f_0 + f_1], \tag{6.23}$$

which corresponds to Simpson's rule.

The numerical integration we used so far is carried out at equally-spaced points $x_0, x_1, ..., x_i, ..., x_n$, and these points are fixed *a priori*. There is no particular reason why we should use the equally-spaced points apart from the fact that it is easy and simple. In fact, we can use any sampling points or integration points as we wish to improve the accuracy of the estimate to the integral. If we use n integration points $(\zeta_i, i = 1, 2, ..., n)$ with a polynomial of $2n-1$ degrees or Legendre polynomial $P_n(x)$, we now have $2n$ unknowns f_i and ζ_i. This means that we can easily construct quadrature formulas, often called Gauss quadrature or Gaussian integration. For both Newton-Cotes quadrature and Gauss quadrature, Figure 6.2 shows the their difference and similarity.

Mathematically, we have the Gauss quadrature

$$J = \int_{-1}^{1} f(\zeta)d\zeta = \sum_{i=1}^{n} w_i f(\zeta_i), \tag{6.24}$$

where ζ_i is determined by the zeros of the Legendre polynomial $P_n(\zeta_i) = 0$ and the weighting coefficient is given by

$$w_i = \frac{2}{(1-\zeta_i^2)[P_n'(\zeta_i)]^2}. \tag{6.25}$$

The error of this quadrature is of the order $O(h^{2n})$. The proof of this formulation is beyond the scope of this book. Readers can find the proof in more advanced mathematical books.

Briefly speaking, Legendre polynomials are obtained by the following generating function or Rodrigue's formula

$$P_n(x) = \frac{1}{2^n n!} \frac{d^n (x^2-1)^n}{dx^n}. \tag{6.26}$$

For example,

$$P_0(x) = 0, \;\; P_1(x) = x, \;\; P_2(x) = \frac{1}{2}(3x^2 - 1), \tag{6.27}$$

$$P_3(x) = \frac{1}{2}(5x^2 - 3x), \quad P_4(x) = \frac{1}{8}(35x^4 - 30x^2 + 3), \tag{6.28}$$

and

$$P_5(x) = \frac{1}{8}(63x^5 - 70x^3 + 15x). \tag{6.29}$$

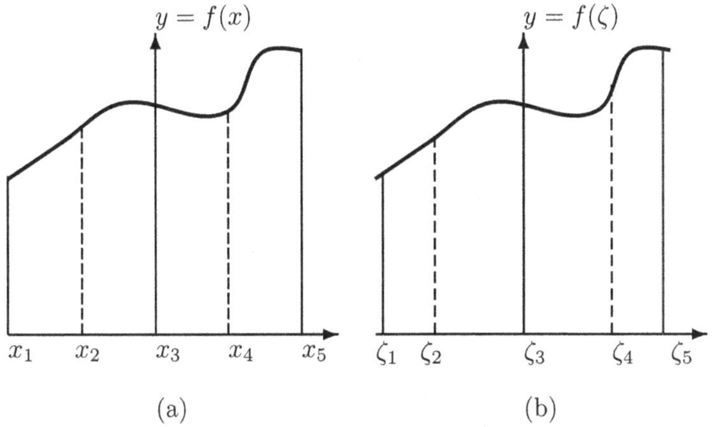

Fig. 6.2 Five integration points: a) Equally spaced with $h = 1/4$; and b) Gauss points with $|\zeta_1 - \zeta_3| = |\zeta_5 - \zeta_3| \approx 0.90617$ and $|\zeta_2 - \zeta_3| = |\zeta_4 - \zeta_3| \approx 0.53847$.

The computation of locations ζ_i of the Gaussian integration points and weighting coefficients w_i is complicated, though straightforward once we know the Legendre polynomials.

For example, when $n = 3$, we have

$$P_3(\zeta) = \frac{1}{2}(5\zeta^3 - 3\zeta) = \frac{\zeta}{2}(5\zeta^2 - 3) = 0, \tag{6.30}$$

which has three solutions

$$\zeta_0 = 0, \quad \zeta_\pm = \pm\sqrt{\frac{3}{5}}. \tag{6.31}$$

Since $P_3'(\zeta) = (15\zeta^2 - 3)/2$, we have

$$w_i = \frac{2}{(1-\zeta^2)[\frac{1}{2}(15\zeta^2 - 3)]^2} = \frac{8}{(1-\zeta^2)(15\zeta^2 - 3)^2}. \tag{6.32}$$

For $\zeta_0 = 0$, we get

$$w_0 = \frac{8}{(1-0^2)(15 \times 0^2 - 3)^2} = \frac{8}{9}. \tag{6.33}$$

Similarly, for $\zeta_\pm = \pm\sqrt{3/5}$, we obtain

$$w_\pm = \frac{8}{[1 - (\pm\sqrt{3/5})^2] \times [15 \times (\pm\sqrt{3/5})^2 - 3]^2} = \frac{5}{9}. \tag{6.34}$$

The coefficients w_i and the integration points are usually listed in tables for various values of n (see Table 6.1).

Table 6.1 Gauss integration points and weighting coefficients.

n	ζ	w
1	0.0000000000	2.000000000
2	±0.5773502692	1.0000000000
3	±0.7745966692	0.5555555556
	0.0000000000	0.8888888889
4	±0.8611363116	0.3478548451
	±0.3399810436	0.6521451549
5	±0.9061798459	0.2369268851
	±0.5384693101	0.4786286705
	0.0000000000	0.5688888889

For multiple integrals

$$J = \int_{-1}^{1}\int_{-1}^{1} f(\zeta, \eta) d\zeta d\eta, \tag{6.35}$$

such Gaussian quadrature can easily be extended by evaluating the integral with η being kept constant first, then evaluating the outer integral. We have

$$J = \int_{-1}^{1} \Big[\int_{-1}^{n} f(\zeta, \eta) d\zeta \Big] d\eta$$

$$= \int_{-1}^{1} \sum_{i=1}^{n} w_i f(\zeta_i, \eta) d\eta = \sum_{i=1}^{n} w_i \int_{-1}^{1} f(\zeta_i, \eta) d\eta$$

$$= \sum_{i=1}^{n} \sum_{j=1}^{n} w_i w_j f(\zeta_i, \eta_j), \tag{6.36}$$

where we have used

$$\int_{-1}^{1} f(\zeta_i, \eta) d\eta = \sum_{j=1}^{n} w_j f(\zeta_i, \eta_j). \tag{6.37}$$

Example 6.2: *To evaluate the integral*

$$I = \frac{2}{\sqrt{\pi}} \int_{-1}^{1} e^{-x^2} dx = \int_{-1}^{1} f(x) dx, \quad f(x) = \frac{2}{\sqrt{\pi}} e^{-x^2},$$

we know that its exact value from Eq. (6.1) is

$$I = 2\,\mathrm{erf}(1) = 1.685401585899...$$

Let us now estimate it using Simpson's rule for three point integration at $x_{-1} = -1$, $x_0 = 0$ *and* $x_1 = 1$, *and we have*

$$I \approx \frac{1}{3}(f_{-1} + 4f_0 + f_1)$$

$$\approx \frac{2}{3\sqrt{\pi}} [e^{-(-1)^2} + 4 \times 1 + e^{-(1)^2}] \approx 1.7812,$$

which differs from its exact value by about 5.6%.

If we use the 3*-point Gauss quadrature at* $x_{\pm 1} = \pm\sqrt{\frac{3}{5}}$ *and* $x_0 = 0$ *with weighting coefficients* $w_{\pm 1} = \frac{5}{9}$ *and* $w_0 = \frac{8}{9}$, *we have*

$$I \approx \sum_{i=-1}^{1} w_i f(x_i)$$

$$\approx \frac{2}{\sqrt{\pi}} [\frac{5}{9} e^{-(-\sqrt{3/5})^2} + \frac{8}{9} \times 1 + \frac{5}{9} e^{-(\sqrt{3/5})^2}] \approx 1.6911,$$

which is a better approximation than 1.7812. In fact, the error of Gauss quadrature is just $(1.6911 - 1.6854)/1.6911 \approx 0.3\%$. *This higher accuracy is the main reason why the Gaussian quadrature is so widely used.*

For triple integrals and other integrals, the Gauss quadrature can be constructed in a similar way. We will see more numerical techniques in the rest of the book.

Chapter 7

Computational Linear Algebra

Linear systems, especially large-scale spare systems, are very commonly in computational sciences. In fact, most numerical methods for solving PDEs such as the finite difference methods and finite element methods often result in large, spare matrices. In this chapter, we introduce the basic concepts and solution methods of linear systems.

7.1 System of Linear Equations

A linear system of m equations for n unknowns $\boldsymbol{u} = (u_1, ..., u_n)^T$

$$\sum_{j=1}^{n} a_{ij} u_j = b_i, \qquad (i = 1, 2, ..., m), \tag{7.1}$$

can be written in the compact form as

$$\begin{pmatrix} a_{11} & a_{12} & ... & a_{1n} \\ a_{21} & a_{22} & ... & a_{2n} \\ \vdots & & \ddots & \\ a_{m1} & a_{m2} & ... & a_{mn} \end{pmatrix} \begin{pmatrix} u_1 \\ u_2 \\ \vdots \\ u_n \end{pmatrix} = \begin{pmatrix} b_1 \\ b_2 \\ \vdots \\ b_n \end{pmatrix}, \tag{7.2}$$

or simply

$$\boldsymbol{A}\boldsymbol{u} = \boldsymbol{b}. \tag{7.3}$$

Here both the matrix $\boldsymbol{A} = [a_{ij}]$ and the column vector $\boldsymbol{b} = (b_1, ..., b_m)^T$ are known. If $m < n$, the system is under-determined as the conditions are not sufficient to guarantee a unique solution set. On the other hand, the system is over-determined if $m > n$ because there are too many conditions and a solution may not exist at all. The unique solution is only possible when $m = n$ or when $\boldsymbol{A}$ is a non-singular square matrix.

The solution of this matrix equation is important to many numerical problems, ranging from the solution of a large system of linear equations to linear mathematical programming, and from data interpolation to finding solutions to finite element problems.

The inverse of $\boldsymbol{A}$ is possible only if $m = n$. If the inverse $\boldsymbol{A}^{-1}$ does not exist, then the linear system is under-determined or there are no unique solutions (or even no solution at all). In order to find the solutions, we multiply both sides by $\boldsymbol{A}^{-1}$,

$$\boldsymbol{A}^{-1}\boldsymbol{A}\boldsymbol{u} = \boldsymbol{A}^{-1}\boldsymbol{b}, \tag{7.4}$$

and we obtain the solution

$$\boldsymbol{u} = \boldsymbol{A}^{-1}\boldsymbol{b}. \tag{7.5}$$

A special case of this equation is when $\boldsymbol{b} = \lambda \boldsymbol{u}$, and this becomes an eigenvalue problem as discussed in Part I.

Mathematically speaking, a linear system can be solved in principle using Cramer's rule,

$$u_i = \frac{\det \boldsymbol{A}_i}{\det \boldsymbol{A}}, \qquad i = 1, 2, ..., n, \tag{7.6}$$

where the matrix $\boldsymbol{A}_i$ is obtained by replacing the i-th column by the column vector $\boldsymbol{b}$.

Example 7.1: *For example, for three linear equations with three unknowns u_1, u_2 and u_3,*

$$a_{11}u_1 + a_{12}u_2 + a_{13}u_3 = b_1,$$

$$a_{21}u_1 + a_{22}u_2 + a_{23}u_3 = b_2,$$

$$a_{31}u_1 + a_{32}u_2 + a_{33}u_3 = b_3,$$

its solution vector is given by the following Cramer's rule

$$u_1 = \frac{1}{\Delta}\begin{vmatrix} b_1 & a_{12} & a_{13} \\ b_2 & a_{22} & a_{23} \\ b_3 & a_{32} & a_{33} \end{vmatrix}, \qquad u_2 = \frac{1}{\Delta}\begin{vmatrix} a_{11} & b_1 & a_{13} \\ a_{21} & b_2 & a_{23} \\ a_{31} & b_3 & a_{33} \end{vmatrix}, \qquad u_3 = \frac{1}{\Delta}\begin{vmatrix} a_{11} & a_{12} & b_1 \\ a_{21} & a_{22} & b_2 \\ a_{31} & a_{32} & b_3 \end{vmatrix},$$

where

$$\Delta = \begin{vmatrix} a_{11} & a_{12} & a_{13} \\ a_{21} & a_{22} & a_{23} \\ a_{31} & a_{32} & a_{33} \end{vmatrix}.$$

Though it is straightforward to extend the rule to any dimensions in theory, this is not an easy task in practice because the calculation of the determinant of a large matrix is not easy. Though Cramer's rule is good for proving theorems, it is not good for numerical implementation. Better methods are to use the inverse matrix and/or iterations.

Finding the inverse $\boldsymbol{A}^{-1}$ of a square $n \times n$ matrix $\boldsymbol{A}$ is not an easy task either, especially when the size of the matrix is large, and it usually requires the algorithm complexity of $O(n^3)$. In fact, many solution methods are designed to avoid the necessity of calculating the inverse $\boldsymbol{A}^{-1}$ if possible.

There are many ways of solving the linear equations, but they fall into two categories: direct algebraic methods and iteration methods. The purpose of the former is to find the solution by elimination, decomposition of matrix, and substitutions, while the latter involves certain iterations to find the approximate solutions. The choice of these methods depends on the characteristics of the matrix $\boldsymbol{A}$, size of the problem, computational time, the type of problem, and the required solution quality.

7.2 Gauss Elimination

The basic idea of Gauss elimination is to transform a square matrix into a triangular matrix by elementary row operations, so that the simplified triangular system can be solved by direct back substitution. For the linear system

$$\begin{pmatrix} a_{11} & a_{12} & a_{13} & \ldots & a_{1n} \\ a_{21} & a_{22} & a_{23} & \ldots & a_{2n} \\ & \vdots & & & \\ a_{n1} & a_{n2} & a_{n3} & \ldots & a_{nn} \end{pmatrix} \begin{pmatrix} u_1 \\ u_2 \\ \vdots \\ u_n \end{pmatrix} = \begin{pmatrix} b_1 \\ b_2 \\ \vdots \\ b_n \end{pmatrix}, \tag{7.7}$$

the aim in the first step is to try to make all the coefficients in the first column (a_{21}, ..., a_{n1}) become zero by elementary row operations, except the first element. This is based on the principle that a linear system will remain the same if its rows are multiplied by some non-zero coefficients, or any two rows are interchanged, or any two (or more) rows are combined through addition and substraction.

To do this, we first divide the first equation by a_{11} (we can always assume $a_{11} \neq 0$; if not, we rearrange the order of the equations to achieve this). We now have

$$\begin{pmatrix} 1 & \frac{a_{12}}{a_{11}} & \frac{a_{13}}{a_{11}} & ... & \frac{a_{1n}}{a_{11}} \\ a_{21} & a_{22} & a_{23} & ... & a_{2n} \\ \vdots & & & & \\ a_{n1} & a_{n2} & a_{n3} & ... & a_{nn} \end{pmatrix} \begin{pmatrix} u_1 \\ u_2 \\ \vdots \\ u_n \end{pmatrix} = \begin{pmatrix} \frac{b_1}{a_{11}} \\ b_2 \\ \vdots \\ b_n \end{pmatrix}. \tag{7.8}$$

Then multiplying the first row by $-a_{21}$ and adding it to the second row, multiplying the first row by $-a_{i1}$ and adding it to the i-th row, we finally have

$$\begin{pmatrix} 1 & \frac{a_{12}}{a_{11}} & \frac{a_{13}}{a_{11}} ... & \frac{a_{1n}}{a_{11}} \\ 0 & a_{22} - \frac{a_{21}a_{12}}{a_{11}} & ... & a_{2n} - \frac{a_{21}a_{1n}}{a_{11}} \\ \vdots & & & \\ 0 & a_{n2} - \frac{a_{n1}a_{12}}{a_{11}} & ... & a_{nn} - \frac{a_{n1}a_{1n}}{a_{11}} \end{pmatrix} \begin{pmatrix} u_1 \\ u_2 \\ \vdots \\ u_n \end{pmatrix} = \begin{pmatrix} \frac{b_1}{a_{11}} \\ b_2 - \frac{a_{21}b_1}{a_{11}} \\ \vdots \\ b_n - \frac{a_{n1}b_n}{a_{11}} \end{pmatrix}.$$

We then repeat the same procedure for the third row to the n-th row, and the final form of the linear system should be in the following generic form

$$\begin{pmatrix} \alpha_{11} & \alpha_{12} & \alpha_{13} & ... & \alpha_{1n} \\ 0 & \alpha_{22} & \alpha_{23} & ... & \alpha_{2n} \\ \vdots & & & \ddots & \\ 0 & 0 & 0 & ... & \alpha_{nn} \end{pmatrix} \begin{pmatrix} u_1 \\ u_2 \\ \vdots \\ u_n \end{pmatrix} = \begin{pmatrix} \beta_1 \\ \beta_2 \\ \vdots \\ \beta_n \end{pmatrix}, \tag{7.9}$$

where $\alpha_{1j} = a_{1j}/a_{11}$, $\alpha_{2j} = a_{2j} - a_{1j}a_{21}/a_{11} (j = 1, 2, ..., n)$, ..., $\beta_1 = b_1/a_{11}$, $\beta_2 = b_2 - a_{21}b_1/a_{11}$ and others. From the above form, we can see that $u_n = \beta_n/\alpha_{nn}$ because there is only one unknown u_n in the n-th row. We can then use the back substitutions to obtain u_{n-1} and up to u_1. Therefore, we have

$$u_n = \frac{\beta_n}{\alpha_{nn}},$$

$$u_i = \frac{1}{\alpha_{ii}}(\beta_i - \sum_{j=i+1}^{n} \alpha_{ij}x_j), \tag{7.10}$$

where $i = n-1, n-2, ..., 1$. Obviously, in our present case, $\alpha_{11} = ... = \alpha_{nn} = 1$. Let us look at an example.

Example 7.2: *For the linear system*

$$\begin{pmatrix} 2 & -1 & 3 & 4 \\ 3 & 2 & -5 & 6 \\ -2 & 1 & 0 & 5 \\ 4 & -5 & -6 & 0 \end{pmatrix} \begin{pmatrix} u_1 \\ u_2 \\ u_3 \\ u_4 \end{pmatrix} = \begin{pmatrix} 21 \\ 9 \\ 12 \\ -3 \end{pmatrix},$$

we first divide the first row by $a_{11} = 2$, *we have*

$$\begin{pmatrix} 1 & -1/2 & 3/2 & 2 \\ 3 & 2 & -5 & 6 \\ -2 & 1 & 0 & 5 \\ 4 & -5 & -6 & 0 \end{pmatrix} \begin{pmatrix} u_1 \\ u_2 \\ u_3 \\ u_4 \end{pmatrix} = \begin{pmatrix} 21/2 \\ 9 \\ 12 \\ -3 \end{pmatrix}.$$

Multiplying the first row by 3 and subtracting it from the second row, and carrying out similar row manipulations for the other rows, we have

$$\begin{pmatrix} 1 & -1/2 & 3/2 & 2 \\ 0 & 7/2 & -19/2 & 0 \\ 0 & 0 & 3 & 9 \\ 0 & -3 & -12 & -8 \end{pmatrix} \begin{pmatrix} u_1 \\ u_2 \\ u_3 \\ u_4 \end{pmatrix} = \begin{pmatrix} 21/2 \\ -45/7 \\ 33 \\ -45 \end{pmatrix}.$$

For the second row, we repeat this procedure again, we have

$$\begin{pmatrix} 1 & -1/2 & 3/2 & 2 \\ 0 & 1 & -19/7 & 0 \\ 0 & 0 & 3 & 9 \\ 0 & 0 & -141/7 & -8 \end{pmatrix} \begin{pmatrix} u_1 \\ u_2 \\ u_3 \\ u_4 \end{pmatrix} = \begin{pmatrix} 21/2 \\ -45/7 \\ 33 \\ -450/7 \end{pmatrix}.$$

Similarly, for the third row, we have

$$\begin{pmatrix} 1 & -1/2 & 3/2 & 2 \\ 0 & 1 & -19/7 & 0 \\ 0 & 0 & 1 & 3 \\ 0 & 0 & 0 & 367/7 \end{pmatrix} \begin{pmatrix} u_1 \\ u_2 \\ u_3 \\ u_4 \end{pmatrix} = \begin{pmatrix} 21/2 \\ -45/7 \\ 11 \\ 1101/7 \end{pmatrix}.$$

The fourth row gives that $u_4 = 3$. *Using the back substitution, we have* $u_3 = 2$ *from the third row. Similarly, we have* $u_2 = -1$ *and* $u_1 = 1$. *So the solution vector is*

$$\begin{pmatrix} u_1 \\ u_2 \\ u_3 \\ u_4 \end{pmatrix} = \begin{pmatrix} 1 \\ -1 \\ 2 \\ 3 \end{pmatrix}.$$

We have seen from the example that there are many floating-point calculations even for the simple system of four linear equations. In fact, the full Gauss elimination is computationally extensive with an algorithmic complexity of $O(2n^3/3)$.

Gauss-Jordan elimination, a variant of Gauss elimination, solves a linear system and, at the same time, can also compute the inverse of a square matrix. The first step is to formulate an augmented matrix from $\boldsymbol{A}$, $\boldsymbol{b}$ and the unit matrix $\boldsymbol{I}$ (with the same size of $\boldsymbol{A}$). That is

$$\boldsymbol{B} = [\boldsymbol{A}|\boldsymbol{b}|\boldsymbol{I}] = \begin{pmatrix} a_{11} \ ... \ a_{1n} \ |b_1| \ 1 \ 0 \ ... \ 0 \\ a_{21} \ ... \ a_{2n} \ |b_2| \ 0 \ 1 \ ... \ 0 \\ \vdots \quad\quad \vdots \quad\quad \vdots \\ a_{n1} \ ... \ a_{nn} \ |b_n| \ 0 \ 0 \ ... \ 1 \end{pmatrix}, \tag{7.11}$$

where the notation $\boldsymbol{A}|\boldsymbol{b}$ denotes the augmented form of two matrix $\boldsymbol{A}$ and $\boldsymbol{b}$. The aim is to reduce $\boldsymbol{B}$ to the following form by elementary row reductions in a way similar to those carried out in Gauss elimination.

$$\begin{pmatrix} 1 \ 0 \ ... \ 0 \ |u_1| \ a'_{11} \ ... \ a'_{1n} \\ 0 \ 1 \ ... \ 0 \ |u_2| \ a'_{21} \ ... \ a'_{2n} \\ \vdots \quad\quad \vdots \quad\quad \vdots \\ 0 \ 0 \ ... \ 1 \ |u_n| \ a'_{n1} \ ... \ a'_{nn} \end{pmatrix} = [\boldsymbol{I}|\boldsymbol{u}|\boldsymbol{A}^{-1}], \tag{7.12}$$

where $\boldsymbol{A}^{-1} = [a'_{ij}]$ is the inverse. This is better demonstrated by an example.

Example 7.3: *In order to solve the following system*

$$\boldsymbol{A}\boldsymbol{u} = \begin{pmatrix} 1 & 2 & 3 \\ -2 & 2 & 5 \\ 4 & 0 & -5 \end{pmatrix} \begin{pmatrix} u_1 \\ u_2 \\ u_3 \end{pmatrix} = \begin{pmatrix} 5 \\ -2 \\ 14 \end{pmatrix} = \boldsymbol{b},$$

we first write it in an augmented form

$$\boldsymbol{B} = \begin{pmatrix} 1 \ \ 2 \ \ 3 \ | \ 5 \ | \ 1 \ 0 \ 0 \\ \\ -2 \ 2 \ \ 5 \ | -2 \ | \ 0 \ 1 \ 0 \\ \\ 4 \ \ 0 \ -5 \ | \ 14 \ | \ 0 \ 0 \ 1 \end{pmatrix}.$$

By elementary row operations, this could be changed into

$$\boldsymbol{B}' = [\boldsymbol{I}|\boldsymbol{u}|\boldsymbol{A}^{-1}] = \begin{pmatrix} 1 \ 0 \ 0 \ | \ 1 \ | \ \frac{5}{7} \ -\frac{5}{7} \ -\frac{2}{7} \\ \\ 0 \ 1 \ 0 \ | \ 5 \ | -\frac{5}{7} \ \frac{17}{14} \ \frac{11}{14} \\ \\ 0 \ 0 \ 1 \ | -2 \ | \ \frac{4}{7} \ -\frac{4}{7} \ -\frac{3}{7} \end{pmatrix},$$

which gives

$$\boldsymbol{u} = \begin{pmatrix} 1 \\ 5 \\ -2 \end{pmatrix}, \qquad \boldsymbol{A}^{-1} = \frac{1}{14} \begin{pmatrix} 10 & -10 & -4 \\ -10 & 17 & 11 \\ 8 & -8 & -6 \end{pmatrix}.$$

We can see that both the solution $\boldsymbol{u}$ and the inverse $\boldsymbol{A}^{-1}$ are simultaneously obtained in the Gauss-Jordan elimination.

The Gauss-Jordan elimination is not quite stable numerically. In order to get better and more stable schemes, a common practice is to use pivoting. Basically, pivoting is a scaling procedure by dividing all the elements in a row by the element with the largest magnitude or norm. If necessary, rows can be exchanged so the largest element is moved so that it becomes the leading coefficient, especially on the diagonal position. This means that all the scaled elements are in the range of $[-1, 1]$. Thus, exceptionally large numbers are removed, which makes the scheme more numerically stable.

An important issue in both Gauss elimination and Gauss-Jordan elimination methods is the non-zero requirement of leading coefficients such as $a_{11} \neq 0$. For a_{11}, it is possible to rearrange the equations to achieve this requirement. However, there is no guarantee that other coefficients such as $a_{22} - a_{21}a_{12}/a_{11}$ should be nonzero. If it is zero, there is a potential difficulty due to the division by zero. In order to avoid this problem, we can use other methods such as the pivoting method and LU factorization or decomposition.

7.3 LU Factorization

Any square matrix $\boldsymbol{A}$ can be written as the product of two triangular matrices in the form

$$\boldsymbol{A} = \boldsymbol{L}\boldsymbol{U}, \tag{7.13}$$

where $\boldsymbol{L}$ and $\boldsymbol{U}$ are the lower and upper triangular matrices, respectively. A lower (upper) triangular matrix has elements only on the diagonal and below (above). That is

$$\boldsymbol{L} = \begin{pmatrix} \beta_{11} & 0 & \dots & 0 \\ \beta_{21} & \beta_{22} & \dots & 0 \\ \vdots & & \ddots & \\ \beta_{n1} & \beta_{n2} & \dots & \beta_{nn} \end{pmatrix}, \tag{7.14}$$

and

$$\boldsymbol{U} = \begin{pmatrix} \alpha_{11} & \dots & \alpha_{1,n-1} & \alpha_{1,n} \\ & \ddots & & \vdots \\ 0 & \dots & \alpha_{n-1,n-1} & \alpha_{n-1,n} \\ 0 & \dots & 0 & \alpha_{nn} \end{pmatrix}. \tag{7.15}$$

The linear system $\mathbf{Au} = \mathbf{b}$ can be written as the following two steps

$$\boldsymbol{A}\boldsymbol{u} = (\boldsymbol{L}\boldsymbol{U})\boldsymbol{u} = \boldsymbol{L}(\boldsymbol{U}\boldsymbol{u}) = \boldsymbol{b}, \tag{7.16}$$

or

$$\boldsymbol{U}\boldsymbol{u} = \boldsymbol{v}, \qquad \boldsymbol{L}\boldsymbol{v} = \boldsymbol{b}, \tag{7.17}$$

which are two linear systems with triangular matrices only, and these systems can be solved by forward and back substitutions. The solutions of v_i are given by

$$v_1 = \frac{b_1}{\beta_{11}}, \qquad v_i = \frac{1}{\beta_{ii}}(b_i - \sum_{j-1}^{i-1} \beta_{ij} v_j), \tag{7.18}$$

where $i = 2, 3, ..., n$. The final solutions u_i are then given by

$$u_n = \frac{v_n}{\alpha_{nn}}, \quad u_i = \frac{1}{\alpha_{ii}}(v_i - \sum_{j=i+1}^{n} \alpha_{ij} u_j), \tag{7.19}$$

where $i = n - 1, ..., 1$.

For triangular matrices such as $\boldsymbol{L}$, there are some interesting properties. The inverse of a lower (upper) triangular matrix is also a lower (upper) triangular matrix. The determinant of the triangular matrix is simply the product of its diagonal entries. That is

$$\det(\boldsymbol{L}) = |\boldsymbol{L}| = \prod_{i=1}^{n} \beta_{ii} = \beta_{11}\beta_{22}...\beta_{nn}. \tag{7.20}$$

More interestingly, the eigenvalues of a triangular matrix are the diagonal entries: β_{11}, β_{22}, ..., β_{nn}. These properties are convenient in determining the stability of an iteration scheme.

But there is another issue here and that is how to decompose a square matrix $\boldsymbol{A} = [a_{ij}]$ into $\boldsymbol{L}$ and $\boldsymbol{U}$. As there are $n(n+1)/2$ coefficients α_{ij} and $n(n+1)/2$ coefficients β_{ij}, so there are $n^2 + n$ unknowns. For the equation (7.13), we know that it could provide only n^2 equations (as there are only n^2 coefficients a_{ij}). They are

$$\sum_{k=1}^{i} \beta_{ik}\alpha_{kj} = a_{ij}, \qquad (i < j), \tag{7.21}$$

$$\sum_{k=1}^{j=i} \beta_{ik}\alpha_{kj} = a_{ij}, \qquad (i = j), \tag{7.22}$$

and

$$\sum_{k=1}^{j} \beta_{ik}\alpha_{kj} = a_{ij}, \qquad (i > j), \tag{7.23}$$

which again forms another system of n equations.

As $n^2 + n > n^2$, there are n free coefficients. Therefore, such factorization or decomposition is not uniquely determined. We have to impose some extra conditions. Fortunately, we can always set either $\alpha_{ii} = 1$ or $\beta_{ii} = 1$ where $i = 1, 2, ..., n$. If we set $\beta_{ii} = 1$, we can use Crout's algorithm to determine α_{ij} and β_{ij}. We have the coefficients for the upper triangular matrix

$$\alpha_{ij} = a_{ij} - \sum_{k=1}^{i-1} \beta_{ik}\alpha_{kj}, \tag{7.24}$$

for $(i = 1, 2, ..., j)$ and $j = 1, 2, ..., n$. For the lower triangular matrix, we have

$$\beta_{ij} = \frac{1}{\alpha_{jj}}(a_{ij} - \sum_{k=1}^{j-1} \beta_{ik}\alpha_{kj}), \tag{7.25}$$

for $i = j + 1, j + 2, ..., n$.

The same issue appears again, that is, all the leading coefficients α_{ii} must be non-zero. For sparse matrices with many zero entries, this could often cause some significant problems numerically. Better methods such as iteration methods should be used in this case.

7.4 Iteration Methods

For a linear system $\mathbf{Au} = \mathbf{b}$, the solution $\mathbf{u} = \mathbf{A}^{-1}\mathbf{b}$ generally involves the inversion of a large matrix. The direct inversion becomes impractical if the matrix is very large (say, if $n > 100,000$). Many efficient algorithms have been developed for solving such systems. Jacobi and Gauss-Seidel iteration methods are just two examples.

7.4.1 *Jacobi Iteration Method*

The basic idea of the Jacobi-type iteration method is to decompose an $n \times n$ square matrix $\boldsymbol{A}$ into three simpler matrices

$$\boldsymbol{A} = \boldsymbol{D} + \boldsymbol{L} + \boldsymbol{U}, \tag{7.26}$$

where $\boldsymbol{D}$ is a diagonal matrix. $\boldsymbol{L}$ and $\boldsymbol{U}$ are the strictly lower and upper triangular matrices, respectively. Here the 'strict' means that the lower (or upper) triangular matrices do not include the diagonal elements. That is say, all the diagonal elements of the triangular matrices are zeros.

It is worth pointing out that here the triangular matrices $\boldsymbol{L}$ and $\boldsymbol{U}$ are different from those in the LU decomposition where it requires matrix products. In comparison with the LU decomposition where $\boldsymbol{LU} = \boldsymbol{A}$, we have used the simple additions here and this makes the decomposition an easier task. Using such decomposition, the linear system $\boldsymbol{Au} = \boldsymbol{b}$ becomes

$$\boldsymbol{Au} = (\boldsymbol{D} + \boldsymbol{L} + \boldsymbol{U})\boldsymbol{u} = \boldsymbol{b}, \tag{7.27}$$

which can be written as an iteration procedure

$$\boldsymbol{Du}^{(n+1)} = \boldsymbol{b} - (\boldsymbol{L} + \boldsymbol{U})\boldsymbol{u}^{(n)}. \tag{7.28}$$

This can be used to calculate the next approximate solution $\boldsymbol{u}^{(n+1)}$ from the current estimate $\boldsymbol{u}^{(n)}$. As the inverse of any diagonal matrix $\boldsymbol{D} = \text{diag}[d_{ii}]$ is easy, we have

$$\boldsymbol{u}^{(n+1)} = \boldsymbol{D}^{-1}[\boldsymbol{b} - (\boldsymbol{L} + \boldsymbol{U})\boldsymbol{u}^{(n)}]. \tag{7.29}$$

Writing in terms of the elements, we have

$$u_i^{(n+1)} = \frac{1}{d_{ii}}[b_i - \sum_{j \neq i} a_{ij} u_j^{(n)}], \tag{7.30}$$

where $d_{ii} = a_{ii}$ are the diagonal elements only.

This iteration usually starts from an initial guess $\boldsymbol{u}^{(0)}$ (say, $\boldsymbol{u}^{(0)} = \boldsymbol{0}$). However, this iteration scheme is only stable under the condition that the square matrix is strictly diagonally dominant. That is to require that

$$|a_{ii}| > \sum_{j=1, j \neq i}^{n} |a_{ij}|, \tag{7.31}$$

for all $i = 1, 2, ..., n$.

In order to show how the iteration works, let us look at an example by solving the following linear system

$$5u_1 + u_2 - 2u_3 = 5,$$
$$u_1 + 4u_2 = -10,$$
$$2u_1 + 2u_2 - 7u_3 = -9. \tag{7.32}$$

We know its exact solution is

$$\boldsymbol{u} = \begin{pmatrix} u_1 \\ u_2 \\ u_3 \end{pmatrix} = \begin{pmatrix} 2 \\ -3 \\ 1 \end{pmatrix}. \tag{7.33}$$

Now let us solve the system by the Jacobi-type iteration method.

Example 7.4: *We first write the above simple system as the compact matrix form* $\boldsymbol{A}\boldsymbol{u} = \boldsymbol{b}$ *or*

$$\begin{pmatrix} 5 & 1 & -2 \\ 1 & 4 & 0 \\ 2 & 2 & -7 \end{pmatrix} \begin{pmatrix} u_1 \\ u_2 \\ u_3 \end{pmatrix} = \begin{pmatrix} 5 \\ -10 \\ -9 \end{pmatrix}.$$

Then we decompose the matrix $\boldsymbol{A}$ *as*

$$\boldsymbol{A} = \begin{pmatrix} 5 & 1 & -2 \\ 1 & 4 & 0 \\ 2 & 2 & -7 \end{pmatrix} = \boldsymbol{D} + (\boldsymbol{L} + \boldsymbol{U})$$

$$= \begin{pmatrix} 5 & 0 & 0 \\ 0 & 4 & 0 \\ 0 & 0 & -7 \end{pmatrix} + \begin{pmatrix} 0 & 0 & 0 \\ 1 & 0 & 0 \\ 2 & 2 & 0 \end{pmatrix} + \begin{pmatrix} 0 & 1 & -2 \\ 0 & 0 & 0 \\ 0 & 0 & 0 \end{pmatrix}.$$

The inverse of $\boldsymbol{D}$ *is simply*

$$\boldsymbol{D}^{-1} = \begin{pmatrix} 1/5 & 0 & 0 \\ 0 & 1/4 & 0 \\ 0 & 0 & -1/7 \end{pmatrix}.$$

The Jacobi-type iteration formula becomes

$$\boldsymbol{u}^{(n+1)} = \boldsymbol{D}^{-1}[\boldsymbol{b} - (\boldsymbol{L} + \boldsymbol{U})\boldsymbol{u}^{(n)}],$$

or

$$\begin{pmatrix} u_1 \\ u_2 \\ u_3 \end{pmatrix}_{n+1} = \begin{pmatrix} 1/5 & 0 & 0 \\ 0 & 1/4 & 0 \\ 0 & 0 & -1/7 \end{pmatrix} \left[\begin{pmatrix} 5 \\ -10 \\ -9 \end{pmatrix} - \begin{pmatrix} 0 & 1 & -2 \\ 1 & 0 & 0 \\ 2 & 2 & 0 \end{pmatrix} \begin{pmatrix} u_1 \\ u_2 \\ u_3 \end{pmatrix}_n \right].$$

If we start from the initial guess $\boldsymbol{u}^{(0)} = \begin{pmatrix} 0 & 0 & 0 \end{pmatrix}^T$, *we have*

$$\boldsymbol{u}^{(1)} \approx \begin{pmatrix} 1 \\ -2.5 \\ 1.2857 \end{pmatrix}, \quad \boldsymbol{u}^{(2)} \approx \begin{pmatrix} 2.0143 \\ -2.7500 \\ 0.8571 \end{pmatrix}, \quad \boldsymbol{u}^{(3)} \approx \begin{pmatrix} 1.8929 \\ 3.0036 \\ 1.0755 \end{pmatrix},$$

$$\boldsymbol{u}^{(4)} \approx \begin{pmatrix} 2.0309 \\ -2.9732 \\ 0.9684 \end{pmatrix}, \quad \boldsymbol{u}^{(5)} \approx \begin{pmatrix} 1.9820 \\ -3.0077 \\ 1.0165 \end{pmatrix}.$$

We can see that after 5 *iterations, the approximate solution is quite close to the true solution* $\boldsymbol{u} = \begin{pmatrix} 2 & -3 & 1 \end{pmatrix}^T$.

However, there is an important issue here. If we interchange the second row (equation) and the third row (equation), the new diagonal matrix becomes

$$\begin{pmatrix} 5 & 0 & 0 \\ 0 & 2 & 0 \\ 0 & 0 & 0 \end{pmatrix},$$

which has no inverse as it is singular. This means the order of the equations is important to ensure that the matrix is diagonally dominant.

Furthermore, if we interchange the first equation (row) and second equation (row), we have an equivalent system

$$\begin{pmatrix} 1 & 4 & 0 \\ 5 & 1 & -2 \\ 2 & 2 & -7 \end{pmatrix} \begin{pmatrix} u_1 \\ u_2 \\ u_3 \end{pmatrix} = \begin{pmatrix} -10 \\ 5 \\ -9 \end{pmatrix}.$$

Now the new decomposition becomes

$$\boldsymbol{A} = \begin{pmatrix} 1 & 4 & 0 \\ 5 & 1 & -2 \\ 2 & 2 & -7 \end{pmatrix} = \begin{pmatrix} 1 & 0 & 0 \\ 0 & 1 & 0 \\ 0 & 0 & -7 \end{pmatrix} + \begin{pmatrix} 0 & 0 & 0 \\ 5 & 0 & 0 \\ 2 & 2 & 0 \end{pmatrix} + \begin{pmatrix} 0 & 4 & 0 \\ 0 & 0 & -2 \\ 0 & 0 & 0 \end{pmatrix},$$

which gives the following iteration formula

$$\begin{pmatrix} u_1 \\ u_2 \\ u_3 \end{pmatrix}_{n+1} = \begin{pmatrix} 1 & 0 & 0 \\ 0 & 1 & 0 \\ 0 & 0 & -\frac{1}{7} \end{pmatrix} \left[\begin{pmatrix} -10 \\ 5 \\ -9 \end{pmatrix} - \begin{pmatrix} 0 & 4 & 0 \\ 5 & 0 & -2 \\ 2 & 2 & 0 \end{pmatrix} \begin{pmatrix} u_1 \\ u_2 \\ u_3 \end{pmatrix}_n \right].$$

Starting from $\boldsymbol{u}^{(0)} = \begin{pmatrix} 0 & 0 & 0 \end{pmatrix}^T$ again, we have

$$\boldsymbol{u}^{(1)} = \begin{pmatrix} -10 \\ 5 \\ 1.2857 \end{pmatrix}, \quad \boldsymbol{u}^{(2)} = \begin{pmatrix} -30 \\ 57.5714 \\ -0.1429 \end{pmatrix}, \quad \boldsymbol{u}^{(3)} = \begin{pmatrix} -240.28 \\ 154.71 \\ 9.16 \end{pmatrix}, \ldots$$

We can see that it diverges. So what is the problem? How can the order of the equations affect the results so significantly?

There are two important criteria for the iterations to converge correctly, and they are: the inverse of $\boldsymbol{D}^{-1}$ must exist, and the spectral radius of the right matrix must be less than 1. The first condition is obvious; if $\boldsymbol{D}^{-1}$ does not exist (say, when any of the diagonal elements is zero), then we cannot carry out the iteration process at all. The second condition requires

$$\rho(\boldsymbol{D}^{-1}) \leq 1, \qquad \rho[\boldsymbol{D}^{-1}(\boldsymbol{L} + \boldsymbol{U})] \leq 1, \tag{7.34}$$

where $\rho(\boldsymbol{A})$ is the spectral radius of the matrix $\boldsymbol{A}$. From the diagonal matrix $\boldsymbol{D}$, its largest absolute eigenvalue is 1. So $\rho(\boldsymbol{D}^{-1}) = \max(|\lambda_i|) = 1$ seems no problem. How about the following matrix?

$$\boldsymbol{N} = \boldsymbol{D}^{-1}(\boldsymbol{L} + \boldsymbol{U}) = \begin{pmatrix} 0 & 4 & 0 \\ 5 & 0 & -2 \\ -2/7 & -2/7 & 0 \end{pmatrix}. \tag{7.35}$$

The three eigenvalues of $\boldsymbol{N}$ are $\lambda_i = 4.590, -4.479, -0.111$. So its spectral radius is $\rho(\boldsymbol{N}) = \max(|\lambda_i|) = 4.59 > 1$, which means that the iteration scheme will diverge.

If we revisit our earlier example, we have

$$\boldsymbol{D}^{-1} = \begin{pmatrix} 1/5 & 0 & 0 \\ 0 & 1/4 & 0 \\ 0 & 0 & -1/7 \end{pmatrix}, \quad \text{eig}(\boldsymbol{D}^{-1}) = \frac{1}{5}, \frac{1}{4}, \frac{-1}{7}, \tag{7.36}$$

and

$$\boldsymbol{N} = \boldsymbol{D}^{-1}(\boldsymbol{L} + \boldsymbol{U}) = \begin{pmatrix} 0 & 1/5 & -2/5 \\ 1/4 & 0 & 0 \\ -2/7 & -2/7 & 0 \end{pmatrix}, \tag{7.37}$$

whose eigenvalues are

$$\lambda_1 = 0.4739, \qquad \lambda_{2,3} = -0.2369 \pm 0.0644i. \tag{7.38}$$

So we have

$$\rho(\boldsymbol{D}^{-1}) = 1/4 < 1, \qquad \rho(\boldsymbol{N}) = 0.4739 < 1. \tag{7.39}$$

That is why the earlier iteration procedure is convergent.

7.4.2 *Gauss-Seidel Iteration*

In the Jacobi-type iterations, we have to store both $\boldsymbol{u}^{(n+1)}$ and $\boldsymbol{u}^{(n)}$ as we will use all the $\boldsymbol{u}_j^{(n)}$ values to compute the values at the next level $t = n+1$, this means that we cannot use the running update when the new approximate has just been computed

$$\boldsymbol{u}_j^{(n+1)} \to \boldsymbol{u}_j^{(n)}, \qquad (j = 1, 2, ...).$$

If the vector size $\boldsymbol{u}$ is large (it usually is), then we can devise other iteration schemes to save memory by using the running update. So only one-vector storage is needed.

The Gauss-Seidel iteration procedure is such a procedure to provide an efficient way of solving the linear matrix equation $\mathbf{A}\mathbf{u} = \mathbf{b}$. It uses the same decomposition as the Jacobi-type iteration by splitting $\boldsymbol{A}$ into

$$\mathbf{A} = \mathbf{L} + \mathbf{D} + \mathbf{U}, \tag{7.40}$$

but the difference from the Jacobi method is that we use $\boldsymbol{L} + \boldsymbol{D}$ instead of $\boldsymbol{D}$ for the inverse so that the running update is possible. The n-th step iteration is updated by

$$(\boldsymbol{L} + \boldsymbol{D})\boldsymbol{u}^{(n+1)} = \boldsymbol{b} - \boldsymbol{U}\boldsymbol{u}^{(n)}, \tag{7.41}$$

or

$$\boldsymbol{u}^{(n+1)} = (\boldsymbol{L} + \boldsymbol{D})^{-1}[\boldsymbol{b} - \boldsymbol{U}\boldsymbol{u}^{(n)}]. \tag{7.42}$$

This procedure, starting from an initial vector $\boldsymbol{u}^{(0)}$, stops if a prescribed criterion is reached.

It is worth pointing out that Gauss-Seidel iteration requires the same criteria of convergence as for the Jacobi-type iteration method. That is to say, the inverse of the matrix must exist, and the largest spectral radius must be less than 1.

7.4.3 *Relaxation Method*

The above Gauss-Seidel iteration method is still slow, and the relaxation method provides a more efficient iteration procedure. A popular method is the successive over-relaxation method which consists of two steps

$$\boldsymbol{v}^{(n)} = (\boldsymbol{L} + \boldsymbol{D} + \boldsymbol{U})\boldsymbol{u}^{(n)} - \boldsymbol{b}, \tag{7.43}$$

and

$$\boldsymbol{u}^{(n+1)} = \boldsymbol{u}^{(n)} - \omega(\boldsymbol{L} + \boldsymbol{D})^{-1}\boldsymbol{v}^{(n)}, \tag{7.44}$$

where $0 < \omega < 2$ is the over-relaxation parameter. If we combine the above equations and rearrange them, we have

$$\boldsymbol{u}^{(n+1)} = (1 - \omega)\boldsymbol{u}^{(n)} + \omega\tilde{\boldsymbol{u}}^{(n)}, \tag{7.45}$$

where $\tilde{\boldsymbol{u}}^{(n)} = (\boldsymbol{L} + \boldsymbol{D})^{-1}(\boldsymbol{b} - \boldsymbol{U}\boldsymbol{u}^{(n)})$ is the standard Gauss-Seidel procedure. Therefore, this method is essentially the weighted average between the previous iteration and the successive Gauss-Seidel iteration. Clearly, if $\omega = 1$, then it reduces to the standard Gauss-Seidel iteration method.

Broadly speaking, a small value of $0 < \omega < 1$ corresponds to under-relaxation with slower convergence, while $1 < \omega < 2$ leads to over-relaxation and faster convergence. It has been proved theoretically that the scheme will not converge if $\omega < 0$ or $\omega > 2$.

7.5 Newton-Raphson Method

Sometimes, the algebraic equations we meet are nonlinear, and direct inversion is not the best technique. In this case, more elaborate techniques should be used.

The nonlinear algebraic equation

$$\mathbf{A}(\mathbf{u})\mathbf{u} = \mathbf{b}(\mathbf{u}), \qquad \text{or} \qquad \mathbf{F}(\mathbf{u}) = \mathbf{A}(\mathbf{u})\mathbf{u} - \mathbf{b}(\mathbf{u}) = \mathbf{0}, \tag{7.46}$$

can be solved using a simple iteration technique

$$\mathbf{A}(\boldsymbol{u}^{(n)})\boldsymbol{u}^{(n+1)} = \boldsymbol{b}(\boldsymbol{u}^{(n)}), \qquad n = 0, 1, 2, ... \tag{7.47}$$

until $||\boldsymbol{u}^{(n+1)} - \boldsymbol{u}^{(n)}||$ is sufficiently small. Iterations require a starting vector $\mathbf{u}^{(0)}$. This method is also referred to as the successive substitution method.

If this simple method does not work, the relaxation method can be used. The relaxation technique first gives a tentative new approximation $\mathbf{u}^*$ from $\boldsymbol{A}(\boldsymbol{u}^{(n)})\boldsymbol{u}^* = \boldsymbol{b}(\boldsymbol{u}^{(n)})$, then we use

$$\boldsymbol{u}^{(n+1)} = \omega \boldsymbol{u}^* + (1-\omega)\boldsymbol{u}^{(n)}, \qquad \omega \in (0, 1], \tag{7.48}$$

where ω is a prescribed relaxation parameter.

The nonlinear equation (7.46) can also be solved using the Newton-Raphson procedure. We approximate $\mathbf{F}(\mathbf{u})$ by a linear function $\boldsymbol{R}(\boldsymbol{u}; \boldsymbol{u}^{(n)})$ in the vicinity of an existing approximation $\boldsymbol{u}^{(n)}$ to $\boldsymbol{u}$:

$$\boldsymbol{R}(\boldsymbol{u}; \boldsymbol{u}^{(n)}) = \boldsymbol{F}(\boldsymbol{u}^{(n)}) + \boldsymbol{J}(\boldsymbol{u}^{(n)})(\boldsymbol{u} - \boldsymbol{u}^{(n)}), \;\; \boldsymbol{J}(\boldsymbol{u}) = \nabla \boldsymbol{F}, \tag{7.49}$$

where $\boldsymbol{J}$ is the Jacobian of $\boldsymbol{F}(\boldsymbol{u}) = (F_1, F_2, ..., F_M)^T$.

For $\boldsymbol{u} = (u_1, u_2, ..., u_M)^T$, we have

$$\boldsymbol{J}_{ij} = \frac{\partial F_i}{\partial u_j}. \tag{7.50}$$

To find the next approximation $\boldsymbol{u}^{(n+1)}$ from $\boldsymbol{R}(\boldsymbol{u}^{(n+1)}; \boldsymbol{u}^{(n)}) = 0$, we can solve the linear system with $\boldsymbol{J}$ as the coefficient matrix

$$\boldsymbol{u}^{(n+1)} = \boldsymbol{u}^{(n)} - \boldsymbol{J}^{-1}\boldsymbol{F}(\boldsymbol{u}^{(n)}), \tag{7.51}$$

under a given termination criterion $||\boldsymbol{u}^{(n+1)} - \boldsymbol{u}^{(n)}|| \leq \epsilon$. Here, $0 < \epsilon \ll 1$ is a very small tolerance.

There are some extensive literature concerning nonlinear iteration methods. Interested readers can refer to more advanced literature.

7.6 QR Decomposition

The QR decomposition employs the orthogonality to decompose an $n \times n$ real matrix $\boldsymbol{A}$. A matrix $\boldsymbol{Q}$ is orthogonal if it satisfies

$$\boldsymbol{Q}^T\boldsymbol{Q} = \boldsymbol{Q}\boldsymbol{Q}^T = I, \tag{7.52}$$

where I is an identity matrix. Since

$$\boldsymbol{A} = \boldsymbol{I}\boldsymbol{A} = (\boldsymbol{Q}\boldsymbol{Q}^T)\boldsymbol{A} = \boldsymbol{Q}(\boldsymbol{Q}^T A) = \boldsymbol{Q}\boldsymbol{R}, \tag{7.53}$$

where

$$\boldsymbol{R} = \boldsymbol{Q}^T\boldsymbol{A}. \tag{7.54}$$

In this decomposition, $\boldsymbol{Q}$ is orthogonal, while $\boldsymbol{R}$ is an upper triangular matrix

$$\boldsymbol{R} = \begin{pmatrix} \alpha_{11} & \alpha_{12} & \dots & \alpha_{1n} \\ 0 & \alpha_{22} & \dots & \alpha_{2n} \\ \vdots & & \ddots & \vdots \\ 0 & 0 & \dots & \alpha_{nn} \end{pmatrix}. \tag{7.55}$$

The advantage of QR decomposition can be used to solve a linear system and to calculate the determinant of $\boldsymbol{A}$. For a linear system,

$$\boldsymbol{A}\boldsymbol{y} = \boldsymbol{b}, \tag{7.56}$$

we have

$$\boldsymbol{A}\boldsymbol{y} = \boldsymbol{Q}\boldsymbol{R}\boldsymbol{y} = \boldsymbol{Q}(\boldsymbol{R}\boldsymbol{y}) = \boldsymbol{Q}\boldsymbol{z} = \boldsymbol{b}, \tag{7.57}$$

which is equivalent to two linear systems

$$\boldsymbol{Q}\boldsymbol{z} = \boldsymbol{b}, \qquad \boldsymbol{R}\boldsymbol{y} = \boldsymbol{z}. \tag{7.58}$$

We first solve $\boldsymbol{Q}\boldsymbol{z} = \boldsymbol{b}$ to get $\boldsymbol{z}$, then solve $\boldsymbol{R}\boldsymbol{y} = \boldsymbol{z}$ to obtain $\boldsymbol{y}$. As $\boldsymbol{Q}$ is an orthogonal matrix, then $\boldsymbol{Q}^{-1} = \boldsymbol{Q}^T$. The solution of $\boldsymbol{z}$ can be obtained by simple transpose and matrix multiplications. That is $\boldsymbol{z} = \boldsymbol{Q}^{-1}\boldsymbol{b} = \boldsymbol{Q}^T\boldsymbol{b}$. Since $\boldsymbol{R}$ is an upper triangular matrix, thus $\boldsymbol{y}$ can be obtained using back substitutions.

From $\boldsymbol{A} = \boldsymbol{Q}\boldsymbol{R}$, we have the determinant

$$\det(\boldsymbol{A}) = \det(\boldsymbol{Q})\det(\boldsymbol{R}). \tag{7.59}$$

Since $\boldsymbol{Q}$ is orthogonal, we have $\det|\boldsymbol{Q}| = 1$. The determinant of a triangular matrix is simply the product of its diagonal elements. Finally, we have

$$\det(\boldsymbol{A}) = \det(\boldsymbol{R}) = \prod_{i=1}^{n} \alpha_{ii}. \tag{7.60}$$

The basic idea of constructing an orthogonal matrix $\boldsymbol{Q}$ is to use the Gram-Schmidt procedure. The first step is to write the matrix $\boldsymbol{A}$ in terms of n column vectors $\boldsymbol{v}_i$. That is

$$\boldsymbol{A} = \begin{pmatrix} a_{11} & a_{12} & \dots & a_{1n} \\ a_{21} & a_{22} & \dots & a_{2n} \\ \vdots & & \ddots & \\ a_{n1} & a_{n2} & \dots & a_{nn} \end{pmatrix} = \begin{pmatrix} \boldsymbol{u}_1 & \boldsymbol{u}_2 & \dots & \boldsymbol{u}_n \end{pmatrix}, \tag{7.61}$$

where

$$\boldsymbol{u}_1 = \begin{pmatrix} a_{11} \\ a_{21} \\ \vdots \\ a_{n1} \end{pmatrix}, \quad \boldsymbol{u}_2 = \begin{pmatrix} a_{12} \\ a_{22} \\ \vdots \\ a_{n2} \end{pmatrix}, \quad \boldsymbol{u}_n = \begin{pmatrix} a_{1n} \\ a_{2n} \\ \vdots \\ a_{nn} \end{pmatrix}. \tag{7.62}$$

Initially, these vectors $\boldsymbol{u}_i$ are in general not orthogonal (perpendicular to each other). So we have to make them orthogonal by the Gram-Schmidt procedure so that the set of unorthogonal $\boldsymbol{u}_i$ will be transformed into a new set of orthogonal $\boldsymbol{v}_i$. First, we let

$$\boldsymbol{v}_1 = \boldsymbol{u}_1 \tag{7.63}$$

as the first new vector. We then define a unit vector associated with this vector as

$$\boldsymbol{e}_1 = \frac{\boldsymbol{v}_1}{\|\boldsymbol{v}_1\|}, \tag{7.64}$$

where $\|\boldsymbol{v}\|$ is the norm of the vector $\boldsymbol{v}$. That is

$$\|\boldsymbol{v}\| = \sqrt{\boldsymbol{v} \cdot \boldsymbol{v}}, \tag{7.65}$$

where $\boldsymbol{v} \cdot \boldsymbol{w}$ is the inner product of two vectors

$$\boldsymbol{v} \cdot \boldsymbol{w} = \boldsymbol{v}^T \boldsymbol{w}. \tag{7.66}$$

For the second vector $\boldsymbol{u}_2$, we have to use projection by inner product to construct a new vector $\boldsymbol{v}_2$ by using

$$\boldsymbol{v}_2 = \boldsymbol{u}_2 - \mathrm{proj}_{\boldsymbol{e}_1}(\boldsymbol{u}_2), \tag{7.67}$$

where the projection operator is defined by

$$\mathrm{proj}_{\boldsymbol{e}_i}(\boldsymbol{u}_j) = \frac{\boldsymbol{e}_i \cdot \boldsymbol{u}_j}{\boldsymbol{e}_i \cdot \boldsymbol{e}_i} \boldsymbol{e}_i. \tag{7.68}$$

Similarly, we can define a unit vector $\boldsymbol{e}_2$ by

$$\boldsymbol{e}_2 = \frac{\boldsymbol{v}_2}{\|\boldsymbol{v}_2\|}. \tag{7.69}$$

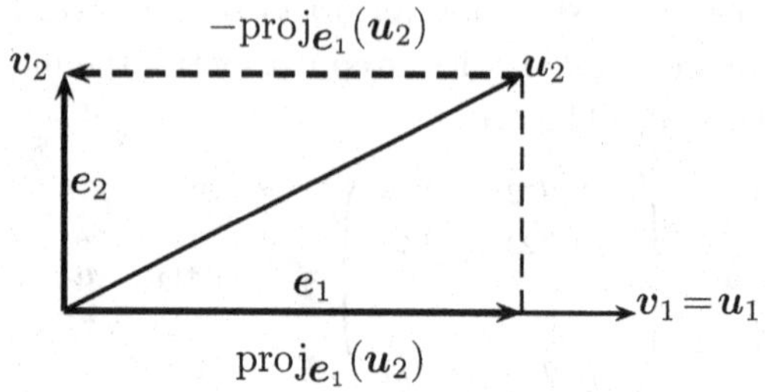

Fig. 7.1 Orthogonalization by projection.

This second step of orthogonalization process is shown in Figure 7.1. We can see that the new unit vectors $\boldsymbol{e}_1$ and $\boldsymbol{e}_2$ are mutually orthogonal. Thus, the new system spanned by such orthogonal unit vectors is an orthonormal system.

Similarly, the other orthogonal vectors are given by

$$\boldsymbol{v}_i = \boldsymbol{u}_i - \sum_{k=1}^{i-1} \text{proj}_{\boldsymbol{e}_k}(\boldsymbol{u}_i), \tag{7.70}$$

so that

$$\boldsymbol{e}_i = \frac{\boldsymbol{v}_i}{\|\boldsymbol{v}_i\|}. \tag{7.71}$$

The $\boldsymbol{Q}$ matrix in the QR factorization is formed by these new orthogonal unit vectors $\boldsymbol{e}_i$. That is

$$\boldsymbol{Q} = \big(\boldsymbol{e}_1\ \boldsymbol{e}_2\ ...\ \boldsymbol{e}_n\big), \tag{7.72}$$

and

$$\boldsymbol{R} = \boldsymbol{Q}^T \boldsymbol{A}. \tag{7.73}$$

This procedure is better demonstrated by an example.

Example 7.5: *We now try to factorize the following matrix*

$$\boldsymbol{A} = \begin{pmatrix} -2 & 2 & 3 \\ 2 & -2 & 2 \\ 0 & 2 & 5 \end{pmatrix},$$

using the above QR decomposition. The first step is to write

$$\boldsymbol{A} = \big(\boldsymbol{u}_1\ \boldsymbol{u}_2\ \boldsymbol{u}_3\big),$$

where

$$\boldsymbol{u}_1 = \begin{pmatrix} -2 \\ 2 \\ 0 \end{pmatrix}, \quad \boldsymbol{u}_2 = \begin{pmatrix} 2 \\ -2 \\ 2 \end{pmatrix}, \quad \boldsymbol{u}_3 = \begin{pmatrix} 3 \\ 2 \\ 5 \end{pmatrix}.$$

The first new orthogonal vector can be chosen as $\boldsymbol{v}_1 = \boldsymbol{u}_1$. *As the length of* $v_1 = \|\boldsymbol{v}_1\| = \sqrt{(-2)^2 + 2^2 + 0^2} = 2\sqrt{2}$, *the unit vector is given by*

$$\boldsymbol{e}_1 = \frac{\boldsymbol{v}_1}{\|\boldsymbol{v}_1\|} = \begin{pmatrix} -\sqrt{2}/2 \\ \sqrt{2}/2 \\ 0 \end{pmatrix}.$$

Since the inner product $\boldsymbol{u}_2 \cdot \boldsymbol{e}_1$ *is*

$$\boldsymbol{e}_1 \cdot \boldsymbol{u}_2 = \boldsymbol{e}_1^T \boldsymbol{u}_2 = \begin{pmatrix} -\sqrt{2}/2 \ \sqrt{2}/2 \ 0 \end{pmatrix} \begin{pmatrix} 2 \\ -2 \\ 2 \end{pmatrix} = -2\sqrt{2},$$

we have

$$\boldsymbol{v}_2 = \boldsymbol{u}_2 - \frac{\boldsymbol{e}_1 \cdot \boldsymbol{u}_2}{\boldsymbol{e}_1 \cdot \boldsymbol{e}_1} \boldsymbol{e}_1 = \boldsymbol{u}_2 - (-2\sqrt{2}) \boldsymbol{e}_1$$

$$= \begin{pmatrix} 2 \\ -2 \\ 2 \end{pmatrix} + 2\sqrt{2} \begin{pmatrix} -\sqrt{2}/2 \\ \sqrt{2}/2 \\ 0 \end{pmatrix} = \begin{pmatrix} 0 \\ 0 \\ 2 \end{pmatrix},$$

where we have used $\boldsymbol{e}_1 \cdot \boldsymbol{e}_1 = 1$. *As* $\|\boldsymbol{v}_2\| = 2$, *we have*

$$\boldsymbol{e}_2 = \frac{\boldsymbol{v}_2}{\|\boldsymbol{v}_2\|} = \begin{pmatrix} 0 \\ 0 \\ 1 \end{pmatrix}.$$

Similarly, since

$$\boldsymbol{e}_1 \cdot \boldsymbol{u}_3 = \begin{pmatrix} -\sqrt{2}/2 \ \sqrt{2}/2 \ 0 \end{pmatrix} \begin{pmatrix} 3 \\ 2 \\ 5 \end{pmatrix} = -\sqrt{2}/2,$$

and

$$\boldsymbol{e}_2 \cdot \boldsymbol{u}_3 = \begin{pmatrix} 0 \ 0 \ 1 \end{pmatrix} \begin{pmatrix} 3 \\ 2 \\ 5 \end{pmatrix} = 5,$$

we have

$$\boldsymbol{v}_3 = \boldsymbol{u}_3 - (-\sqrt{2}/2) \boldsymbol{e}_1 - 5 \boldsymbol{e}_2$$

$$= \begin{pmatrix} 3 \\ 2 \\ 5 \end{pmatrix} + \frac{\sqrt{2}}{2} \begin{pmatrix} -\sqrt{2}/2 \\ \sqrt{2}/2 \\ 0 \end{pmatrix} - 5 \begin{pmatrix} 0 \\ 0 \\ 1 \end{pmatrix} = \begin{pmatrix} 5/2 \\ 5/2 \\ 0 \end{pmatrix}.$$

As the length of $\boldsymbol{v}_3$ is $\|\boldsymbol{v}_3\| = 5\sqrt{2}/2$, we have

$$\boldsymbol{e}_3 = \frac{\boldsymbol{v}_3}{\|\boldsymbol{v}_3\|} = \begin{pmatrix} \sqrt{2}/2 \\ \sqrt{2}/2 \\ 0 \end{pmatrix}.$$

So the matrix $\boldsymbol{Q}$ in the QR decomposition is formed by all these unit vectors that are mutually orthogonal, and we have

$$\boldsymbol{Q} = \begin{pmatrix} \boldsymbol{e}_1 \ \boldsymbol{e}_2 \ \boldsymbol{e}_3 \end{pmatrix} = \begin{pmatrix} -\sqrt{2}/2 & 0 & \sqrt{2}/2 \\ \sqrt{2}/2 & 0 & \sqrt{2}/2 \\ 0 & 1 & 0 \end{pmatrix}.$$

Furthermore, we have

$$\boldsymbol{R} = \boldsymbol{Q}^T \boldsymbol{A} = \begin{pmatrix} -\sqrt{2}/2 & \sqrt{2}/2 & 0 \\ 0 & 0 & 1 \\ \sqrt{2}/2 & \sqrt{2}/2 & 0 \end{pmatrix} \begin{pmatrix} -2 & 2 & 3 \\ 2 & -2 & 2 \\ -2 & 5 \end{pmatrix}$$

$$= \begin{pmatrix} 2\sqrt{2} & -2\sqrt{2} & -\sqrt{2}/2 \\ 0 & 2 & 5 \\ 0 & 0 & 5\sqrt{2}/2 \end{pmatrix}.$$

Therefore, the final QR decomposition $\boldsymbol{A} = \boldsymbol{QR}$ becomes

$$\boldsymbol{Q} = \begin{pmatrix} -\sqrt{2}/2 & 0 & \sqrt{2}/2 \\ \sqrt{2}/2 & 0 & \sqrt{2}/2 \\ 0 & 1 & 0 \end{pmatrix}, \quad \boldsymbol{R} = \begin{pmatrix} 2\sqrt{2} & -2\sqrt{2} & -\sqrt{2}/2 \\ 0 & 2 & 5 \\ 0 & 0 & 5\sqrt{2}/2 \end{pmatrix}.$$

It is straightforward to verify that $\boldsymbol{Q}$ is an orthogonal matrix because

$$\boldsymbol{Q}^T\boldsymbol{Q} = \boldsymbol{Q}\boldsymbol{Q}^T = \begin{pmatrix} 1 & 0 & 0 \\ 0 & 1 & 0 \\ 0 & 0 & 1 \end{pmatrix}.$$

Indeed, $\boldsymbol{R}$ is an upper triangular matrix.

Furthermore, the determinant of $\boldsymbol{A}$ is

$$\det(\boldsymbol{A}) = 20,$$

while the determinant of the upper triangular matrix $\boldsymbol{R}$ is simply

$$\det(\boldsymbol{R}) = 2\sqrt{2} \times 2 \times \frac{5\sqrt{2}}{2} = 20,$$

so we have $\det(\boldsymbol{A}) = \det(\boldsymbol{R}) = 20.$

There are many other powerful methods for computing QR decompositon, including Householder's reflection method and the Givens rotation method.

Although we have used the square matrices in our discussion here, QR decomposition can be carried out in general for any $m \times n$ matrix where $m \geq n$. In this case, $\boldsymbol{Q}$ is $m \times m$ and $\boldsymbol{R}$ is $m \times n$. In addition, there is another decomposition $\boldsymbol{A} = \boldsymbol{QL}$, called QL decomposition, which is similar to QR decomposition. The only difference here is that the matrix $\boldsymbol{L}$ is a lower triangular matrix, and the procedure is essentially the same as the that for QR decomposition.

7.7 Conjugate Gradient Method

The method of conjugate gradient can be used to solve the following linear system

$$\boldsymbol{A}\boldsymbol{u} = \boldsymbol{b}, \tag{7.74}$$

where $\boldsymbol{A}$ is often a symmetric positive definite matrix. The above system is equivalent to minimising the following function $f(\boldsymbol{u})$

$$f(\boldsymbol{u}) = \frac{1}{2}\boldsymbol{u}^T\boldsymbol{A}\boldsymbol{u} - \boldsymbol{b}^T\boldsymbol{u} + \boldsymbol{v}, \tag{7.75}$$

where $\boldsymbol{v}$ is a constant and can be taken to be zero. We can easily see that $\nabla f(\boldsymbol{u}) = 0$ leads to $\boldsymbol{A}\boldsymbol{u} = \boldsymbol{b}$. The theory behind these iterative methods is closely related to the Krylov subspace $\mathcal{K}$ spanned by $\boldsymbol{A}$ and $\boldsymbol{b}$ as defined by

$$\mathcal{K}_n(\boldsymbol{A}, \boldsymbol{b}) = \{\boldsymbol{I}\boldsymbol{b}, \boldsymbol{A}\boldsymbol{b}, \boldsymbol{A}^2\boldsymbol{b}, ..., \boldsymbol{A}^{n-1}\boldsymbol{b}\}, \tag{7.76}$$

where $\boldsymbol{A}^0 = \boldsymbol{I}$.

If we use a iterative procedure to obtain the estimate $\boldsymbol{u}_n$ at n-th iteration, the residual is given by

$$\boldsymbol{r}_n = \boldsymbol{b} - \boldsymbol{A}\boldsymbol{u}_n, \tag{7.77}$$

which is essentially the negative gradient $\nabla f(\boldsymbol{u}_n)$. The search direction vector in the conjugate gradient method is subsequently determined by

$$\boldsymbol{d}_{n+1} = \boldsymbol{r}_n - \frac{\boldsymbol{d}_n^T\boldsymbol{A}\boldsymbol{r}_n}{\boldsymbol{d}_n^T\boldsymbol{A}\boldsymbol{d}_n}\boldsymbol{d}_n. \tag{7.78}$$

The solution often starts with an initial guess $\boldsymbol{u}_0$ at $n = 0$, and proceeds iteratively. The above steps can compactly be written as

$$\boldsymbol{u}_{n+1} = \boldsymbol{u}_n + \alpha_n\boldsymbol{d}_n, \quad \boldsymbol{r}_{n+1} = \boldsymbol{r}_n - \alpha_n\boldsymbol{A}\boldsymbol{d}_n, \tag{7.79}$$

and

$$\boldsymbol{d}_{n+1} = \boldsymbol{r}_{n+1} + \beta_n \boldsymbol{d}_n, \tag{7.80}$$

where

$$\alpha_n = \frac{\boldsymbol{r}_n^T \boldsymbol{r}_n}{\boldsymbol{d}_n^T \boldsymbol{A} \boldsymbol{d}_n}, \qquad \beta_n = \frac{\boldsymbol{r}_{n+1}^T \boldsymbol{r}_{n+1}}{\boldsymbol{r}_n^T \boldsymbol{r}_n}. \tag{7.81}$$

Iterations stop when a prescribed accuracy is reached. This can easily be programmed in any programming language, especially Matlab.

The conjugate gradient method is a powerful method with a diverse range of applications. In fact, it was voted as one of the top 10 most popular methods in scientific computing.

Chapter 8

Interpolation

Interpolation forms an important part of numerical algorithms which are widely used in modern scientific computing and computational geometry. We will briefly review the standard interpolation methods such as the spline interpolation and the Bézier curve.

8.1 Spline Interpolation

The spline interpolation is to construct a function, called spline function, of degree m for given $n+1$ known values y_i at $n+1$ data points $x_i (i = 0, 1, 2, ..., n)$. These given points are organized in an increasing order so that

$$x_0 < x_1 < ... < x_n. \tag{8.1}$$

The values y_i at the given data points are often called knot values. The major requirement is that the constructed spline function $S(x)$ should be continuous and can produce the exact values at the data points. The spline function can be constructed in a piecewise manner so that each spline function $S_i(x)$ is valid in each data interval $x \in [x_i, x_{i+1}]$.

8.1.1 *Linear Spline Functions*

In the simplest case of three points with given values (y_{i-1}, y_i, y_{i+1}) at three distinct points x_{i-1}, x_i and x_{i+1}, the simplest spline is the linear functions, which can be constructed in each interval $[x_i, x_{i+1}]$ so that we have

$$S_i(x) = y_i + \frac{(y_{i+1} - y_i)}{(x_{i+1} - x_i)}(x - x_i), \quad x \in [x_i, x_{i+1}], \tag{8.2}$$

which corresponds to the piecewise line segments in Fig. 8.1.

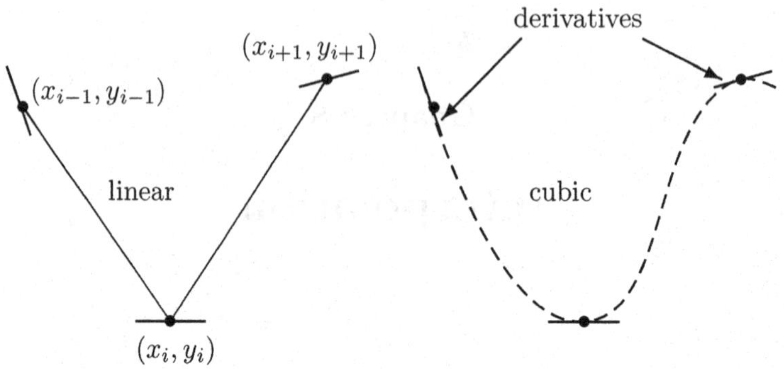

Fig. 8.1 Spline construction for given values (y_{i-1}, y_i, y_{i+1}) and their derivatives $(y'_{i-1}, y'_i, y'_{i+1})$ at three points x_{i-1}, x_i and x_{i+1}.

For two consecutive intervals, the functions S_{i-1} and S_i should be continuous, that is to say, $S_i(x_i) = S_{i-1}(x_i) = y_i$. As $i = 0, 2, ..., n-1$, there are n such spline functions. In order to systematically construct the spline interpolation for $n+1$ points, we can write

$$y = f(x) = \sum_{i=0}^{n} y_i \theta(x), \tag{8.3}$$

where $\theta_i(x)$ are the linear elementary basis functions.

$$\theta_i(x) = \begin{cases} \frac{x-x_i}{h_i}, & x \in [x_i, x_{i+1}] \\ \\ \frac{x_i-x}{h_{i-1}}, & x \in [x_{i-1}, x_i], \end{cases} \tag{8.4}$$

where $h_i = (x_{i+1} - x_i)$ and $h_{i-1} = (x_i - x_{i-1})$. The linear spline function $S_i(x)$ in the interval $x \in [x_{i-1}, x_i]$ is expressed as

$$S_i(x) = y_{i-1}\theta_{i-1} + y_i\theta_i, \tag{8.5}$$

which can be represented geometrically as the dashed and dotted lines in Fig. 8.2. We can see that $S_i(x)$ is continuous, but not necessary smooth. The first derivatives of the linear spline function are not continuous.

8.1.2 *Cubic Spline Functions*

The linear spline interpolation is continuous, but not smooth as there is a discontinuity in the first derivatives. Now suppose the three derivatives y'_{i-1}, y'_i, y'_{i+1} are also given at these three points (x_{i-1}, y_{i-1}), (x_i, y_i) and

(x_{i+1}, y_{i+1}), can we construct a class of better and smoother spline functions (such as the dashed curve shown in Fig. 8.1) so that their values and derivatives meet the given conditions? The answer is yes, that is the cubic spline function. You may wonder why cubic?

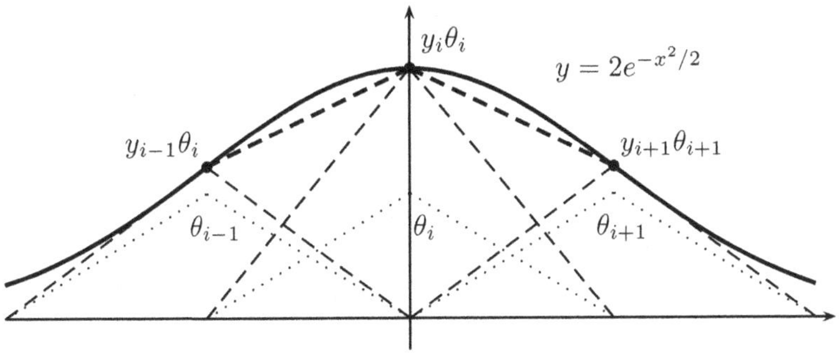

Fig. 8.2 Linear spline construction from $\theta(x)$.

In the interval $[x_i, x_{i+1}]$, we now have four conditions: two function values y_i, y_{i+1}, and two derivatives y'_i, y'_{i+1}. Obviously, linear function is not enough, how about quadratic function $S(x) = ax^2 + bx + c$? The function is relatively smooth and the first derivative could meet these requirements, but the second derivative $S''(x) = a$ is constant, and if we require the second-derivative also continuous, this means that either the second derivative is constant everywhere (thus a is same everywhere) or there is a discontinuity. Furthermore, we have only three unknown a, b, c with four conditions, so in general they are over-determined, and not all conditions will be met. Thus, we need a cubic function.

If we use the following generic cubic function

$$S_i(x) = \alpha_i(x - x_i)^3 + \beta_i(x - x_i)^2 + \gamma_i(x - x_i) + \delta_i, \tag{8.6}$$

where $i = 1, 2, ..., n$, then the spline function is twice continuous differentiable. Its first derivative is

$$S'_i(x) = 3\alpha_i(x - x_i)^2 + 2\beta_i(x - x_i) + \gamma_i. \tag{8.7}$$

The four conditions become

$$S_i(x_i) = y_i, \qquad S_i(x_{i+1}) = y_{i+1}, \tag{8.8}$$

and

$$S'_i(x_i) = y'_i, \qquad S'_i(x_{i+1}) = y'_{i+1}. \tag{8.9}$$

Thus, there are four equations and four unknown $\alpha_i, \beta_i, \gamma_i$ and δ_i, so all the four unknowns can be uniquely determined. The general requirements among different intervals are $S_i(x), S_i'(x)$, and $S_i''(x)$ should be continuous. We have

$$S_i(x_i) = y_i, S_i(x_{i+1}) = y_{i+1}, \qquad i = 0, 2, ..., n-1, \tag{8.10}$$

$$S_i'(x_{i+1}) = S_{i+1}'(x_{i+1}) = y'(x_{i+1}), \;\; i = 0, 1, ..., n-2, \tag{8.11}$$

and

$$S_i''(x_{i+1}) = S_{i+1}''(x_{i+1}), \qquad i = 0, 1, ..., n-2. \tag{8.12}$$

In the n intervals, we have $4n$ unknowns, but we have $4n-2$ conditions: $n+1$ from $y_i (i = 0, 1, ..., n-1)$; $2(n-1)$ from y_i' and S_i' $(i = 0, 1, ..., n-2)$; $n-1$ from S_i'', $(i = 0, 1, ..., n-2)$, so we need 2 more conditions.

The two extra conditions are at the two end points $i = 0$ and $i = n$. The clamped boundary conditions are to set

$$S_0'(x_0) = y_0', \qquad S_{n-1}'(x_n) = y_n', \tag{8.13}$$

which are usually given. If the derivatives at the end points are not given, we can use the natural or free boundary conditions:

$$S_0''(x_0) = 0, \qquad S_{n-1}''(x_n) = 0. \tag{8.14}$$

In order to find the spline functions $S_i(x)$, it is conventional to rewrite them in terms of their second derivatives $\xi_i = S_i''(x)$, $(i = 0, 1, 2, ..., n-1)$, and we have

$$\xi_i = 6\alpha_i(x - x_i) + 2\beta_i. \tag{8.15}$$

At $x = x_i$, we have

$$\xi_i(x_i) = 2\beta_i, \tag{8.16}$$

or

$$\beta_i = \frac{\xi_i}{2}. \tag{8.17}$$

Using the continuity of ξ_i at $x = x_{i+1}$, we have

$$\xi_i(x_{i+1}) = \xi_{i+1}(x_{i+1}) = 6\alpha_i h_i + 2\beta_i, \;\; h_i = x_{i+1} - x_i. \tag{8.18}$$

Combining with Eq. (8.17), we have

$$\alpha_i = \frac{\xi_{i+1} - \xi_i}{6h_i}. \tag{8.19}$$

Since $S_i(x_i) = y_i$ at $x = x_{i+1}$, we have

$$\delta_i = y_i. \tag{8.20}$$

Substituting α_i, β_i and δ_i into $S_i(x_{i+1}) = y_{i+1}$ at $x = x_{i+1}$ and after some re-arrangements, we have

$$\gamma_i = \frac{(y_{i+1} - y_i)}{h_i} - \frac{h_i(2\xi_i + \xi_{i+1})}{6}. \tag{8.21}$$

Now substituting these coefficients into Eq. (8.3), we can express S_i in terms of the second derivatives ξ_i, and we have

$$S_i(x) = \frac{1}{6h_i}[\xi_{i+1}(x - x_i)^3 + \xi_i(x_{i+1} - x)^3]$$

$$+ [\frac{y_{i+1}}{h_i} - \frac{h_i}{6}\xi_{i+1}](x - x_i) + [\frac{y_i}{h_i} - \frac{h_i}{6}\xi_i](x_{i+1} - x). \tag{8.22}$$

This is a cubic polynomial, and the only thing left is to find the coefficients ξ_i. Using the continuity conditions: $S_i(x_i) = S_{i-1}(x_i) = y_i$ and $S_i'(x_i) = S_{i-1}'(x_i)$ at $x = x_i$; $S_i(x_{i+1}) = S_{i+1}(x_{i+1}) = y_{i+1}$ and $S_i'(x_{i+1}) = S_{i+1}'(x_{i+1})$ at $x = x_{i+1}$, we can rewrite the above equation as

$$h_{i-1}\xi_{i-1} + 2(h_{i-1} + h_i)\xi_i + h_i\xi_{i+1} = 6[\frac{(y_{i+1} - y_i)}{h_i} - \frac{(y_i - y_{i-1})}{h_{i-1}}], \tag{8.23}$$

where $i = 1, 2, ..., n-1$. Writing them in a matrix form, we have

$$\begin{pmatrix} 2(h_0 + h_1) & h_1 & \dots & 0 \\ h_1 & 2(h_1 + h_2) & \dots & 0 \\ \vdots & & \ddots & h_{n-2} \\ 0 & \dots & h_{n-2} & 2(h_{n-2} + h_{n-1}) \end{pmatrix} \begin{pmatrix} \xi_1 \\ \xi_2 \\ \vdots \\ \xi_{n-1} \end{pmatrix}$$

$$= \begin{pmatrix} 6(\frac{y_2 - y_1}{h_1} - \frac{y_1 - y_0}{h_0}) \\ \vdots \\ 6(\frac{y_n - y_{n-1}}{h_{n-1}} - \frac{y_{n-1} - y_{n-2}}{h_{n-2}}) \end{pmatrix}. \tag{8.24}$$

Since $\xi_0 = 0$ and $\xi_n = 0$ are given from the natural boundary conditions, this linear system will uniquely determine $\xi_1, ..., \xi_{n-1}$. For any given set of data, we should solve the linear system to get ξ_i, then to compute $S_i(x)$.

In the case of equal spacing $h_0 = h_i = h_{n-1} = h$, the above equation becomes

$$\begin{pmatrix} 4 & 1 & \dots & 0 \\ 1 & 4 & \dots & 0 \\ \vdots & & \ddots & 1 \\ 0 & \dots & 1 & 4 \end{pmatrix} \begin{pmatrix} \xi_1 \\ \xi_2 \\ \vdots \\ \xi_{n-1} \end{pmatrix} = \frac{6}{h^2} \begin{pmatrix} y_2 - 2y_1 + y_0 \\ y_3 - 2y_2 + y_1 \\ \vdots \\ y_n - 2y_{n-1} + y_{n-2} \end{pmatrix}. \tag{8.25}$$

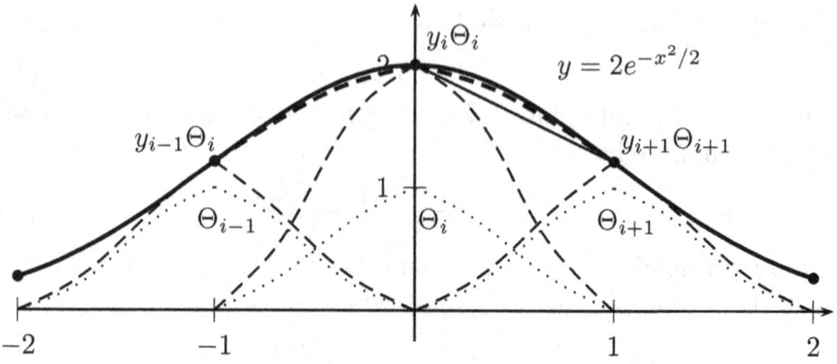

Fig. 8.3 Cubic spline construction from $\Theta(x)$.

Example 8.1: *For the function*

$$y = 2e^{-\frac{x^2}{2}},$$

we now try to approximate it using cubic spline functions constructed from five points $x_0 = -2, x_1 = -1, x_2 = 0,\ x_3 = 1, x_4 = 2$. *We known that* $y_0 = 0.2707, y_1 = 1.2130, y_2 = 2.0,\ y_3 = 1.2130, y_4 = 0.2707$. *Since these points are equally-spaced with* $h = 1$, *Eq. (8.25) then becomes*

$$\begin{pmatrix} 4 & 1 & 0 \\ 1 & 4 & 1 \\ 0 & 1 & 4 \end{pmatrix} \begin{pmatrix} \xi_1 \\ \xi_2 \\ \xi_3 \end{pmatrix} = \begin{pmatrix} -0.9327 \\ -9.443 \\ -0.9327 \end{pmatrix},$$

whose solution is

$$\begin{pmatrix} \xi_1 \\ \xi_2 \\ \xi_3 \end{pmatrix} = \begin{pmatrix} 0.4080 \\ -2.565 \\ 0.4080 \end{pmatrix}.$$

At the two end points, we use the natural boundary condition $\xi_0 = 0$ *and* $\xi_4 = 0$. *Now substituting* $\xi_0, \xi_1, \xi_2, \xi_3, \xi_4$ *into Eq. (8.22) for* $i = 0, 1, 2, 3$, *we have*

$$S_0 = \frac{0.4080}{6}(x+2)^3 + 0.4275(x+2),$$

$$S_1 = -0.4955(x+1)^3 + 1.282(x+1),$$

$$S_2 = 0.4955(x-1)^3 - 1.282(x-1),$$

and

$$S_3 = -\frac{0.4080}{6}(x-2)^3 - 0.4275(x-2).$$

These spline functions are plotted in Fig. 8.3 as the heavy dashed curve. We can see that cubic spline curves almost fall on the exact curve (solid) of the original function $y = 2\exp(-x^2/2)$.

Alternatively, similar to the elementary basis function $\theta_i(x)$ for linear spline, we can find the corresponding basis function $\Theta_i(x)$ for cubic function. In a given interval $[x_i, x_{i+1}]$, if we assume that $\Theta_i(x_i) = 0$, and $\Theta_i'(x_i) = 0$, we get

$$\Theta_i(x) = (x - x_i)^2[\alpha(x - x_i) + \beta]. \tag{8.26}$$

As we require that $\Theta_i(x)$ reaches the maximum $\Theta_i = 1$ at $x = x_{i+1}$, it leads to

$$\Theta_i'(x_{i+1}) = 3\alpha h_i^2 + 2\beta h_i = 0, \tag{8.27}$$

and

$$\Theta_i(x_{i+1}) = h_i^2(\alpha h_i + \beta) = 1. \tag{8.28}$$

The solutions are

$$\alpha = -\frac{2}{h_i^3}, \qquad \beta = \frac{3}{h_i^2}. \tag{8.29}$$

Therefore, we get

$$\Theta_i = (x - x_i)^2[\frac{3}{h_i^2} - \frac{2(x - x_i)}{h_i^3}]. \tag{8.30}$$

The cubic spline function can in general be written as

$$S(x) = \sum_{i=0}^{n-1} y_i \Theta_i(x), \tag{8.31}$$

which can be geometrically represented as the dashed and dotted curves in Fig. 8.3 where, as an example, $y = 2\exp(-x^2/2)$ is approximated using cubic spline functions.

8.2 Lagrange Interpolating Polynomials

We have seen that the construction of spline functions is tedious. Lagrange polynomials provide a systematic way to construct interpolation functions.

For any given n points $(x_i, y_i), (i = 1, 2, ..., n)$, there is a Lagrange interpolating polynomial $P(x)$ of degree $k \leq (n - 1)$ which passes through all n points. That is

$$P(x) = \sum_{i=1}^{n} P_i(x) y_i, \tag{8.32}$$

where

$$P_i = \prod_{j=1, j\neq i}^{n} \frac{(x - x_j)}{(x_i - x_j)}. \tag{8.33}$$

For example, for $n = 4$, we have

$$P(x) = \frac{(x-x_2)(x-x_3)(x-x_4)y_1}{(x_1-x_2)(x_1-x_3)(x_1-x_4)} + \frac{(x-x_1)(x-x_3)(x-x_4)y_2}{(x_2-x_1)(x_2-x_3)(x_2-x_4)}$$
$$+ \frac{(x-x_1)(x-x_2)(x-x_4)y_3}{(x_3-x_1)(x_3-x_2)(x_3-x_4)} + \frac{(x-x_1)(x-x_2)(x-x_3)y_4}{(x_4-x_1)(x_4-x_2)(x_4-x_3)}. \tag{8.34}$$

Example 8.2: *For equally-spaced five points* $(x_i, y_i) = P1(0,1)$, $P2(1,1)$, $P3(2,2)$, $P4(3,-0.5)$, *and* $P5(4,1)$, *the Lagrange polynomial becomes*

$$P(x) = \frac{(x-1)(x-2)(x-3)(x-4)}{24} y_1 - \frac{x(x-2)(x-3)(x-4)}{6} y_2$$
$$+ \frac{x(x-1)(x-3)(x-4)}{4} y_3 - \frac{x(x-1)(x-2)(x-4)}{6} y_4$$
$$+ \frac{x(x-1)(x-2)(x-3)y_5}{24},$$

which is equivalent to

$$P(x) = \frac{1}{2}x^4 - \frac{15}{4}x^3 + \frac{47}{6}x^2 - \frac{20}{3}x + 1.$$

This polynomial and the five points are plotted in Figure 8.4. If we use any consecutive four points, we can construct Lagrange polynomials of degree $k = 3$. *We have*

$$\phi(x) = -\frac{3}{4}x^3 + \frac{11}{4}x^2 - 2x + 1,$$

for $P1, P2, P3$ *and* $P4$; *and*

$$\psi(x) = \frac{5}{4}x^3 - \frac{37}{4}x^2 + 20x - 11,$$

for $P2, P3, P4$ *and* $P5$. *These two polynomials are also plotted in the same figure. We can see that Lagrange polynomials can have strong oscillations, depending on the degrees of polynomials.*

The disadvantage of Lagrange polynomials is that when n increases, the order of the polynomials also increases, and this leads to greater oscillations between data points. For equally-spaced points, the Lagrange interpolation oscillates around the true function. However, the advantages of Lagrange polynomials are that they are unique and rigorous and thus they become handy in mathematical proofs. In addition, they form the basic formulation for shape functions in finite element analysis discussed in later chapters and they are also widely used in signal processing including audio-video analysis and seismic wave processing.

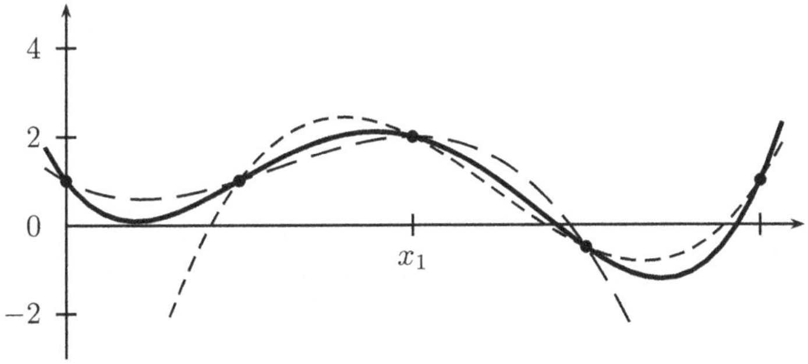

Fig. 8.4 Lagrange polynomials of degree $k = 4$ for 5 given points, and degree $k = 3$ for consecutive four points.

8.3 Bézier Curve

We now know that linear and quadratic spline functions are not quite smooth, but splines of higher degrees are not straightforward to construct. There is an alternative way to construct smooth interpolation functions, that is to use Bézier curves (see Fig. 8.5). These interpolation curves are smooth and can be easily extended to higher dimensions to construct surfaces and volumes. Therefore, this interpolation method is widely used in engineering, computer graphics, and computational sciences.

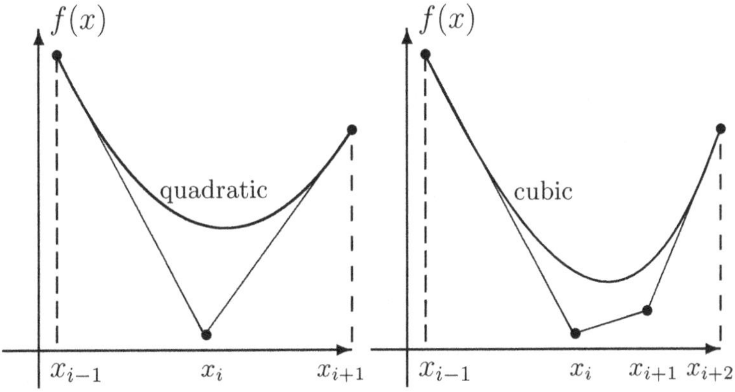

Fig. 8.5 Bézier interpolation: quadratic and cubic.

The quadratic Bézier curve for any three points: $P_0(x_{i-1},\ y_{i-1})$, $P_1(x_i, y_i)$, $P_2(x_{i+1}, y_{i+1})$ as shown in Fig. 8.5, can be constructed using a parameter $t \in [0, 1]$

$$P(t) = (1-t)^2 P_0 + 2t(1-t)P_1 + t^2 P_2, \tag{8.35}$$

which is equivalent to

$$x(t) = (1-t)^2 x_{i-1} + 2t(1-t)x_i + t^2 x_{i+1}, \tag{8.36}$$

and

$$y(t) = (1-t)^2 y_{i-1} + 2t(1-t)y_i + t^3 y_{i+1}. \tag{8.37}$$

Clearly, $t = 0$ corresponds to (x_{i-1}, y_{i-1}) (one end point), while $t = 1$ gives (x_{i+1}, y_{i+1}) (the other end point). The only unusual feature is that the curve does not go through the point (x_i, y_i). This might be a disadvantage as it is not an exact interpolation, however, it becomes an advantage as the curve is very smooth and tangential to both end points. This characteristic is also true for Bézier curves of higher degrees.

For four points, the cubic Bézier curve leads to

$$x(t) = (1-t)^3 x_{i-1} + 3t(1-t)^2 x_i + 3t^2(1-t)x_{i+1} + t^3 x_{i+2}, \tag{8.38}$$

and

$$y(t) = (1-t)^3 y_{i-1} + 3t(1-t)^2 y_i + 3t^2(1-t)y_{i+1} + t^3 y_{i+2}, \tag{8.39}$$

where $t \in [0, 1]$. It is straightforward to extend to any degree n using coefficients of binomial expansion $[(1-t)+t]^n$.

Cubic Bézier curves are very versatile in approximating various curves, and some examples are shown in Fig. 8.6.

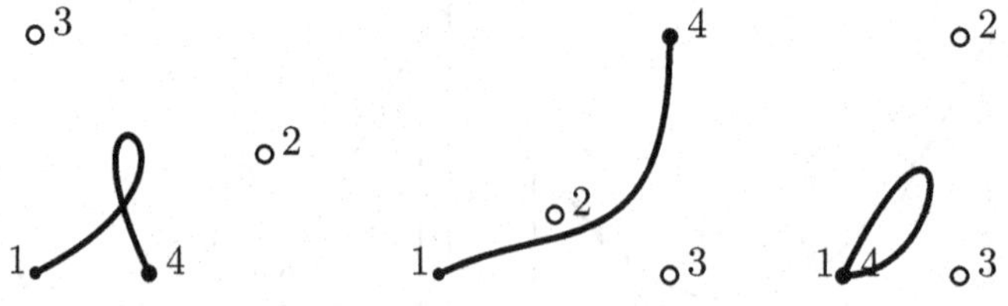

Fig. 8.6 Flexibility of Bézier curves.

With these fundamentals of mathematics and numerical techniques, we are now ready to solve various problems involving PDEs and optimization. In Part III, we will first introduce numerical methods for partial differential equations, and in Part IV, we will study the conventional methods that are widely used in mathematical programming. In Part V, we will study some widely used stochastic methods, and finally in Part VI, we introduce some nature-inspired metaheuristic methods for computational intelligence and optimization.

Part III

Numerical Methods of PDEs

Chapter 9

Finite Difference Methods for ODEs

The finite difference method is one of the most popular methods that are used commonly in computer simulations. It has the advantage of simplicity and clarity, especially in 1-D configuration and other cases with regular geometry. The finite difference method essentially transforms an ordinary differential equation into a coupled set of algebraic equations by replacing the continuous derivatives with finite difference approximations on a grid of mesh or node points that span the domain of interest based on the Taylor series expansions. In general, the boundary conditions and boundary nodes need certain special treatment.

9.1 Integration of ODEs

The second-order or higher-order ordinary differential equations can be written as a first-order system of ODEs. Since the technique for solving a system is essentially the same as that for solving a single equation

$$\frac{dy}{dx} = f(x, y), \tag{9.1}$$

we shall focus on the first-order equation in the rest of this section. In principle, the solution can be obtained by direct integration,

$$y(x) = y_0 + \int_{x_0}^{x} f(x, y(x))dx, \tag{9.2}$$

but in practice it is usually impossible to do the integration analytically as it requires the solution of $y(x)$ to evaluate the right-hand side. Thus, some approximations shall be utilized. Numerical integration is the most common technique for obtaining approximate solutions. There are various integration schemes with different orders of accuracy and convergent rates.

These schemes include the simple Euler scheme, Runge-Kutta method, Relaxation method, and many others.

9.2 Euler Scheme

Using the notation $h = \Delta x = x_{n+1} - x_n$, $y_n = y(x_n)$, $x_n = x_0 + n\Delta x$ ($n = 0, 1, 2, ..., N$), and $' = d/dx$ for convenience, then the explicit Euler scheme can simply be written as

$$y_{n+1} = y_n + \int_{x_n}^{x_{n+1}} f(x, y)dx \approx y_n + hf(x_n, y_n). \tag{9.3}$$

This is a forward difference method as it is equivalent to the approximation of the first derivative

$$y'_n = \frac{y_{n+1} - y_n}{\Delta x}. \tag{9.4}$$

The order of accuracy can be estimated using the Taylor expansion

$$y_{n+1} = y_n + hy'|_n + \frac{h^2}{2} y''|_n + ...$$

$$\approx y_n + hf(x_n, y_n) + O(h^2). \tag{9.5}$$

Thus, the Euler method is first-order accurate.

For any numerical algorithms, the algorithm must be stable in order to reach convergent solutions. Thus, stability is an important issue in numerical analysis. Defining δy as the discrepancy between the actual numerical solution and the true solution of the Euler finite difference equation, we have

$$\delta y_{n+1} = [1 + hf'(y)] = \xi \delta y_n. \tag{9.6}$$

In order to avoid the discrepancy to grow, it requires the following stability condition $|\xi| \leq 1$. The stability restricts the size of interval h, which is usually small.

One alternative that can use larger h is the implicit Euler scheme, and this scheme approximates the derivative by a backward difference $y'_n = (y_n - y_{n-1})/h$ and the right-hand side of equation (9.2) is evaluated at the new y_{n+1} location. Now the scheme can be written as

$$y_{n+1} = y_n + hf(x_{n+1}, y_{n+1}). \tag{9.7}$$

The stability condition becomes

$$\delta y_{n+1} = \xi \delta y_n = \frac{\delta y_n}{1 - hf'(y)}, \tag{9.8}$$

which is always stable if $f'(y) = \frac{\partial f}{\partial y} \le 0$. This means that any step size is acceptable. However, the step size cannot be too large as the accuracy reduces as the step size increases.

Another practical issue is that, for most problems such as nonlinear ODEs, the evaluation of y' and $f'(y)$ requires the value of y_{n+1} which is unknown. Thus, an iteration procedure is needed to march to a new value y_{n+1}, and the iteration starts with a guess value which is usually taken to be zero for most cases. The implicit scheme generally gives better stability.

Example 9.1: *To solve the equation*

$$\frac{dy}{dx} = f(y) = e^{-y} - y,$$

we use the explicit Euler scheme, and we have

$$y_{n+1} \approx y_n + hf(y_n) = y_n + h(e^{-y_n} - y_n).$$

Suppose the discrepancy between true solution y_n^* and the numerical y_n is δy_n so that $y_n^* = y_n + \delta y_n$, then the true solution satisfies

$$y_{n+1}^* = y_n^* + hf(y_n^*).$$

Since $f(y_n^*) = f(y_n) + \frac{df}{dy}\delta y_n$, the above equation becomes

$$y_{n+1} + \delta y_{n+1} = y_n + \delta y_n + h[f(y_n) + f'(y_n)\delta y_n].$$

Together with the Euler scheme, we have

$$\delta y_{n+1} = \delta y_n + f'\delta y_n.$$

Suppose that $\delta y_n \propto \xi^n$, then we have

$$\xi^{n+1} = \xi^n + hf'\xi^n, \qquad or \qquad \xi = 1 + hf'.$$

In order for the scheme to be stable (or $\xi^n \to 0$), it requires that

$$|\xi| \le 1, \qquad or \qquad -1 \le 1 + hf' = 1 - h(e^{-y_n} + 1) \le 1.$$

The stability condition becomes $0 \le h \le \frac{2}{e^{-y_n}+1}$.

9.3 Leap-Frog Method

The leap-frog scheme is the central difference

$$y_n' = \frac{y_{n+1} - y_{n-1}}{2\Delta x}, \tag{9.9}$$

which leads to

$$y_{n+1} = y_{n-1} + 2hf(x_n, y_n). \tag{9.10}$$

The central difference method is second-order accurate. In a similar way as equation (9.6), the leap-frog method becomes

$$\delta y_{n+1} = \delta y_{n-1} + 2hf'(y)\delta y_n, \tag{9.11}$$

or

$$\delta y_{n+1} = \xi^2 \delta y_{n-1}, \tag{9.12}$$

where $\xi^2 = 1 + 2hf'(y)\xi$. This scheme is stable only if $|\xi| \leq 1$, and a special case is $|\xi| = 1$ when $f'(y)$ is purely imaginary. Therefore, the central scheme is not necessarily a better scheme than the forward scheme.

9.4 Runge-Kutta Method

We have so far seen that stability of the Euler method and the central difference method is limited. The Runge-Kutta method uses a trial step to the midpoint of the interval by central difference and combines with the forward difference at two steps

$$\hat{y}_{n+1/2} = y_n + \frac{h}{2} f(x_n, y_n), \tag{9.13}$$

$$y_{n+1} = y_n + hf(x_{n+1/2}, \hat{y}_{n+1/2}). \tag{9.14}$$

This scheme is second-order accurate with higher stability compared with previous simple schemes. One can view this scheme as a predictor-corrector method. In fact, we can use multisteps to devise higher-order methods if the right combinations are used to eliminate the error terms order by order. The popular classical Runge-Kutta method can be written as

$$a = hf(x_n, y_n),$$

$$b = hf(x_n + h/2, y_n + a/2),$$

$$c = hf(x_n + h/2, y_n + b/2),$$

$$d = hf(x_n + h, y_n + c),$$

$$y_{n+1} = y_n + \frac{a + 2(b + c) + d}{6}, \tag{9.15}$$

which is fourth-order accurate.

Example 9.2: *Let us solve the following nonlinear equation numerically*

$$\frac{dy}{dx} + y^2 = -1, \qquad x \in [0, 2]$$

with the initial condition

$$y(0) = 1.$$

We know that it has an analytical solution

$$y(x) = -\tan(x - \frac{\pi}{4}).$$

On the interval $[0, 2]$*, let us first solve the equation using the Euler scheme for* $h = 0.5$*. There are five points* $x_i = ih (i = 0, 1, 2, 3, 4)$*. As* $dy/dx = f(y) = -1 - y^2$*, we have the Euler scheme*

$$y_{n+1} = y_n + hf(y) = y_n - h - hy_n^2.$$

From the initial condition $y_0 = 1$*, we now have*

$$y_1 = y_0 - h - hy_0^2 = 1 - 0.5 - 0.5 \times 1^2 = 0,$$

$$y_2 \approx -0.5, \qquad y_3 \approx -1.125, \qquad y_4 \approx -2.2578.$$

These are significantly different (about 30%*) from the exact solutions*

$$y_0^* = 1, \; y_1^* \approx 0.2934079, \; y_2^* = -0.21795809,$$

$$y_3^* = -0.86756212, \; y_4^* = -2.68770693.$$

Now let us use the Runge-Kutta method to solve the same equation to see if it is better. Since $f(x_n, y_n) = -1 - y_n^2$*, we have*

$$a = hf(x_n, y_n) = -h(1 + y_n^2), \qquad b = -h[1 + (y_n + \frac{a}{2})^2],$$

$$c = -h[1 + (y_n + \frac{b}{2})^2], \qquad d = -h[1 + (y_n + c)^2],$$

and

$$y_{n+1} = y_n + \frac{a + 2(b + c) + d}{6}.$$

From $y_0 = 1$ *and* $h = 0.5$*, we have*

$$y_1 \approx 0.29043, \qquad y_2 \approx -0.22062,$$

$$y_3 = -0.87185, \qquad y_4 \approx -2.67667.$$

These values are within about 1% *of the analytical solutions* y_n^**. We can see that even with the same step size, the Runge-Kutta method is much more efficient and accurate than the Euler scheme.*

Generally speaking, higher-order schemes are better than lower order schemes, but not always.

9.5 Shooting Methods

So far, the numerical integration of an ordinary differential equations is always possible because the problems of ODEs we have solved are mainly initial value problems. That is say, for a given ODE, the conditions we provided are always at $t = 0$ (or $x = 0$), usually in terms of $y(0)$ or $y'(0)$ and their combinations. The integration starts from this initial time (or a single point) and all the solutions at other times (or locations) can be obtained.

Now let us try to solve the following problem in the known interval $[0, \tau]$ where $\tau > 0$ is a given constant.

$$\frac{d^2y(t)}{dt^2} = f(t, y, y'), \qquad t \in [0, \tau], \tag{9.16}$$

with the boundary conditions

$$y(0) = \alpha, \qquad y(\tau) = \beta, \tag{9.17}$$

where $y' \equiv dy/dt$, and both α and β are constant. If we start the integration from $t = 0$, then condition $y(0) = \alpha$ is not sufficient to determine a unique solution. The same is true if we start the integration from $t = \tau$. It seems that the direct numerical integration does not work. In fact, this problem is a two-point boundary problem because its boundary conditions are provided at two points or different times.

Shooting methods are a class of methods for solving such two-point boundary conditions. In general, the boundary conditions can be provided at more than two points, and the boundary conditions are often in functions either linear $ay + by' = c$ or nonlinear $\psi(y, y') = 0$. The shooting method described below will work in both cases; however, for simplicity, we will use the simplest conditions given in Eq. (9.17).

As a single condition $y(0) = \alpha$ is not enough to ensure a unique solution if we start the integration from $t = 0$, we will assume an additional condition $y'(0) = s$ where s is a parameter which will remain constant during the integration. In general, the discrepancy between the value $y(\tau, s)$ (for a given s) and the prescribed boundary value $y(\tau) = \beta$ will be a function of s. We have

$$\Pi(s) = y(\tau, s) - \beta. \tag{9.18}$$

The aim is now to minimize $\Pi(s)$, and the true solution will be obtained when $s = s_*$ which is the root of

$$\Pi(s_*) = 0. \tag{9.19}$$

For higher-order ODEs or a system of first-order ODEs, the discrepancy Π should become a discrepancy vector. In principle, the root(s) s_* can be found using the standard root-finding algorithms such as the Newton-Raphson method discussed earlier in the book. We can see that this method can deal with boundary conditions in any form, even nonlinear or inexplicit functions. In addition, the numerical integration can be carried out using any of the efficient schemes such as the Runge-Kutta method.

In practice, the basic step of a shooting method is to start from a guess value $s = s_0$ and integrate the equation to get a trial solution, then this trial solution is tested to see if it satisfies the other condition $y(\tau) = \beta$ at $t = \tau$. If the condition at the other end is not satisfied (usually not), we adjust the parameter s and carry out the integration again. This iterative procedure continues by varying the values of the parameter s until the solution satisfies both boundary conditions or within a prescribed accuracy. Here, the parameter $y'(0) = s$ acts as a velocity if the solution $y(t)$ (or $y(x)$ in a spatial coordinate) is considered as a trajectory (see Figure 9.1). This is similar to shooting a bullet, aiming to hit the target at $y(\tau) = \beta$ (hence, the name of the shooting method).

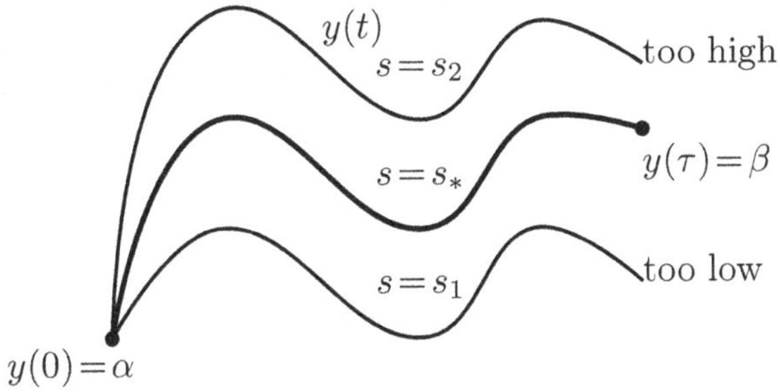

Fig. 9.1 Schematic representation of the shooting method with various values of the parameter s.

Example 9.3: *We now try to solve the following two-point boundary problem*

$$y''(t) - 2y'(t) + y(t) = t,$$

with boundary conditions at

$$y(0) = 0, \qquad y(1) = 1.$$

The analytical solution can be obtained using the standard technique of ODEs with constant coefficients. We have

$$y(t) = t + 2(1 - e^t) + \frac{2(e-1)}{e} te^t.$$

Suppose we do not know the exact solution and try to solve this problem using the standard shooting method. We first rewrite the above second-order ODE as a linear system by letting $u_1 = y(t)$ *and* $u_2 = du_1/dt = y'(t)$, *and we have*

$$\dot{\boldsymbol{u}} = \frac{d}{dt}\begin{pmatrix} u_1 \\ u_2 \end{pmatrix} = \begin{pmatrix} u_2 \\ t + 2u_2 - u_1 \end{pmatrix}.$$

This is a system of first-order ODEs, which can be integrated directly if we use a guess value for $y'(0) = s$ *at* $t = 0$. *That is to say, the boundary value problem is now converted into an initial value problem with the initial values as*

$$\boldsymbol{u}(\boldsymbol{0}) = \begin{pmatrix} u_1(0) \\ u_2(0) \end{pmatrix} = \begin{pmatrix} 0 \\ s \end{pmatrix}.$$

For simplicity, let us use the bisection method to find the appropriate value s *by starting from the initial guess* $s = 0$. *For* $s = 0$, *we can integrate the system using the Runge-Kutta method with* $h = 0.25$ *in a similar manner to the previous example. After some numerical calculations, we have* $u_1(1) = y(1) \approx 0.2817$ *which is too low for the given boundary condition* $y(1) = 1$. *We then try this again using* $s = 1$, *we have* $u_1(1) = 3$ *which is too high. Now we try* $s = (0+1)/2$, *we get* $u_1(1) \approx 1.6409$ *which is still too high. We then try* $s = (0 + 0.5)/2 = 0.25$, *we get* $u_1(1) \approx 0.961$ *which is lower than* $y(1) = 1$. *So we tried* $s = (0.25 + 0.5)/2$ *again, we get* $u_1(1) \approx 1.301$ *(too high). We continue in this way until we reach* $s = 0.2656$ *after 7 iterations. We can then continue the iterations until a prescribed accuracy.*

If we use the guess $y'(0) = s$ *to solve the equation analytically, we have*

$$y(t) = 2 + t - 2e^t + (1 + s)te^t.$$

In order to satisfy $y(1) = 1$ *at* $t = 1$, *we have the solution* $s_* = 1 - 2e^{-1}$. *The solution after 7 iterations corresponds to* $s = 0.2656$ *which is quite close (within about 0.5%) to the true value of* $y'(0) = 1 - 2e^{-1} \approx 0.264241117657$.

This is a relatively simple example, but it does demonstrate how the shooting method works. In fact, nonlinear equations with more complex boundary conditions can be solved using shooting methods. In the standard shooting method we discussed here, we start the integration (or shooting) from one boundary and tries to match the target at the other boundary. Sometimes, it might be difficult to do so. In this case, it is possible or even more efficient to shoot from both sides (boundaries) and try to match a common target at a convenient mid location. This forms the shooting-to-a-fitting-point method. There are other variations of shooting methods as well, and readers can refer to more advanced literature.

In addition, the two-point boundary problem can be solved using the implicit methods to be discussed in the next chapter. The advantage of the shooting method over the implicit method is that it can easily deal with nonlinear equations such as

$$y'''(t) + \sin[y(t)] + y^2(t) = \cos(t), \tag{9.20}$$

with the boundary conditions, say,

$$y(0) = 1, \qquad y(\pi) + \alpha y'(\pi) = \beta. \tag{9.21}$$

Furthermore, these types of problems can also be solved using more advanced methods such as boundary element methods and finite element methods.

Chapter 10

Finite Difference Methods for PDEs

The numerical solution of partial differential equations is more complicated than that of ODEs because it involves time and space variables and the geometry of the domain of interest. Usually, boundary conditions are more complex and the solution domain can be irregular. In addition, nonlinear problems are very common in a wide range of applications. We start with the simplest first-order equations and then move onto more complicated cases.

10.1 Hyperbolic Equations

For simplicity, we first look at the one-dimensional scalar equation of hyperbolic type,

$$\frac{\partial u}{\partial t} + c\frac{\partial u}{\partial x} = 0, \tag{10.1}$$

where c is a constant or the velocity of advection. By using the forward Euler scheme for time and central scheme for space, we have

$$\frac{u_j^{n+1} - u_j^n}{\Delta t} + c[\frac{u_{j+1}^n - u_{j-1}^n}{2h}] = 0, \tag{10.2}$$

where $t = n\Delta t, n = 0, 1, 2, ..., x = x_0 + jh, j = 0, 1, 2, ...,$ and $h = \Delta x$. In order to see how this method behaves numerically, we use the von Neumann stability analysis.

Assuming the independent solutions or eigenmodes, also called Fourier modes, in spatial coordinate x in the form of $u_j^n = \xi^n e^{ikhj}$ where k is the equivalent wavenumber, and substituting into equation (10.2), we have

$$\xi = 1 - i\frac{c\Delta t}{h}\sin(kh). \tag{10.3}$$

The stability criterion $|\xi| \le 1$ requires

$$(\frac{c\Delta t}{h})^2 \sin^2 kh \le 0. \tag{10.4}$$

However, this inequality is impossible to satisfy and this scheme is thus unconditionally unstable.

To avoid the difficulty of instability, we can use other schemes such as the upwind scheme and Lax scheme. For the upwind scheme, the equation becomes

$$\frac{u_j^{n+1} - u_j^n}{\Delta t} + c[\frac{u_j^n - u_{j-1}^n}{h}] = 0, \tag{10.5}$$

whose stability condition is

$$|\xi| = \left|1 - \frac{c\Delta t}{h}[1 - \cos(kh) + i\sin(kh)]\right| \le 1, \tag{10.6}$$

which is equivalent to

$$0 < \frac{c\Delta t}{h} \le 1. \tag{10.7}$$

This the well-known Courant-Friedrichs-Lewy stability condition, often referred to as the Courant stability condition. Thus, the upwind scheme is conditionally stable.

The wave propagation of the first-order hyperbolic equation can be demonstrated by the following simple case,

$$u_t + u_x = 0, \qquad 0 \le x \le L, \tag{10.8}$$

with an initial condition

$$u(x,0) = \frac{1}{2}e^{-[(x-\frac{L}{4})/L]^2} + e^{-[(x-\frac{L}{2})/L]^2}, \tag{10.9}$$

and boundary conditions $u(0,t) = u(L,t) = 0$.

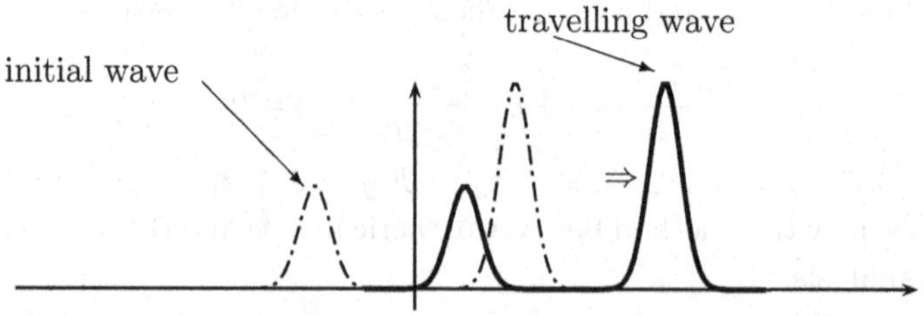

Fig. 10.1 First-order hyperbolic equation and its travelling wave solution $u_t + u_x = 0$.

Figure 10.1 shows the wave propagation where the dashed curve corresponds to the initial wave profile, while the solid curve corresponds to the travelling wave. We can see that the wave profile does not change with time but moves with a constant velocity.

Higher-order equations such as the second-order wave equation can be written as a system of hyperbolic equations and then be solved using numerical integration. They can also be solved by direct discretization using any finite difference scheme. The wave equation

$$\frac{\partial^2 u}{\partial t^2} = c^2 \frac{\partial^2 u}{\partial x^2}, \tag{10.10}$$

consists of second derivatives. If we approximate the first derivatives at each time step n using

$$u_i' = \frac{u_{i+1}^n - u_i^n}{\Delta x}, \qquad u_{i-1}' = \frac{u_i^n - u_{i-1}^n}{\Delta x}, \tag{10.11}$$

then we can use the following approximation for the second derivative

$$u_i'' = \frac{u_i' - u_{i-1}'}{\Delta x} = \frac{u_{i+1}^n - 2u_i^n + u_{i-1}^n}{(\Delta x)^2}. \tag{10.12}$$

This is in fact a central difference scheme of second-order accuracy. If we use the similar scheme for time-stepping, then we get a central difference scheme in both time and space.

Thus, the numerical scheme for this equation becomes

$$\frac{u_i^{n+1} - 2u_i^n + u_i^{n-1}}{(\Delta t)^2} = c^2 \frac{u_{i+1}^n - 2u_i^n + u_{i-1}^n}{(\Delta x)^2}. \tag{10.13}$$

This is a two-level scheme with a second order accuracy. The idea of solving this difference equation is to express (or to solve) u_i^{n+1} at time step $t = n+1$ in terms of the known values or data u_i^n and u_i^{n-1} at two previous time steps $t = n$ and $t = n-1$.

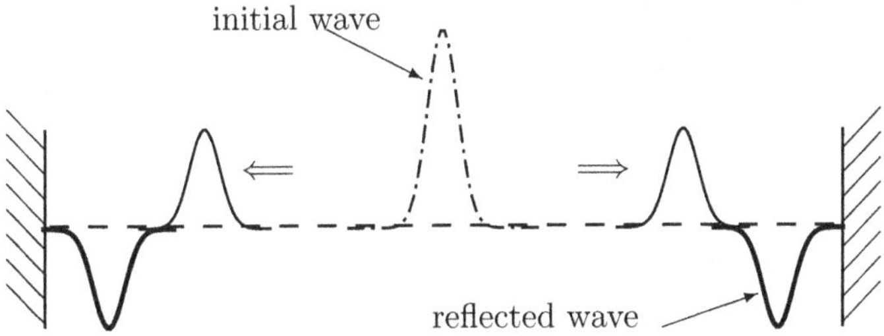

Fig. 10.2 Travelling wave solution of the wave equation: $u_{tt} - c^2 u_{xx} = 0$.

It leaves as an exercise to implement the above method using any programming language for the case of $i = 1, 2, ..., 100$. Solving the wave equation (10.10) with the initial condition

$$u(x, 0) = e^{-x^2}, \tag{10.14}$$

together with the wave reflection boundary conditions at both ends $u(-L,t) = u(L,t) = 0$, we have the solution shown in Figure 10.2. We can see that the initial profile is split into two travelling waves: one travels to the left and one to the right.

10.2 Parabolic Equation

For the parabolic equation such as the diffusion or heat conduction equation

$$\frac{\partial u}{\partial t} = \frac{\partial}{\partial x}(D\frac{\partial u}{\partial x}), \tag{10.15}$$

a simple Euler method for the time derivative and centered second-order approximations for space derivatives lead to

$$u_j^{n+1} = u_j^n + \frac{D\Delta t}{h^2}(u_{j+1}^n - 2u_j^n + u_{j-1}^n). \tag{10.16}$$

From the application of von Neumann stability analysis by assuming $u_j^n = \xi^n e^{ikhj}$, the above equation becomes

$$\xi = 1 - \frac{4D\Delta t}{h^2}\sin^2(\frac{kh}{2}). \tag{10.17}$$

The stability requirement $|\xi| \leq 1$ leads to the constraint on the timestep,

$$\Delta t \leq \frac{h^2}{2D}. \tag{10.18}$$

This scheme is thus conditionally stable.

For simplicity, we consider a 1-D heat conduction equation $u_t = \kappa u_{xx}$ with an initial condition

$$u(x,0) = [H(x - L/5) - H(x)]$$

where $H(x)$ is a Heaviside function.

$$H(x) = 1, \quad x \geq 0, \qquad H(x) = 0 \quad x < 0.$$

The evolution of the temperature profile is shown in Figure 10.3 where the initial profile is plotted as a dashed curve. It leaves as an exercise to show such a profile using any programming language such as Matlab or Java.

We can see that the profile is gradually smoothed out as time increases and this is the typical behaviour of the diffusive system. The time-stepping scheme we used limits the step size of time as larger time steps will make the scheme unstable. There are many ways to improve this, and one of most widely used schemes is the implicit scheme.

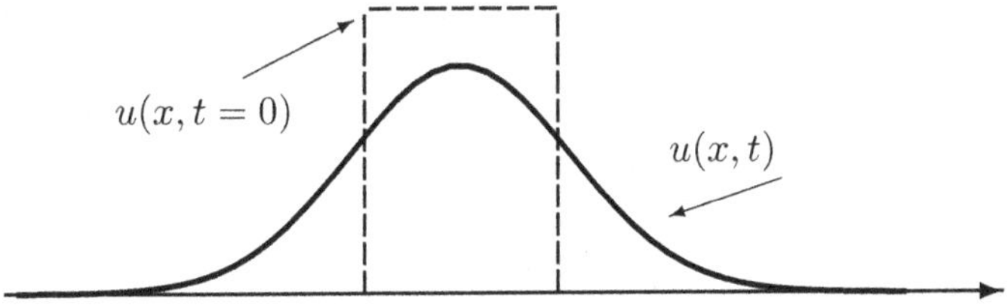

Fig. 10.3 The 1-D time-dependent diffusion equation: $u_t - \kappa u_{xx} = 0$.

To avoid the limitation due to very small time steps, we now use an implicit scheme for time derivative differencing, and thus we have

$$u_j^{n+1} - u_j^n = \frac{D\Delta t}{h^2}(u_{j+1}^{n+1} + 2u_j^{n+1} + u_{j-1}^{n+1}). \tag{10.19}$$

Applying the stability analysis, we have

$$\xi = \frac{1}{1 + \frac{4D\Delta t}{h^2}\sin^2\frac{kh}{2}}, \tag{10.20}$$

whose norm is always less than unity ($|\xi| \leq 1$). This means the implicit scheme is unconditionally stable for any size of time steps. That is why implicit methods are more desirable in simulations. However, there is one disadvantage of this method, which requires more programming skills and the inverse of a large matrix is usually needed in implicit schemes. To avoid the inverse of a large, sparse matrix, some iteration methods have also to be used for such implicit schemes.

10.3 Elliptical Equation

In the parabolic equation, if the time derivative is zero or u does not change with time $u_t = 0$, then we reach a steady-state problem that is governed by the elliptic equation. For the steady state heat conduction problem, we generally have the Poisson problem,

$$\nabla \cdot [\kappa(u, x, y, t)\nabla u] = f. \tag{10.21}$$

If κ is a constant, this becomes

$$\nabla^2 u = q, \qquad q = \frac{f}{\kappa}. \tag{10.22}$$

There are many methods available to solve this problem, such as the boundary integral method, the relaxation method, and the multigrid method.

Two major ones are the long-time approximation of the transient parabolic diffusion equations, the other includes the iteration method.

The so-called long-time approximation method is essentially based on the fact that the parabolic equation

$$\frac{\partial u}{\partial t} + \kappa \nabla^2 u = f, \tag{10.23}$$

evolves with a typical scale of $\sqrt{\kappa t}$. If $\sqrt{\kappa t} \gg 1$, the system is approaching its steady state. Assuming $t \to \infty$ and $\kappa \gg 1$, we then have

$$\nabla^2 u = \frac{f}{\kappa} - \frac{1}{\kappa} u_t \to 0. \tag{10.24}$$

In the case of $\kappa = \text{const}$, it degenerates into the above steady-state equation (10.21) because $u_t \to 0$ as $t \to \infty$. This approximation becomes better if $\kappa \gg 1$. Thus, the usual numerical methods for solving parabolic equations are valid. However, other methods may obtain the results more quickly.

The iteration method uses the second-order scheme for space derivatives, and equation (10.22) in the 2-D case becomes

$$\frac{u_{i+1,j} - 2u_{i,j} + u_{i-1,j}}{(\Delta x)^2} + \frac{u_{i,j+1} - 2u_{i,j} + u_{i,j-1}}{(\Delta y)^2} = q. \tag{10.25}$$

If we use $\Delta x = \Delta y = h$, then the above equation simply becomes

$$(u_{i,j+1} + u_{i,j-1} + u_{i+1,j} + u_{i-1,j}) - 4u_{i,j} = h^2 q, \tag{10.26}$$

which can be written as

$$\mathbf{A}\mathbf{u} = \mathbf{b}. \tag{10.27}$$

This equation can be solved using the various methods such as the Gauss-Seidel iteration method, as discussed earlier in Part II.

Example 10.1: *The 1D steady state solution of a heat transfer problem through a thin plate can be approximated as*

$$\frac{d^2 u}{dx^2} = q, \qquad 0 \le x \le w,$$

where w is the thickness and q is a constant. The boundary conditions are

$$u(x = 0, t) = u_0,$$

and

$$u(x = w, t) = 0.$$

Integrating the equation twice and using the boundary conditions, we get the exact solution

$$u = \frac{q}{2}x^2 - \frac{qw}{2}x + u_0(1 - \frac{x}{w}).$$

Let us now solve this problem using the finite difference method. For simplicity, we can assume $q = 1$, $u_0 = 1$ *and* $w = 1$. *We also use five points* $i = 0, 1, ..., 4$ *so that* $h = \Delta x = w/4 = 0.25$. *Now the second-order scheme becomes*

$$\frac{u_{i+1} - 2u_i + u_{i-1}}{h^2} = q,$$

or

$$u_{i-1} - 2u_i + u_{i+1} = h^2 q.$$

For $i = 1, 2, 3$, *we have*

$$u_0 - 2u_1 + u_2 = h^2 q,$$

$$u_1 - 2u_2 + u_3 = h^2 q,$$

$$u_2 - 2u_3 + u_4 = h^2 q.$$

At $x_0 = 0$, *we have*

$$u_0 = 1.$$

At $x_4 = w = 1$, *we have*

$$u_4 = 0.$$

Combining these into a matrix equation $\boldsymbol{A}\boldsymbol{u} = \boldsymbol{b}$, *we have*

$$\begin{pmatrix} 1 & 0 & 0 & 0 & 0 \\ 1 & -2 & 1 & 0 & 0 \\ 0 & 1 & -2 & 1 & 0 \\ 0 & 0 & 1 & -2 & 1 \\ 0 & 0 & 0 & 0 & 1 \end{pmatrix} \begin{pmatrix} u_0 \\ u_1 \\ u_2 \\ u_3 \\ u_4 \end{pmatrix} = \begin{pmatrix} 1 \\ 0.0625 \\ 0.0625 \\ 0.0625 \\ 0 \end{pmatrix}.$$

The solution can be obtained by

$$\boldsymbol{u} = \boldsymbol{A}^{-1}\boldsymbol{b} = \begin{pmatrix} 1.0000 \\ 0.6563 \\ 0.3750 \\ 0.1562 \\ 0.0000 \end{pmatrix},$$

which is almost the same as the exact solution

$$\boldsymbol{u}_{\rm true} = \begin{pmatrix} 1 \\ 0.65625 \\ 0.375 \\ 0.15625 \\ 0 \end{pmatrix}.$$

Obviously, the above illustrative examples are very small scale. For large-scale problems or problems with complex geometry, the number of grid or nodal points can be very high. In this case, the inverse of matrices should not be used. As we mentioned in earlier chapters, iteration methods for solving large, sparse systems should be recommended.

10.4 Spectral Methods

The accuracy of any finite difference scheme depends on many factors such as the way of approximating the relevant derivatives and thus their accuracy can be described by $O(h)$ (first-order scheme) or $O(h^2)$ (second-order scheme) or $O(h^s)$ in general. In order to design higher-order schemes, we have to use elaborate approximations for derivatives.

There are other alternatives which provide higher-order accuracy for spatial derivatives. Among these techniques, spectral methods are widely used. The essence of spectral methods (and their variants) is to approximate the solutions by a linear combination of some basis functions, very similar to the widely used finite element methods. In the finite element methods, the basis functions are piecewise and often local. In contrast, basis functions in spectral methods are smooth and often global. This may lead to some advantages of spectral methods in approximating smooth solutions more efficiently.

Spectral methods often transform a partial differential equation into a set of ordinary differential equations which can in turn be solved using a wide range of techniques for solving ODEs. For example, now we consider the diffusion problem discussed earlier

$$\frac{\partial u}{\partial t} = D\frac{\partial^2 u}{\partial x^2}, \qquad x \in [0, L]. \tag{10.28}$$

In spectral methods, the solution $u(x, t)$ can be approximated by

$$u(x,t) \approx u_N = \sum_{k=0}^{N-1} \alpha_k(t)\psi_k(x), \tag{10.29}$$

where the coefficients $\alpha_k(t)$ are amplitudes which only depend on time t. The basis functions $\psi_k(x)$ represent spectral modes which may vary with spatial coordinates x, not time t. N is the number of terms in the approximation, and is often related to the number of spectral modes. The basis functions can be any orthogonal set of functions such as Chebyshev or Legendre polynomials. The simplest basis functions are probably the trigonometric functions as they are related to Fast Fourier Transforms (FFT) which can be calculated very efficiently and quickly with modern computers.

If we substitute the approximation (10.29) into (10.28), we have

$$\frac{\partial}{\partial t}[\sum_{k=0}^{N-1} \alpha_k(t)\psi_k(x)] = D\frac{\partial}{\partial x}[\sum_{k=0}^{N-1} \alpha_k(t)\psi_k(x)], \tag{10.30}$$

which leads to

$$\sum_{k=0}^{N-1} \frac{d\alpha_k(t)}{dt}\psi_k(x) = D\sum_{k=0}^{N-1} \alpha_k(t)\frac{d^2\psi_k(x)}{dx^2}. \tag{10.31}$$

If we use the basis functions

$$\psi_k(x) = e^{-i\omega_k x}, \qquad \omega_k = \frac{2\pi k}{L}, \quad (k = 1, 2, ...), \tag{10.32}$$

we then have

$$\frac{d^2\psi_k(x)}{dx^2} = -\omega_k^2 e^{-i\omega_k x}. \tag{10.33}$$

Now we finally get

$$\sum_{k=0}^{N-1} \frac{d\alpha_k(t)}{dt} e^{-i\omega_k x} = -\sum_{k=0}^{N-1} D\omega_k^2 \alpha_k(t) e^{-i\omega_k x}. \tag{10.34}$$

Multiplying both sides by $e^{-i\omega_j x}$ and integrating from 0 to L together with the orthogonality condition

$$\int_0^L e^{-i\omega_k x} e^{-i\omega_j x} dx = \begin{cases} 0 \text{ if } (k \neq j) \\ L \text{ if } (k = j) \end{cases}, \tag{10.35}$$

we have the following set of ordinary differential equations

$$\frac{d\alpha_k(t)}{dt} = -D\omega_k^2 \alpha_k(t), \qquad (k = 1, 2, ..., N). \tag{10.36}$$

This set of equations can be solved using standard time-stepping methods such as the forward Euler scheme. We can see that the spectral methods essentially transform the PDE problem into a set of ODE problems. We can then use the advantages of FFT techniques and other efficient ODE schemes to solve the original problem.

A significant advantage of spectral methods is that the spatial accuracy is very high as we have not used any approximation for the spatial derivatives in the original equation, except for the truncated errors in (10.29). However, spectral methods do have some disadvantages. For example, boundary conditions cannot be applied naturally as we have to transform them in some way so that they can be related to the coefficients α_k. Also for solutions with sharp local features, spectral methods are not the best options. There are other methods with improved accuracy, including pseudospectral methods, spectral element methods and their relevant variants. Interested readers can refer to more specialised books.

10.5 Pattern Formation

One of the most studied nonlinear reaction-diffusion equations in the 2-D case is the Kolmogorov-Petrovskii-Piskunov (KPP) equation

$$\frac{\partial u}{\partial t} = D(\frac{\partial^2 u}{\partial x^2} + \frac{\partial^2 u}{\partial y^2}) + \gamma q(u), \tag{10.37}$$

and

$$q(u) = u(1 - u). \tag{10.38}$$

The KPP equation can describe a huge number of physical, chemical and biological phenomena. The most interesting feature of this nonlinear equation is its ability to generate beautiful patterns. We can solve it using the finite difference scheme by applying the periodic boundary conditions and using a random initial condition $u = \text{random}(n, n)$ where n is the size of the grid.

Figure 10.4 shows the pattern formation of the above equation on a 200 $\times$ 200 grid for $D = 0.2$ and $\gamma = 0.5$. We can see that rings and ribbon-like patterns are formed, arising naturally from random initial conditions. The landscape surface shows the variations in the location and values of the field $u(x, y)$. The following simple 15-line Matlab program can be used to solve this nonlinear system.

```
% Pattern formation:  a 15 line matlab program
% PDE form: u_t=D*(u_{xx}+u_{yy})+gamma*q(u)
% where q(u)='u.*(1-u)';
% The solution of this PDE is obtained by the
% finite difference method, assuming dx=dy=dt=1.
% Written by Xin-She Yang (Cambridge University)
```

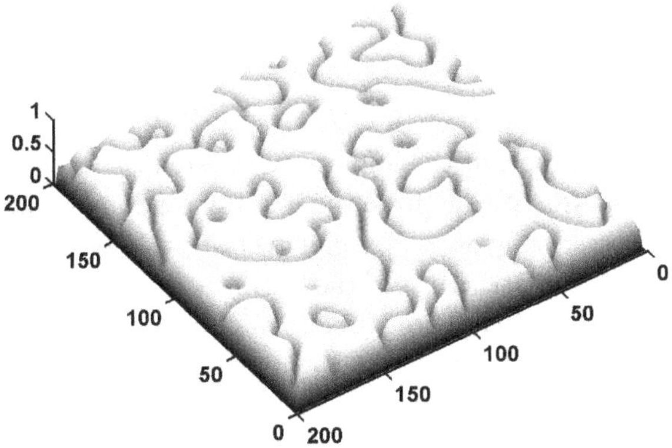

Fig. 10.4 2-D pattern formation for $D = 0.2$ and $\gamma = 0.5$.

```
% Usage: pattern(200)   or  simply >pattern
% ---------------------------------------------------
function pattern(n)                         % line 1
% Input the domain size n
if nargin<1, n=200; end                     % line 2
% ---------------------------------------------------
% Initialize parameters
% ---- time=100, D=0.2; gamma=0.5; --------------
time=100; D=0.2; gamma=0.5;                 % line 3
% ---- Set initial values of u randomly ---------
u=rand(n,n);  grad=u*0;                     % line 4
% Vectorization/index for u(i,j) and the loop ---
I = 2:n-1; J = 2:n-1;                       % line 5
% ---- Time stepping ----------------------------
for step=1:time,                            % line 6
% Laplace gradient of the equation      % line 7
 grad(I,J)= u(I,J-1)+u(I,J+1)+u(I-1,J)+u(I+1,J);
 u =(1-4*D)*u+D*grad+gamma*u.*(1-u);    % line 8
% ----- Show results ----------------------------
 pcolor(u);  shading interp;                % line 9
% ----- Coloring and showing colorbar -----------
 colorbar; colormap jet;                    % line 10
 drawnow;                                   % line 11
end                                         % line 12
```

```
% ----- Topology of the final surface ----------
surf(u);                                 % line 13
shading interp;                          % line 14
view([-25 70]);                          % line 15
```

If you use this program to simulate the system, you will see that the pattern emerges automatically from the initially random background. Once the pattern is formed, it evolves gradually with time, but the characteristics such as the shape and structure of the patterns do not change much with time. The beautiful patterns are stable.

10.6 Cellular Automata

The modelling of a physical system does not necessarily have to be carried out in terms of differential equations or any other equation-based relationship. In fact, it is possible to characterise the behaviour of a system by using rules. This rule-based approach provides a powerful alternative to equation-based mathematical models. A cellular automaton (CA) is such a rule-based computing machine which can be used to simulate a dynamical system or a physical system. This kind of system is discrete in both space and time.

The innovative concept of a cellular automaton (CA) was first proposed by von Newmann in the early 1950s and the systematic studies were pioneered by Wolfram from the 1980s. There has been a substantial amount of research in these areas over the past decades. Here, we will introduce some of the important concepts.

The simplest cellular automaton is one-dimensional, as shown in Figure 10.5. On a one-dimensional grid that consists of N consecutive cells, each cell i $(= 1, 2, ., N)$ may be at any of the finite number of states, k. At each time step, t, the next state of a cell is determined by its present state and the states of its local neighbours. Generally speaking, the state u_i at $t + 1$ is a function of its $2r + 1$ neighbours with r cells on the left of the cell concerned and r cells on its right. The parameter r is often referred to as the radius of the neighbourhood.

The number of possible permutations for k finite states and a radius of r is $p = k^{2r+1}$, thus the number of all possible rules to generate the state of cells at the next time step is k^p, which is usually very large. For example, $r = 2$, $k = 5$, then $p = 125$ and the number of possible rules is $5^{125} \approx 2.35 \times 10^{87}$, which is much larger than the number of stars in the

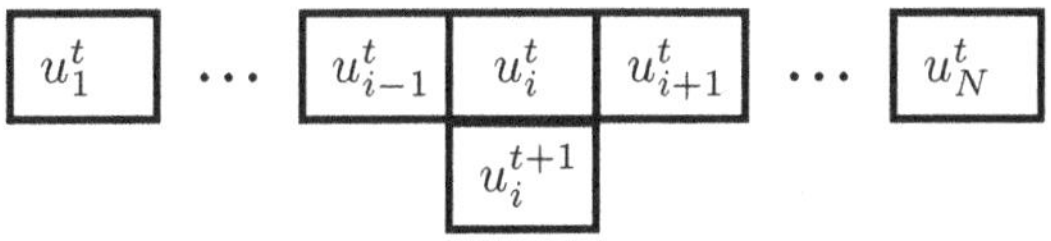

Fig. 10.5 Schematic representation of a one-dimensional cellular automaton.

whole universe.

The rule determining the new state is often referred to as the transition rule or updating rule. In principle, the state of a cell at the next time step can be any function of the states of some neighbour cells, and the function can be linear or nonlinear. There is a subclass of the possible rules in which the new state only depends on the sum of the states in a neighbourhood, and this will simplify the rules significantly. For this type of updating rule, the number of possible permutations for k states and $2r+1$ neighbours is simply $k(2r+1)$, and thus the number of all possible rules is $k^{k(2r+1)}$.

For the same parameter $k = 5$, $r = 2$, then the number of possible rules is $5^{15} = 3.05 \times 10^{10}$, which is much smaller compared with 5^{125}. This subclass of sum-rule or totalistic rule is especially important in the popular cellular automata such as Conway's game of life, and in the cellular automaton implementation of partial differential equations.

Cellular automata can be formulated in higher dimensions such as 2-D and 3-D. The dynamics and complexity of cellular automata are extremely rich. Stephan Wolfram's pioneering research and mathematical analysis of cellular automata led to his famous classification of one-dimensional cellular automata. There are four basic classes: homogeneous, periodic, aperiodic, and complex.

By extending the discretization procedure for differential equations to derivation of automaton rules for cellular automata, we can now formulate the cellular automata from their corresponding partial differential equations. First, let us start with the reaction-diffusion equation in the two-dimensional configuration

$$\frac{\partial u}{\partial t} = D(\frac{\partial^2 u}{\partial x^2} + \frac{\partial^2 u}{\partial y^2}) + f(u), \tag{10.39}$$

where $u(x, y, t)$ is the state variable that evolves with time in a 2-D domain, and the function $f(u)$ can be either linear or nonlinear. D is a constant depending on the properties of diffusion. The discretization of this equation can be written as

$$\frac{u_{i,j}^{n+1} - u_{i,j}^{n}}{\Delta t} = f(u_{i,j}^{n})$$

$$+D[\frac{u_{i+1,j}^{n} - 2u_{i,j}^{n} + u_{i-1,j}^{n}}{(\Delta x)^2} + \frac{u_{i,j+1}^{n} - 2u_{i,j}^{n} + u_{i,j-1}^{n}}{(\Delta y)^2}]. \tag{10.40}$$

By choosing $\Delta t = \Delta x = \Delta y = 1$, we have

$$u_{i,j}^{n+1} = D[u_{i+1,j}^{n} + u_{i-1,j}^{n} + u_{i,j+1}^{n} + u_{i,j-1}^{n}] + f(u_{i,j}^{n}) + (1 - 4D)u_{i,j}^{n},$$

which can be written as the generic form

$$u_{i,j}^{t+1} = \sum_{\alpha,\beta=-r}^{r} A_{\alpha,\beta} u_{i+\alpha,j+\beta}^{t} + f(u_{i,j}^{t}), \tag{10.41}$$

where the summation is over the $4r + 1$ neighbourhood. This is a finite state 2-D cellular automaton with the coefficients $A_{\alpha,\beta}$ being determined from the discretization of the governing equations, and for this special case, we have $A_{-1,0} = A_{+1,0} = A_{0,-1} = A_{0,+1} = D$, $A_{0,0} = 1 - 4D$, and $r = 1$.

For the 1-D linear wave equation,

$$\frac{\partial^2 u}{\partial t^2} = c^2 \frac{\partial^2 u}{\partial x^2}, \tag{10.42}$$

where c is the wave speed. The simplest central difference scheme leads to

$$\frac{u_i^{n+1} - 2u_i^n + u_i^{n-1}}{(\Delta t)^2} = c^2 \frac{u_{i+1}^n - 2u_i^n + u_{i-1}^n}{(\Delta x)^2}. \tag{10.43}$$

By choosing $\Delta t = \Delta x = 1$ and $t = n$, we have

$$u_i^{t+1} = [u_{i+1}^t + u_{i-1}^t + 2(1 - c^2)u_i^t] - u_i^{t-1}. \tag{10.44}$$

This can be written in the generic form

$$u_i^{t+1} + u_i^{t-1} = f(u^t), \tag{10.45}$$

which is reversible under certain conditions. This property comes from the reversibility of the wave equation because it is invariant under the transformation: $t \to -t$.

There are some interesting connections between cellular automata and PDEs although this is not always straightforward and sometimes may be very difficult to see their connections. The derivation of updating rules for cellular automata from their corresponding partial differential equations are relatively straightforward by using the finite differencing schemes, while the formulation of differential equations from cellular automata is usually difficult and non-unique. More studies are highly needed in these areas. In addition, either rule-based systems or equation-based systems can have complex pattern formation under appropriate conditions, and these spatiotemporal patterns can simulate many phenomena in nature and biological applications.

Chapter 11

Finite Volume Method

The finite difference method discussed in the previous chapters approximates the ordinary differential equations and partial differential equations using Taylor series expansions, resulting in a system of algebraic equations. The finite volume method resembles the finite difference method in certain ways but the starting point is the integral formulation of the problem. It uses the integral form of the partial differential equations in terms of conservation laws, then approximates the surface and boundary integrals in the control volumes. This becomes convenient for problems involving flow or flux boundaries.

11.1 Concept of the Finite Volume

For a hyperbolic equation that is valid in the domain Ω with boundary $\partial\Omega$,

$$\frac{\partial u}{\partial t} - \nabla \cdot (\kappa \nabla u) = q, \tag{11.1}$$

or written in terms of a flux function $\mathbf{F} = \mathbf{F}(u) = -\kappa \nabla u$, we have

$$\frac{\partial u}{\partial t} + \nabla \cdot \mathbf{F} = q. \tag{11.2}$$

The integral form of this equation becomes

$$\int_\Omega \frac{\partial u}{\partial t} d\Omega + \int_\Omega \nabla \cdot \mathbf{F} d\Omega = \int_\Omega q d\Omega. \tag{11.3}$$

If the integral form is decomposed into many small control volumes, or finite volumes, we have $\Omega = \bigcup_{i=1}^{N} \Omega_i$ and $\Omega_i \bigcap \Omega_j = \emptyset$. By defining the control volume cell average or mean value

$$u_i = \frac{1}{V_i} \int_{\Omega_i} u d\Omega_i, \qquad q_i = \frac{1}{V_i} \int_{\Omega_i} q d\Omega_i, \tag{11.4}$$

where $V_i = |\Omega_i|$ is the volume of the small control volume Ω_i, the above equation can be written as

$$\frac{\partial u_i}{\partial t} + \sum_{i=1}^{N} \frac{1}{V_i} \int_{\Omega_i} \nabla \cdot \mathbf{F}(u_i) d\Omega_i = q_i. \tag{11.5}$$

By using the divergence theorem

$$\int_V \nabla \cdot \mathbf{F} dV = \int_\Gamma \mathbf{F} \cdot \mathbf{n} dA, \tag{11.6}$$

we have

$$\frac{\partial u_i}{\partial t} + \sum_{i=1}^{N} \frac{1}{V_i} \int_{\Gamma_i} \mathbf{F} \cdot \mathbf{dS} = q_i, \tag{11.7}$$

where $\mathbf{dS} = \boldsymbol{n} dA$ is the surface element and $\boldsymbol{n}$ is the outward pointing unit vector on the surface Γ_i enclosing the finite volume Ω_i. The integration can be approximated using various numerical integration schemes. In the simplest 1-D case with $h = \Delta x$, the integration

$$u_i = \frac{1}{h} \int_{(i-1/2)h}^{(i+1/2)h} u dx, \tag{11.8}$$

is a vertex-centred finite volume scheme. In the following sections, we will discuss the three major types of partial differential equations (elliptic, parabolic and hyperbolic) and their finite volume discretization schemes.

11.2 Elliptic Equations

Laplace's equation is one of the most studied elliptic equations

$$\nabla^2 u(x,y) = 0, \qquad (x,y) \in \Omega, \tag{11.9}$$

its integral form can be written as

$$\int_\Omega \nabla^2 u d\Omega = \int_\Gamma \frac{\partial u}{\partial \boldsymbol{n}} \cdot \mathbf{dS} = 0. \tag{11.10}$$

For the simple regular grid points $(i\Delta x, j\Delta y)$, the control volume in this case is a cell centred at $(i\Delta x, j\Delta y)$ with a size of Δx (along x-axis) and Δy (along y-axis), and the boundary integral on any cell consists of four parts integrated on each of the four sides. By using the simple approximation $\frac{\partial u}{\partial n}$ with $\frac{\partial u}{\partial x} = (u_{i+1,j} - u_{i,j})/\Delta x$ and $\frac{\partial u}{\partial y} = (u_{i,j+1} - u_{i,j})/\Delta y$, we have

$$\int_{\Omega_{i,j}} \frac{\partial u}{\partial n} d\Omega = \frac{\Delta y}{\Delta x}(u_{i+1,j} + u_{i-1,j} - 2u_{i,j})$$

$$+\frac{\Delta x}{\Delta y}(u_{i,j+1}+u_{i,j-1}-2u_{i,j})=0. \tag{11.11}$$

Dividing both sides by $\Delta x \Delta y$, and letting $\Delta x = \Delta y = h$, we obtain

$$(u_{i+1,j}+u_{i,j+1}+u_{i-1,j}+u_{i,j-1})-4u_{i,j}=0, \tag{11.12}$$

which resembles the formulas in the finite difference methods in many ways. In fact, this is exactly the Laplace operator for a 5-point differencing scheme.

11.3 Parabolic Equations

For the case of a heat conduction problem

$$\frac{\partial u}{\partial t}=k\frac{\partial^2 u}{\partial x^2}+q(u,x,t), \tag{11.13}$$

we have its integral form

$$\int_t\int_\Omega(\frac{\partial u}{\partial t}-k\frac{\partial^2 u}{\partial x^2}-q)dxdt=0. \tag{11.14}$$

If we use the control volume from $(i-1/2)h$ to $(i+1/2)h$ where $h=\Delta x$, and with time from step n to $n+1$, we have

$$\int_{n\Delta t}^{(n+1)\Delta t}\int_{(i-1/2)h}^{(i+1/2)h}(\frac{\partial u}{\partial t}-k\frac{\partial^2 u}{\partial x^2}-q)dxdt=0. \tag{11.15}$$

By using the mid-point approximation

$$\int_a^b \psi(x)dx=\psi[\frac{(a+b)}{2}](b-a), \tag{11.16}$$

and the DuFort-Frankel scheme where we first approximate the gradient

$$\frac{\partial^2 u}{\partial x^2}=\frac{u_{i+1}^n-2u_i^n+u_{i-1}^n}{h^2}, \tag{11.17}$$

then replace $-2u_i^n$ with $-(u_j^{n+1}+u_j^{n-1})$, we have

$$\frac{u_i^{n+1}-u_i^{n-1}}{2\Delta t}=\frac{[(u_{i+1}^n-(u_i^{n+1}+u_i^{n-1})+u_{i-1}^n)]}{h^2}+q_i^n, \tag{11.18}$$

where we have used the central scheme for time as well. This is exactly the DuFort-Frankel explicit scheme in the finite difference method; however, the starting point is different. In addition, the finite volume scheme is more versatile in dealing with irregular geometry and more natural in applying boundary conditions. Similar stability analysis will lead to the condition $|\xi|<1$, which is always true and thus the Dufort-Frankel scheme is unconditionally stable for all Δt and Δx. However, it may show oscillatory behavior under certain conditions, and its implementation should be carried out carefully.

11.4 Hyperbolic Equations

For the hyperbolic equation of the conservation law in the one-dimensional case

$$\frac{\partial u}{\partial t} + \frac{\partial \Psi(u)}{\partial x} = 0, \tag{11.19}$$

we have its integral form in the fixed domain

$$\int_{x_a}^{x_b} \frac{\partial u}{\partial t} dx = \frac{\partial}{\partial t} \int_{x_a}^{x_b} u dx = -\{\Psi[u(x_b)] - \Psi[u(x_a)]\} = 0.$$

If we use the mid-point rule u^* to approximate the integral, we have

$$(x_b - x_a)\frac{\partial u^*}{\partial t} = -\{\Psi[u(x_b)] - \Psi[u(x_a)]\}. \tag{11.20}$$

By choosing the control volume $[(i-1/2)\Delta x, (i+1/2)\Delta x]$ centred at the mesh point $x_i = i\Delta x = ih$ with the approximation $u_i \approx u_i^*$ in each interval, and using the forward differencing scheme for the time derivative, we have

$$u_i^{n+1} - u_i^n = -\frac{\Delta t}{h}[\Psi(x_{i+1/2}) - \Psi(x_{i-1/2})]. \tag{11.21}$$

By further approximation of the flux $\Psi(x_{i+1/2}) \approx \Psi(x_i)$, we have the upward scheme

$$u_i^{n+1} - u_i^n = -\frac{\Delta t}{h}[\Psi(u_i) - \Psi(u_{i-1})], \tag{11.22}$$

which is conditionally stable as we know this in the finite difference method. For the simplest flux $\Psi(u) = cu$, we have

$$u^{n+1} = u_i^n - \frac{c\Delta t}{h}(u_i^n - u_{i-1}^n), \tag{11.23}$$

and its stability requires that

$$0 < \frac{c\Delta t}{h} \leq 1. \tag{11.24}$$

The literature above the finite volume method is vast. For example, the so-called computational fluid dynamics (CFD) dedicates solely to the solution of the Navier-Stokes equations and applications. Interested readers can refer to more advanced literature.

Chapter 12

Finite Element Method

The basic idea of finite element analysis is to divide the domain of interest into many small blocks or elements. This is equivalent to imaginarily cutting a solid structure such as a building or bridge into many pieces or elements. These small blocks are characterized by nodes, edges and surfaces, and the whole domain can be considered as if these blocks or elements are glued together at these nodes and along the element boundaries. In this way, we essentially transform a continuum system with infinite degrees of freedom into a discrete system with finite degrees of freedom.

12.1 Finite Element Formulation

As we know that most continuum systems are governed by differential equations, the major advantage of this finite-element approach is that the differential equation for a continuum system is transformed into a set of simultaneous algebraic equations for the discrete system with a finite number of elements.

12.1.1 *Weak Formulation*

Many problems are modelled in terms of partial differential equations, which can generally be written as

$$\mathcal{L}(u) = 0, \qquad \boldsymbol{x} \in \Omega, \tag{12.1}$$

where $\mathcal{L}$ is a differential operator, often linear. For example, Laplace's equation $\nabla^2 u = 0$ is equivalent to $\mathcal{L}u = 0$ and $\mathcal{L} = \nabla^2$. This problem is usually completed with the essential boundary condition (or prescribed values $\bar{u}$), $\mathcal{E}(u) = u - \bar{u} = 0$ for $\boldsymbol{x} \in \partial\Omega_E$, and natural boundary conditions

$\mathcal{B}(u) = 0$ for $\boldsymbol{x} \in \partial\Omega_N$ where $\mathcal{B}$ is a known function. Natural boundary conditions are usually concerned with flux or force.

Assuming that the true solution u can be approximated by u_h over a finite element mesh with an averaged element size or mean distance h between two adjacent nodes, the above equation can be approximated as

$$\mathcal{L}(u_h) \approx 0. \tag{12.2}$$

The ultimate goal is to construct a method of computing u_h such that the error $|u_h - u|$ is minimized. Generally speaking, the residual $R(u_1, ..., u_M, \boldsymbol{x}) = \mathcal{L}(u_h(\boldsymbol{x}))$ varies with space and time. There are several methods of minimizing R. Depending on the scheme of minimization and the choice of shape functions, various methods can be formulated. These include the weighted residual method, the method of least squares, the Galerkin method and others.

Multiplying both sides of equation (12.2) by a test function or a proper weighting function w_i, integrating over the domain and using associated boundary conditions, we can write the general weak formulation as

$$\int_\Omega \mathcal{L}(u_h) w_i d\Omega + \int_{\partial\Omega_N} \mathcal{B}(u_h)\bar{w}_i d\Gamma + \int_{\partial\Omega_E} \mathcal{E}(u_h)\tilde{w}_i d\Gamma_E \approx 0,$$

where $(i = 1, 2, ..., M)$, and $\bar{w}_i$ and $\tilde{w}_i$ are the values of w_i on the natural and essential boundaries. If we can approximate the solution u_h by the expansion in term of shape function N_i

$$u_h(u, t) = \sum_{i=1}^{M} u_i(t) N_i(x) = \sum_{j=1}^{M} u_j N_j, \tag{12.3}$$

it requires that $N_i = 0$ on $\partial\Omega_E$ so that we can choose $\tilde{w}_i = 0$ on $\partial\Omega_E$. Thus, only the natural boundary conditions are included since the essential boundary conditions are automatically satisfied. In addition, there is no such limitation on the choice of w_i and $\bar{w}_i$. If we choose $\bar{w}_i = -w_i$ on the natural boundary so as to simplify the formulation, we have

$$\int_\Omega \mathcal{L}(u_h) w_i d\Omega \approx \int_{\partial\Omega_N} \mathcal{B}(u_h) w_i d\Gamma. \tag{12.4}$$

12.1.2 *Galerkin Method*

There are many different ways to choose the test functions w_i and shape functions N_i. One of the most popular methods is the Galerkin method where the test functions are the same as the shape functions, or

$$w_i = N_i. \tag{12.5}$$

In this special case, the formulation simply becomes

$$\int_{\Omega} \mathcal{L}(u_h) N_i d\Omega \approx \int_{\partial\Omega_N} \mathcal{B}(u_h) N_i d\Gamma. \tag{12.6}$$

The discretization of this equation will usually lead to an algebraic matrix equation.

On the other hand, if we use the Dirac delta function as the test functions $w_i = \delta(\boldsymbol{x} - \boldsymbol{x}_i)$, the method is called the collocation method which uses the interesting properties of the Dirac function

$$\int_{\Omega} f(\boldsymbol{x})\delta(\boldsymbol{x} - \boldsymbol{x}_i) d\Omega = f(\boldsymbol{x}_i), \tag{12.7}$$

together with $\delta(\boldsymbol{x} - \boldsymbol{x}_i) = 1$ at $\boldsymbol{x} = \boldsymbol{x}_i$ and $\delta(\boldsymbol{x} - \boldsymbol{x}_i) = 0$ at $\boldsymbol{x} \neq \boldsymbol{x}_i$.

12.1.3 *Shape Functions*

The main aim of the finite element method is to find an approximate solution $u_h(\boldsymbol{x}, t)$ for the exact solution u on some nodal points,

$$u_h(\boldsymbol{x}, t) = \sum_{i=1}^{M} u_i(t) N_i(\boldsymbol{x}) \tag{12.8}$$

where u_i are unknown coefficients or the values of u at the discrete nodal point i. Functions N_i $(i = 1, 2, ..., M)$ are linearly independent functions that vanish on the part of the essential boundary. At any node i, we have $N_i = 1$, and $N_i = 0$ at any other nodes, or

$$\sum_{i=1}^{M} N_i = 1, \qquad N_i(\boldsymbol{x}_j) = \delta_{ij}. \tag{12.9}$$

The functions $N_i(\boldsymbol{x})$ are referred to as basis functions, trial functions or more often shape functions in the literature of finite element methods.

12.1.3.1 *Linear Shape Functions*

For the simplest 1-D element with two nodes i and j, the linear shape functions (shown in Figure 12.1) can be written as

$$N_i = \frac{x_j - x}{L} = \frac{1 - \xi}{2}, \qquad N_j = \xi = \frac{x - x_i}{L} = \frac{1 + \xi}{2}, \tag{12.10}$$

where ξ is the natural coordinate

$$\xi = \frac{x - x_o}{L/2}, \qquad L = |x_j - x_i|, \qquad x_o = \frac{x_i + x_j}{2}, \tag{12.11}$$

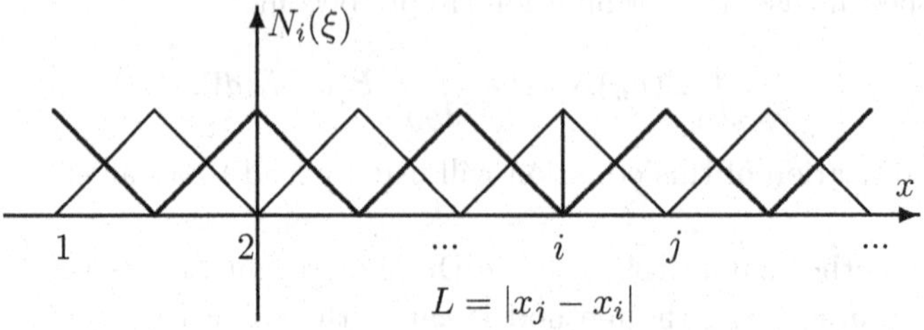

Fig. 12.1 The 1-D linear shape functions.

where x_o is the midpoint of the element, and $\xi_i = -1$ at $x = x_i$ and $\xi_j = 1$ at $x = x_j$.

A linear shape function spans only two adjacent nodes i and j, and it requires two coefficients in the generic form

$$N(\xi) = a + b\xi. \tag{12.12}$$

12.1.3.2 *Quadratic Shape Functions*

Suppose we want to get higher-order approximations, we can use, say, the quadratic shape functions which span three adjacent nodes i, j, and k. Three coefficients need to be determined

$$N(\xi) = a + b\xi + c\xi^2. \tag{12.13}$$

Using the conditions $\xi_i = -1$ at $x = x_i$ and $\xi_j = 1$ at $x = x_j$, and the known displacements u_i, u_j and u_k, we have

$$u_i = a + b(-1) + c(-1)^2, \tag{12.14}$$

$$u_j = a, \tag{12.15}$$

and

$$u_k = a + b(1) + c(1)^2, \tag{12.16}$$

whose solutions are

$$\begin{pmatrix} a \\ b \\ c \end{pmatrix} = \begin{pmatrix} u_j \\ \frac{1}{2}(u_i - 2u_j + u_k) \\ \frac{1}{2}(u_k - u_i) \end{pmatrix}. \tag{12.17}$$

Substituting this into equation (12.13), we have

$$u = \frac{\xi(\xi-1)}{2} u_i + (1 - \xi^2) u_j + \frac{\xi(\xi+1)}{2} u_k, \tag{12.18}$$

which is equivalent to

$$u = N_i u_i + N_j u_j + N_k u_k, \tag{12.19}$$

where

$$\boldsymbol{N} = [N_i, N_j, N_k] = [\frac{\xi(\xi-1)}{2}, (1-\xi^2), \frac{\xi(\xi+1)}{2}]. \tag{12.20}$$

12.1.3.3 *Lagrange Polynomials*

The essence of the shape functions is the interpolation, and the interpolation functions can be many different types. Lagrange polynomials are popularly used to construct shape functions. The $n-1$ order Lagrange polynomials require n nodes, and the associated shape functions can generally be written as

$$N_i(\xi) = \Pi_{j=1, j\neq i}^{n} \frac{(\xi - \xi_j)}{(\xi_i - \xi_j)}$$

$$= \frac{(\xi-\xi_1)...(\xi-\xi_{i-1})(\xi-\xi_{i+1})...(\xi-\xi_n)}{(\xi_i-\xi_1)...(\xi_i-\xi_{i-1})(\xi_i-\xi_{i+1})...(\xi_i-\xi_n)}, \tag{12.21}$$

where ξ_j means that the value of ξ at node j. For $n = 3$, it is straightforward to validate that

$$N_1(\xi) = \frac{\xi(\xi-1)}{2}, \quad N_2(\xi) = 1 - \xi^2, \quad N_3(\xi) = \frac{\xi(\xi+1)}{2}.$$

This method of formulating shape functions can be easily extended to the 2D and 3D cases as well as isoparametric elements. The derivative of $N_i(x)$ with respect to ξ is given by

$$N_i'(\xi) = \sum_{k=1, k\neq i}^{n} \frac{1}{(\xi_i - \xi_j)} \Pi_{j=1, j\neq i}^{n} \frac{(\xi - \xi_j)}{(\xi_i - \xi_j)}. \tag{12.22}$$

12.1.3.4 *2D Shape Functions*

The shape functions we discussed earlier are 1D shape functions (with one independent variables x or ξ) for 1D elements. For 2D elements such as quadrilateral elements, their corresponding shape functions are functions of two independent variables: x and y, or ξ and η. Using the natural coordinates ξ and η shown in Fig. 12.2, we can construct various shape functions.

For a bilinear quadrilateral (Q4) element, we use bilinear approximations for the displacement field u and v. If we use

$$u = \alpha_0 + \alpha_1 x + \alpha_2 y + \alpha_3 xy, \tag{12.23}$$

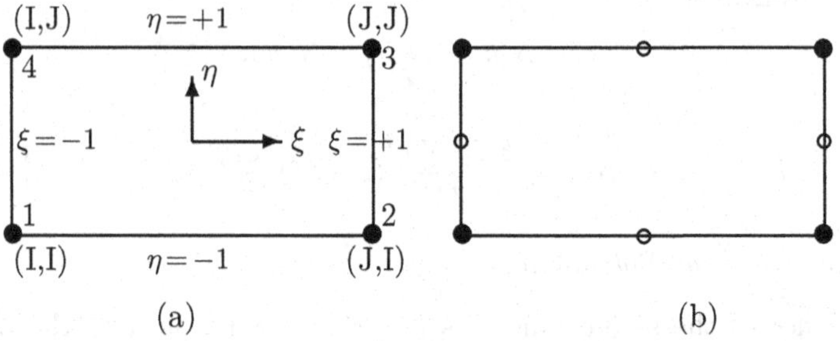

Fig. 12.2 (a) A bilinear quadrilateral element; (b) A quadratic quadrilateral element.

$$v = \beta_0 + \beta_1 x + \beta_2 y + \beta_3 xy, \tag{12.24}$$

and express them in terms of shape function N_i

$$u = \sum N_i u_i, \qquad v = \sum N_j v_j, \tag{12.25}$$

we can derive the shape functions by following the similar procedure as discussed above. We have

$$N_1 = \frac{(1-\xi)(1-\eta)}{4}, \qquad N_2 = \frac{(1+\xi)(1-\eta)}{4}, \tag{12.26}$$

$$N_3 = \frac{(1+\xi)(1+\eta)}{4}, \qquad N_4 = \frac{(1-\xi)(1+\eta)}{4}. \tag{12.27}$$

From the 1-D linear shape functions

$$N_I^{(2)}(\xi) = \frac{(1-\xi)}{2}, \qquad N_J^{(2)}(\xi) = \frac{(1+\xi)}{2}, \tag{12.28}$$

for a 2-node element (along x) where the superscript '(2)' means 2 nodes, we can also write another set of linear shape functions for a 2-node element in the y-direction. We have

$$N_I^{(2)}(\eta) = \frac{(1-\eta)}{2}, \qquad N_J^{(2)}(\eta) = \frac{(1+\eta)}{2}. \tag{12.29}$$

If we label the nodes by a pair (I, J) in 2D coordinates, we have

$$N_i(\xi,\eta) = N_{IJ} = N_I^{(2)} N_J^{(2)}. \tag{12.30}$$

We can see that

$$N_1(\xi,\eta) = N_I^{(2)}(\xi) N_I^{(2)}(\eta), \quad N_2(\xi,\eta) = N_J^{(2)}(\xi) N_I^{(2)}(\eta),$$

and

$$N_3(\xi,\eta) = N_J^{(2)}(\xi) N_J^{(2)}(\eta), \quad N_4(\xi,\eta) = N_I^{(2)}(\xi) N_J^{(2)}(\eta).$$

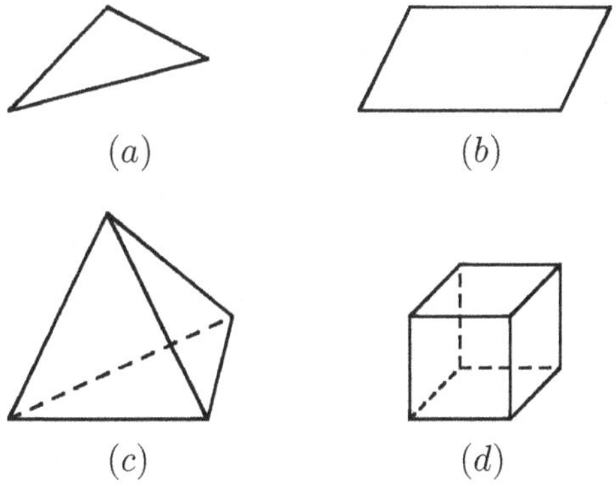

Fig. 12.3 Common elements: (a) triangular element; (b) quadrilateral element; (c) tetrahedron; and (d) hexahedron.

In fact, higher-order shape functions for 2D and 3D elements can be systematically derived in this manner. Figure 12.3 shows several common elements.

If we approximate the displacement field using higher-order approximations, then we are dealing with the quadratic quadrilateral (Q8) element because we have to use eight nodes (4 finite element nodes and 4 midpoints). In this case, the shape functions are much more complicated; for example, the shape function N_2 becomes

$$N_2 = \frac{(1-\xi)(1-\eta)}{4} - \frac{1}{4}[(1-\xi^2)(1-\eta) + (1+\xi)(1-\eta^2)]. \quad (12.31)$$

12.2 Derivatives and Integration

12.2.1 *Derivatives*

Using the assumptions that $u_i(t)$ does not depend on space and $N_i(\boldsymbol{x})$ does not depend on time, the derivatives of u can be approximated as

$$\frac{\partial u}{\partial \boldsymbol{x}} \approx \frac{\partial u_h}{\partial \boldsymbol{x}} = \sum_{i=1}^{M} u_i(t) N'(\boldsymbol{x}),$$

$$\dot{u} \approx \frac{\partial u_h}{\partial t} = \sum_{i=1}^{M} \dot{u}_i N(\boldsymbol{x}), \quad (12.32)$$

where we have used the notations: $' = d/d\boldsymbol{x}$ and $\dot{} = \frac{\partial}{\partial t}$. The derivatives of the shape functions N_i can be calculated in the similar manner as in

Eq. (12.22). For a linear shape function, its first derivative is constant in each interval or element, which may lead to discontinuity in u'. For a quadratic shape function, its first derivative varies linearly (in a piecewise manner), but its second derivative u'' is a constant, which may be discontinuous. If we want a continuous and smooth u', we have to use higher-order shape functions. Higher-order derivatives are then calculated in a similar way.

12.2.2 *Gauss Quadrature*

In the finite element analysis, the calculations of stiffness matrix and application of boundary conditions such as Eq. (12.6) involve the integration over elements. Such numerical integration is often carried out in terms of natural coordinates ξ and η, and the Gauss integration or Gauss quadrature as discussed in earlier chapters is usually used for evaluating integrals numerically. Gauss quadrature has relatively high accuracy. For example, the n-point Gauss quadrature for one-dimensional integrals

$$\mathcal{I} = \int_{-1}^{1} \psi(\xi) d\xi \approx \sum_{i=1}^{n} w_i \psi_i. \tag{12.33}$$

For the case of $n = 3$, we have

$$\int_{-1}^{1} \psi(\xi) d\xi \approx \sum_{i=1}^{3} w_i \psi_i = \frac{1}{9}[8\psi_2 + 5(\psi_1 + \psi_3)], \tag{12.34}$$

which is schematically shown in Figure 12.4.

For two-dimensional integrals, we use n^2-point Gauss quadrature of order n, and we have

$$\mathcal{I} = \int_{-1}^{1} \int_{-1}^{1} \psi(\xi, \eta) d\xi d\eta = \sum_{i=1}^{n} \sum_{j=1}^{n} w_i w_j \psi_{ij}, \tag{12.35}$$

where $\psi_{ij} = \psi(\xi_i, \eta_j)$. In the case of $n = 3$, we have 9 points (shown in Figure 12.4), and the quadrature becomes

$$\mathcal{I} = \int_{-1}^{1} \int_{-1}^{1} \psi(\xi, \eta) \approx \sum_{i=1}^{3} \sum_{j=1}^{3} w_i w_j \psi_{i+3*(j-1)}(\xi_i, \eta_j)$$

$$= \frac{1}{81}[25(\psi_1 + \psi_3 + \psi_7 + \psi_9) + 64\psi_5 + 40(\psi_2 + \psi_4 + \psi_6 + \psi_8)]. \tag{12.36}$$

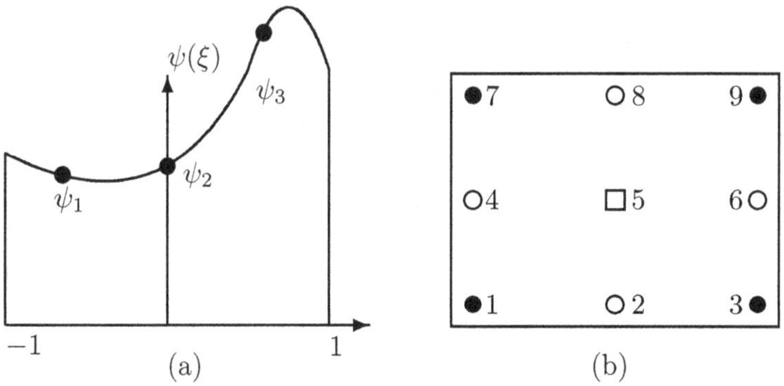

Fig. 12.4 Gauss quadrature: a) 1-D integration with $|\xi_1 - \xi_2| = |\xi_2 - \xi_3| = \sqrt{3/5}$; and b) 2-D 9-point integration over a quadrilateral element with point 3 at $(\xi_3, \eta_1) = (\sqrt{3/5}, -\sqrt{3/5})$ and point 9 at $(\xi_3, \eta_3) = (\sqrt{3/5}, \sqrt{3/5})$.

12.3 Poisson's Equation

Heat transfer problems are very common in engineering and computational modelling. The geometry in most applications is irregular. Thus, finite element methods are especially useful in this case.

The steady-state heat transfer is governed by the heat conduction equation or Poisson's equation

$$\nabla \cdot (k \nabla u) + Q = 0, \tag{12.37}$$

with the essential boundary condition

$$u = \bar{u}, \qquad \boldsymbol{x} \in \partial \Omega_E, \tag{12.38}$$

and the natural boundary condition

$$k \frac{\partial u}{\partial n} - q = 0, \qquad \boldsymbol{x} \in \partial \Omega_N. \tag{12.39}$$

Multiplying both sides of equation (12.37) by the shape function N_i and using the formulation similar to the formulation (12.6) in terms of $u \approx u_h$, we have

$$\int_\Omega [\nabla \cdot (k \nabla u) + Q] N_i d\Omega - \int_{\partial \Omega_N} [k \frac{\partial u}{\partial n} - q] N_i d\Gamma = 0. \tag{12.40}$$

Integrating by parts and using Green's theorem, we have

$$- \int_\Omega (\nabla u_h \cdot k \cdot \nabla N_i) d\Omega + \int_{\partial \Omega} k \frac{\partial u_h}{\partial n} N_i d\Gamma$$

$$+\int_{\Omega} QN_i d\Omega - \int_{\partial\Omega_N} [k\frac{\partial u_h}{\partial n} - q]N_i d\Gamma = 0. \tag{12.41}$$

Since $N_i = 0$ on $\partial\Omega_E$, thus we have

$$\int_{\partial\Omega} [\,]N_i d\Gamma = \int_{\partial\Omega_N} [\,]N_i d\Gamma, \tag{12.42}$$

where $[\,]$ is any integrand. Therefore, the above weak formulation becomes

$$\int_{\Omega} (\nabla u_h \cdot k \cdot \nabla N_i) d\Omega - \int_{\Omega} QN_i d\Omega - \int_{\partial\Omega_N} qN_i d\Gamma = 0. \tag{12.43}$$

Substituting

$$u_h = \sum_{j=1}^{M} u_j N_j(\boldsymbol{x}),$$

into the equation, we have

$$\sum_{j=1}^{M} [\int_{\Omega} (k\nabla N_i \cdot \nabla N_j) d\Omega] u_j - \int_{\Omega} QN_i d\Omega - \int_{\partial\Omega_N} qN_i d\Gamma = 0. \tag{12.44}$$

This can be written in the compact matrix form

$$\sum_{j=1}^{M} K_{ij} U_j = f_i, \qquad \mathbf{KU} = \mathbf{f}, \tag{12.45}$$

where $\mathbf{K} = [K_{ij}], (i, j = 1, 2, ..., M)$, $\mathbf{U}^T = (u_1, u_2, ..., u_M)$, and $\mathbf{f}^T = (f_1, f_2, ..., f_M)$. That is,

$$K_{ij} = \int_{\Omega} k\nabla N_i \nabla N_j d\Omega, \tag{12.46}$$

$$f_i = \int_{\Omega} QN_i d\Omega + \int_{\partial\Omega_N} qN_i d\Gamma. \tag{12.47}$$

As a simple example, we consider the 1-D steady-state heat conduction problem,

$$u''(x) + Q(x) = 0,$$

with boundary conditions

$$u(0) = \beta, \qquad u'(1) = q.$$

The assembly of the global system matrix for the example with 4 elements and five nodes is shown below. For each element with i and j nodes, we have

$$N_i = 1 - \xi, \qquad N_j = \xi, \qquad \xi = \frac{x}{L}, \qquad L = h_e,$$

$$K_{ij}^{(e)} = [\int_0^L kN_i'N_j'dx] = \frac{k}{h_e}\begin{pmatrix} 1 & -1 \\ -1 & 1 \end{pmatrix},$$

$$f_i^{(e)} = \frac{Qh_e}{2}\begin{pmatrix} 1 \\ 1 \end{pmatrix},$$

so that, for example in elements 1 and 2, these can extend to all nodes (with $h_i = x_{i+1} - x_i, i = 1, 2, 3, 4$),

$$K^{(1)} = \begin{pmatrix} k/h_1 & -k/h_1 & 0 & 0 & 0 \\ -k/h_1 & k/h_1 & 0 & 0 & 0 \\ 0 & 0 & 0 & 0 & 0 \\ 0 & 0 & 0 & 0 & 0 \\ 0 & 0 & 0 & 0 & 0 \end{pmatrix}, \quad f^{(1)} = \frac{Q}{2}\begin{pmatrix} h_1 \\ h_1 \\ 0 \\ 0 \\ 0 \end{pmatrix},$$

$$K^{(2)} = \begin{pmatrix} 0 & 0 & 0 & 0 & 0 \\ 0 & k/h_2 & -k/h_2 & 0 & 0 \\ 0 & -k/h_2 & k/h_2 & 0 & 0 \\ 0 & 0 & 0 & 0 & 0 \\ 0 & 0 & 0 & 0 & 0 \end{pmatrix}, \quad f^{(2)} = \frac{Q}{2}\begin{pmatrix} 0 \\ h_2 \\ h_2 \\ 0 \\ 0 \end{pmatrix},$$

and so on. Now the global system matrix is obtained by adding all the stiffness matrices (of each element), which leads to

$$K = \begin{pmatrix} k/h_1 & -k/h_1 & 0 & 0 & 0 \\ -k/h_1 & \frac{k}{h_1} + \frac{k}{h_2} & -k/h_2 & 0 & 0 \\ 0 & -k/h_2 & \frac{k}{h_2} + \frac{k}{h_3} & -k/h_3 & 0 \\ 0 & 0 & -k/h_3 & \frac{k}{h_3} + \frac{k}{h_4} & -k/h_4 \\ 0 & 0 & 0 & -k/h_4 & k/h_4 \end{pmatrix},$$

$$\mathbf{U} = \begin{pmatrix} u_1 \\ u_2 \\ u_3 \\ u_4 \\ u_5 \end{pmatrix}, \qquad \boldsymbol{f} = \begin{pmatrix} Qh_1/2 \\ Q(h_1 + h_2)/2 \\ Q(h_2 + h_3)/2 \\ Q(h_3 + h_4)/2 \\ Qh_4/2 + q \end{pmatrix},$$

where the last row of $\boldsymbol{f}$ has already included the natural boundary condition at $u'(1) = q$.

We now use the direct application method for the essential boundary conditions. We can replace the first equation $\sum_{j=1}^{5} K_{1j}u_j = f_1$ with $u_1 = \beta$, so that the first row becomes $K_{1j} = (1\ 0\ 0\ 0\ 0)$ and $f_1 = \beta$. Thus, we have

$$\boldsymbol{K} = \begin{pmatrix} 1 & 0 & 0 & 0 & 0 \\ -k/h_1 & \frac{k}{h_1} + \frac{k}{h_2} & -k/h_2 & 0 & 0 \\ 0 & -k/h_2 & \frac{k}{h_2} + \frac{k}{h_3} & -k/h_3 & 0 \\ 0 & 0 & -k/h_3 & \frac{k}{h_3} + \frac{k}{h_4} & -k/h_4 \\ 0 & 0 & 0 & -k/h_4 & k/h_4 \end{pmatrix},$$

$$\boldsymbol{U} = \begin{pmatrix} u_1 \\ u_2 \\ u_3 \\ u_4 \\ u_5 \end{pmatrix}, \qquad \boldsymbol{f} = \begin{pmatrix} \beta \\ Q(h_1 + h_2)/2 \\ Q(h_2 + h_3)/2 \\ Q(h_3 + h_4)/2 \\ Qh_4/2 + q \end{pmatrix}.$$

Now let us look at a simpler case as an example.

Example 12.1: *For the case of* $k = 1, Q = -1, h_1 = ... = h_4 = 0.25$, $\beta = 1$ *and* $q = -0.25$, *we have*

$$\boldsymbol{K} = \begin{pmatrix} 1 & 0 & 0 & 0 & 0 \\ -4 & 8 & -4 & 0 & 0 \\ 0 & -4 & 8 & -4 & 0 \\ 0 & 0 & -4 & 8 & -4 \\ 0 & 0 & 0 & -4 & 4 \end{pmatrix}, \qquad \boldsymbol{f} = \begin{pmatrix} 1 \\ -0.25 \\ -0.25 \\ -0.25 \\ -0.375 \end{pmatrix}.$$

Hence, the solution is

$$\boldsymbol{U} = \boldsymbol{K}^{-1}\boldsymbol{f} = \begin{pmatrix} 1.00 & 0.71875 & 0.50 & 0.34375 & 0.25 \end{pmatrix}^T.$$

We know that the analytical solution for this problem is

$$u(x) = -\frac{Q}{2}x^2 + (Q + q)x + \beta$$

$$= \frac{1}{2}x^2 - 1.25x + 1,$$

therefore, the exact values of u *at the same five nodes are*

$$\boldsymbol{U}_{\text{true}} = \begin{pmatrix} 1 & 23/32 & 0.5 & 11/32 & 0.25 \end{pmatrix}^T.$$

We can see that the numerical solutions obtained by the finite element method are almost the same as the true values.

12.4 Transient Problems

The problems we have discussed so far are static or time-independent because the time dimension is not involved. For time-dependent problems, the standard finite element formulation first produces an ordinary differential equation for matrices rather than algebraic matrix equations. Therefore, besides the standard finite element formulations, extra time-stepping schemes should be used in a similar manner to that in finite difference methods.

As the weak formulation uses the Green theorem that involves the spatial derivatives, the time derivatives can be considered as the source term. Thus, one simple and yet instructive way to extend the finite element formulation to include the time dimension is to replace Q in equation (12.37) with $Q - \alpha u_t - \beta u_{tt} = Q - \alpha\dot{u} - \beta\ddot{u}$ so that we have

$$\nabla \cdot (k\nabla u) + (Q - \alpha\dot{u} - \beta\ddot{u}) = 0. \tag{12.48}$$

The boundary conditions and initial conditions are $u(\boldsymbol{x}, 0) = \phi(\boldsymbol{x})$, $u = \boldsymbol{u}, \boldsymbol{x} \in \partial\Omega_E$, and $k\frac{\partial u}{\partial n} - q = 0, \boldsymbol{x} \in \partial\Omega_N$. Using integration by parts and the expansion $u_h = \sum_{j=1}^{M} u_j N_j$, we have

$$\sum_{j=1}^{M}[\int_{\Omega}(k\nabla N_i \nabla N_j)d\Omega] + \sum_{j=1}^{M}\int_{\Omega}[(N_i\alpha N_j)\dot{u}_j + (N_i\beta N_j)\ddot{u}_j]d\Omega$$

$$- \int_{\Omega} N_i Q d\Omega - \int_{\partial\Omega_N} N_i q d\Gamma = 0, \tag{12.49}$$

which can be written in a compact form as

$$\boldsymbol{M}\ddot{\boldsymbol{u}} + \boldsymbol{C}\dot{\boldsymbol{u}} + \boldsymbol{K}\boldsymbol{u} = \boldsymbol{f}, \tag{12.50}$$

where

$$K_{ij} = \int_{\Omega}[(k\nabla N_i \nabla N_j)]d\Omega, \tag{12.51}$$

$$f_i = \int_{\Omega} N_i Q d\Omega + \int_{\partial\Omega_N} N_i q d\Gamma, \tag{12.52}$$

and

$$C_{ij} = \int_{\Omega} N_i \alpha N_j d\Omega, \qquad M_{ij} = \int_{\Omega} N_i \beta N_j d\Omega. \tag{12.53}$$

The matrices $\boldsymbol{K}, \boldsymbol{M}$, and $\boldsymbol{C}$ are symmetric, that is to say, $K_{ij} = K_{ji}, M_{ij} = M_{ji}, C_{ij} = C_{ji}$ due to the interchangeability of the orders in the product of the integrand k, N_i and N_j (i.e., $\nabla N_i \cdot k \cdot \nabla N_j = k \nabla N_i \nabla N_j$, $N_i \alpha N_j = N_j \alpha N_i = \alpha N_i N_j$, etc.). The matrix $\mathbf{C} = [C_{ij}]$ is the damping matrix similar to the damping coefficient of damped oscillations. $\mathbf{M} = [M_{ij}]$ is the general mass matrix due to a similar role acting as an equivalent mass in dynamics.

In addition, before the boundary conditions are imposed, the stiffness matrix is usually singular, which may imply many solutions. Only after the proper boundary conditions have been enforced, the stiffness matrix will be nonsingular, thus unique solutions may be obtained.

Example 12.2: *For the wave equation ($\boldsymbol{C} = 0$), we have*

$$\boldsymbol{M}\ddot{\boldsymbol{u}} + \boldsymbol{K}\boldsymbol{u} = \boldsymbol{f}.$$

Using

$$\ddot{\boldsymbol{u}} = \frac{\boldsymbol{u}^{n+1} - 2\boldsymbol{u}^n + \boldsymbol{u}^{n-1}}{(\Delta t)^2},$$

we have

$$\boldsymbol{u}^{n+1} = \boldsymbol{M}^{-1}\boldsymbol{f}(\Delta t)^2 + [2\boldsymbol{I} - (\Delta t)^2 \boldsymbol{M}^{-1}\boldsymbol{K}]\boldsymbol{u}^n - \boldsymbol{u}^{n-1},$$

where $\mathbf{I}$ *is an identity matrix. In principle, this can be solved by iterative methods starting from an initial vector* $\boldsymbol{u}^0$.

The literature of the finite element methods and their variants is extremely vast, from general textbooks to very specialized topics in engineering and computational sciences. Therefore, we will not discuss this class of methods any further, and interested readers can refer to more advanced textbooks.

Part IV

Mathematical Programming

Chapter 13

Mathematical Optimization

Optimization is everywhere, from business transactions and engineering design to planning your holidays and daily travel routes. Business organisations have to maximize their profits and minimize the costs. Engineering design has to maximize the performance of the designed product while minimizing the manufacturing cost at the same time. Even when we plan holidays we want to maximize the enjoyment and minimize the cost. Therefore, the studies of optimization are of both scientific interest and practical implications and subsequently the methodology can potentially have many applications.

13.1 Optimization

Whatever the real-world applications may be, it is usually possible to formulate an optimization problem in a generic form. All optimization problems with explicit objectives can in general be expressed as a nonlinearly constrained optimization problem

$$\underset{\boldsymbol{x}\in\Re^d}{\text{maximize/minimize}} \, f(\boldsymbol{x}), \quad \boldsymbol{x}=(x_1, x_2, ..., x_d)^T \in \Re^d,$$

$$\text{subject to } \phi_j(\boldsymbol{x})=0, \quad (j=1,2,...,M),$$

$$\psi_k(\boldsymbol{x}) \geq 0, \quad (k=1,...,N), \tag{13.1}$$

where $f(\boldsymbol{x})$, $\phi_i(\boldsymbol{x})$ and $\psi_j(\boldsymbol{x})$ are scalar functions of the design vector $\boldsymbol{x}$. Here the components x_i of $\boldsymbol{x}=(x_1,...,x_d)^T$ are called design or decision variables, and they can be either continuous, or discrete or mixed of these two. The vector $\boldsymbol{x}$ is often called the decision vector which varies in a d-dimensional space $\Re^d$. It is worth pointing out that we use a column vector here for $\boldsymbol{x}$ (thus with a transpose T). We can also use a row vector $\boldsymbol{x} = (x_1, ..., x_d)$ and the results will be the same. Different textbooks may use slightly different formulations, once we are aware of such minor variations, it causes no difficulty or confusion.

In addition, the function $f(\boldsymbol{x})$ is called the objective function or cost function. In addition, $\phi_i(\boldsymbol{x})$ are constraints in terms of M equalities, and $\psi_j(\boldsymbol{x})$ are constraints written as N inequalities. So there are $M + N$ constraints in total. The optimization problem formulated here is a nonlinear constrained problem.

The space spanned by the decision variables is called the search space $\Re^d$, while the space formed by the values of the objective function is called the solution space. The optimization problem essentially maps the $\Re^d$ domain or space of decision variables into a solution space $\Re$ (or the real axis in general).

The objective function $f(\boldsymbol{x})$ can be either linear or nonlinear. If the constraints ϕ_i and ψ_j are all linear, it becomes a linearly constrained problem. Furthermore, when ϕ_i, ψ_j and the objective function $f(\boldsymbol{x})$ are all linear, then it becomes a linear programming problem. If the objective is at most quadratic with linear constraints, then it is called quadratic programming. If all the values of the decision variables can be integers, then this type of linear programming is called integer programming or integer linear programming.

Linear programming is very important in applications and has been well studied. However, there is still no generic method for solving nonlinear programming in general, though some important progress has been made in the last few decades. It is worth pointing out that the term *programming* here means *planning*, it has nothing to do with computer programming, and the similarity in wording is purely coincidental.

On the other hand, if no constraints are specified so that x_i can take any values in the real axis (or any integers), the optimization problem is referred to as an unconstrained optimization problem.

As a very simple example of optimization problems without any constraints, we discuss the search of the maxima or minima of a function. For example, to find the maximum of a univariate function $f(x)$

$$f(x) = x^2 e^{-x^2}, \qquad -\infty < x < \infty, \tag{13.2}$$

is a simple unconstrained problem, while the following problem is a simple constrained minimization problem

$$f(x_1, x_2) = x_1^2 + x_1 x_2 + x_2^2, \qquad (x_1, x_2) \in \Re^2, \tag{13.3}$$

subject to

$$x_1 \geq 1, \qquad x_2 - 2 = 0. \tag{13.4}$$

It is worth pointing out that the objective are explicitly known in all the optimization problems to be discussed in this book. However, in reality, it is often difficult to quantify what we want to achieve, but we still try to optimize certain things such as the degree of enjoyment or a quality service on holiday. In other cases, it might be impossible to write the objective function in any explicit form mathematically.

Example 13.1: *To find the minimum of $f(x) = x^2 e^{-x^2}$, we have the stationary condition $f'(x) = 0$ or*

$$f'(x) = 2x \times e^{-x^2} + x^2 \times (-2x)e^{-x^2} = 2(x - x^3)e^{-x^2} = 0.$$

As $e^{-x^2} > 0$, we have

$$x(1 - x^2) = 0,$$

or

$$x = 0, \qquad x = \pm 1.$$

The second derivative is given by

$$f''(x) = 2e^{-x^2}(1 - 5x^2 + 2x^4),$$

which is an even function with respective to x.

So at $x = \pm 1$, $f''(\pm 1) = 2[1 - 5(\pm 1)^2 + 2(\pm 1)^4]e^{-(\pm 1)^2} = -4e^{-1} < 0$. Thus, there are two maxima that occur at $x_* = \pm 1$ with $f_{\max} = e^{-1}$. At $x = 0$, we have $f''(0) = 2 > 0$, thus the minimum of $f(x)$ occurs at $x_* = 0$ with $f_{\min}(0) = 0$.

Whatever the objective is, we have to evaluate it many times. In most cases, the evaluations of the objective functions consume a substantial amount of computational power (which costs money) and design time. Any efficient algorithm that can reduce the number of objective evaluations will save both time and money.

13.2 Optimality Criteria

In mathematical programming, there are many important concepts, and we will first introduce three related concepts: feasible solutions, the strong local optimum and weak local optimum.

A point $\boldsymbol{x}$ which satisfies all the constraints is called a feasible point and thus is a feasible solution to the problem. The set of all feasible points is called the feasible region. A point $\boldsymbol{x}_*$ is called a strong local maximum of the nonlinearly constrained optimization problem if $f(\boldsymbol{x})$ is defined in a δ-neigbourhood $N(\boldsymbol{x}_*, \delta)$ and satisfies $f(\boldsymbol{x}_*) > f(\boldsymbol{u})$ for $\forall \boldsymbol{u} \in N(\boldsymbol{x}_*, \delta)$ where $\delta > 0$ and $\boldsymbol{u} \neq \boldsymbol{x}_*$. If $\boldsymbol{x}_*$ is not a strong local maximum, the inclusion of equality in the condition $f(\boldsymbol{x}_*) \geq f(\boldsymbol{u})$ for $\forall \boldsymbol{u} \in N(\boldsymbol{x}_*, \delta)$ defines the point $\boldsymbol{x}_*$ as a weak local maximum (see Fig. 13.1). The local minima can be defined in the similar manner when $>$ and $\geq$ are replaced by $<$ and $\leq$, respectively.

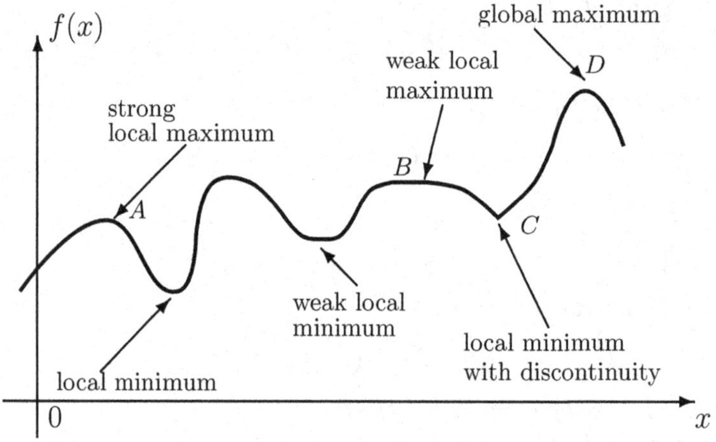

Fig. 13.1 Strong and weak maxima and minima.

Figure 13.1 shows various local maxima and minima. Point A is a strong local maximum, while point B is a weak local maximum because there are many (in fact infinite) different values of $\boldsymbol{x}$ which will lead to the same value of $f(\boldsymbol{x}_*)$. Point D is the global maximum. However, point C is a strong local minimum, but it has a discontinuity in $f'(\boldsymbol{x}_*)$. So the stationary condition for this point $f'(\boldsymbol{x}_*) = 0$ is not valid. We will not deal with these types of minima or maxima in detail. In our present discussion, we will assume that both $f(\boldsymbol{x})$ and $f'(\boldsymbol{x})$ are always continuous or $f(\boldsymbol{x})$ is everywhere twice-continuously differentiable.

For example, the minimum of $f(x) = x^2$ at $x = 0$ is a strong local minimum. The minimum of $g(x, y) = (x - y)^2 + (x - y)^2$ at $x = y = 0$ is a weak local minimum because $g(x, y) = 0$ along the line $x = y$ so that $g(x, y = x) = 0 = g(0, 0)$.

13.3 Unconstrained Optimization

13.3.1 *Univariate Functions*

The simplest optimization problem without any constraints is probably the search for the maxima or minima of a univariate function $f(x)$. For unconstrained optimization problems, the optimality occurs at either boundary or more often at the critical points given by the stationary condition $f'(x) = 0$.

However, this stationary condition is just a necessary condition, but it is not a sufficient condition. If $f'(x_*) = 0$ and $f''(x_*) > 0$, it is a local minimum. Conversely, if $f'(x_*) = 0$ and $f''(x_*) < 0$, then it is a local maximum. However, if $f'(x_*) = 0$ but $f''(x)$ is indefinite (both positive and negative) when $x \to x_*$, then x_* corresponds to a saddle point. For example, $f(x) = x^3$ has a saddle point $x_* = 0$ because $f'(0) = 0$ but f'' changes sign from $f''(0+) > 0$ to $f''(0-) < 0$.

Example 13.2: *For example, in order to find the maximum or minimum of a univariate function* $f(x)$

$$f(x) = xe^{-x^2}, \qquad -\infty < x < \infty, \tag{13.5}$$

we have to find first the stationary point x_* *when the first derivative* $f'(x)$ *is zero. That is*

$$\frac{df(x_*)}{dx_*} = e^{-x_*^2} - 2x_*^2 e^{-x_*^2} = 0. \tag{13.6}$$

Since $\exp(-x_*^2) \neq 0$, *we have*

$$x_* = \pm\frac{\sqrt{2}}{2}. \tag{13.7}$$

From the basic calculus we know that the maximum requires $f''(x_*) \leq 0$ *while the minimum requires* $f''(x_*) \geq 0$. *At* $x_* = \sqrt{2}/2$, *we have*

$$f''(x_*) = (4x_*^2 - 6)x_* e^{-x_*^2} = -2\sqrt{2}e^{-1/2} < 0, \tag{13.8}$$

so this point corresponds to the maximum $f(x_*) = \frac{1}{2}e^{-1/2}$.

Similarly, at $x_* = -\sqrt{2}/2$, $f''(x_*) = 2\sqrt{2}e^{-1/2} > 0$, *we have the minimum* $f(x_*) = -\frac{1}{2}e^{-1/2}$.

A maximum of a function $f(x)$ can be converted into a minimum of $A - f(x)$ where A is usually a large positive number (though $A = 0$ will do). For example, we know the maximum of $f(x) = e^{-x^2}, x \in (-\infty, \infty)$ is 1 at $x_* = 0$. This problem can be converted to a minimum problem $-f(x)$. For this reason, the optimization problems can be expressed as either minima or maxima depending on the context and convenience of finding the solutions.

In fact, in the optimization literature, some books formulate all the optimization problems in terms of maximization, while others write these problems in terms of minimization, though they are in essence dealing with the same topics.

13.3.2 *Multivariate Functions*

To find the maximum or minimum of a multivariate function $f(\boldsymbol{x})$ where $\boldsymbol{x} = (x_1, ..., x_d)^T$, we can express it as an univariate optimization problem concerning a design vector $\boldsymbol{x}$

$$\underset{\boldsymbol{x} \in \Re^d}{\text{minimize/maximize}} f(\boldsymbol{x}). \tag{13.9}$$

For a general function $f(\boldsymbol{x})$, we can expand it using Taylor series about a point $\boldsymbol{x} = \boldsymbol{x}_*$ so that $\boldsymbol{x} = \boldsymbol{x}_* + \epsilon \boldsymbol{u}$

$$f(\boldsymbol{x} + \epsilon \boldsymbol{u}) = f(\boldsymbol{x}_*) + \epsilon \boldsymbol{u} \boldsymbol{G}(\boldsymbol{x}_*) + \frac{1}{2}\epsilon^2 \boldsymbol{u}^T \boldsymbol{H}(\boldsymbol{x}_* + \epsilon \boldsymbol{u})\boldsymbol{u} + ..., \tag{13.10}$$

where $\boldsymbol{G}$ and $\boldsymbol{H}$ are its gradient vector and Hessian matrix, respectively. ϵ is a small parameter, and $\boldsymbol{u}$ is a vector. For example, for a generic quadratic function

$$f(\boldsymbol{x}) = \frac{1}{2}\boldsymbol{x}^T \boldsymbol{A} \boldsymbol{x} + \boldsymbol{k}^T \boldsymbol{x} + \boldsymbol{b},$$

where $\boldsymbol{A}$ is a constant square matrix, $\boldsymbol{k}$ is the gradient vector, and $\boldsymbol{b}$ is a vector constant, we have

$$f(\boldsymbol{x}_* + \epsilon \boldsymbol{u}) = f(\boldsymbol{x}_*) + \epsilon \boldsymbol{u}^T \boldsymbol{k} + \frac{1}{2}\epsilon^2 \boldsymbol{u}^T \boldsymbol{A} \boldsymbol{u} + ..., \tag{13.11}$$

where

$$f(\boldsymbol{x}_*) = \frac{1}{2}\boldsymbol{x}_*^T \boldsymbol{A} \boldsymbol{x}_* + \boldsymbol{k}^T \boldsymbol{x}_* + \boldsymbol{b}. \tag{13.12}$$

Thus, in order to study the local behaviour of a quadratic function, we only need to study $\boldsymbol{G}$ and $\boldsymbol{H}$. In addition, for simplicity, we can take $\boldsymbol{b} = \boldsymbol{0}$ as it is a constant vector anyway.

At a stationary point $\boldsymbol{x}_*$, the first derivatives are zero or $\boldsymbol{G}(\boldsymbol{x}_*) = 0$. Therefore, equation (13.10) becomes

$$f(\boldsymbol{x}_* + \epsilon \boldsymbol{u}) \approx f(\boldsymbol{x}_*) + \frac{1}{2}\epsilon^2 \boldsymbol{u}^T \boldsymbol{H} \boldsymbol{u}. \tag{13.13}$$

If $\boldsymbol{H} = \boldsymbol{A}$, then

$$\boldsymbol{A}\boldsymbol{v} = \lambda \boldsymbol{v} \tag{13.14}$$

forms an eigenvalue problem. For an $n \times n$ matrix $\boldsymbol{A}$, there will be n eigenvalues $\lambda_j (j = 1, ..., n)$ with n corresponding eigenvectors $\boldsymbol{v}$. As we have seen earlier that $\boldsymbol{A}$ is symmetric, these eigenvectors are orthonormal. That is,

$$\boldsymbol{v}_i^T \boldsymbol{v}_j = \delta_{ij}. \tag{13.15}$$

Near any stationary point $\boldsymbol{x}_*$, if we take $\boldsymbol{u}_j = \boldsymbol{v}_j$ as the local coordinate systems, we then have

$$f(\boldsymbol{x}_* + \epsilon \boldsymbol{v}_j) = f(\boldsymbol{x}_*) + \frac{1}{2}\epsilon^2 \lambda_j, \tag{13.16}$$

which means that the variations of $f(\boldsymbol{x})$, when $\boldsymbol{x}$ moves away from the stationary point $\boldsymbol{x}_*$ along the direction $\boldsymbol{v}_j$, are characterized by the eigenvalues. If $\lambda_j > 0$, $|\epsilon| > 0$ will lead to $|\Delta f| = |f(\boldsymbol{x}) - f(\boldsymbol{x}_*)| > 0$. In other words, $f(\boldsymbol{x})$ will increase as $|\epsilon|$ increases. Conversely, if $\lambda_j < 0$, $f(\boldsymbol{x})$ will decrease as $|\epsilon| > 0$ increases. Obviously, in the special case $\lambda_j = 0$, the function $f(\boldsymbol{x})$ will remain constant along the corresponding direction of $\boldsymbol{v}_j$.

Example 13.3: *We know that function*

$$f(x, y) = xy,$$

has a saddle point at $(0, 0)$. *It increases along the* $x = y$ *direction and decreases along* $x = -y$ *direction. From the above analysis, we know that* $\boldsymbol{x}_* = (x_*, y_*)^T = (0, 0)^T$ *and* $f(x_*, y_*) = 0$. *We now have*

$$f(\boldsymbol{x}_* + \epsilon \boldsymbol{u}) \approx f(\boldsymbol{x}_*) + \frac{1}{2}\epsilon^2 \boldsymbol{u}^T \boldsymbol{A} \boldsymbol{u},$$

where

$$\boldsymbol{A} = \nabla^2 f(\boldsymbol{x}_*) = \begin{pmatrix} \frac{\partial^2 f}{\partial x^2} & \frac{\partial^2 f}{\partial x \partial y} \\ \frac{\partial^2 f}{\partial x \partial y} & \frac{\partial^2 f}{\partial y^2} \end{pmatrix} = \begin{pmatrix} 0 & 1 \\ 1 & 0 \end{pmatrix}.$$

The eigenvalue problem is simply

$$\boldsymbol{A} v = \lambda_j \boldsymbol{v}_j, \quad (j = 1, 2),$$

or

$$\begin{vmatrix} -\lambda_j & 1 \\ 1 & -\lambda_j \end{vmatrix} = 0,$$

whose solutions are

$$\lambda_j = \pm 1.$$

For $\lambda_1 = 1$*, the corresponding eigenvector is*

$$\boldsymbol{v}_1 = \begin{pmatrix} \sqrt{2}/2 \\ \sqrt{2}/2 \end{pmatrix}.$$

Similarly, for $\lambda_2 = -1$*, the eigenvector is*

$$\boldsymbol{v}_2 = \begin{pmatrix} \sqrt{2}/2 \\ -\sqrt{2}/2 \end{pmatrix}.$$

Since A *is symmetric,* $\boldsymbol{v}_1$ *and* $\boldsymbol{v}_2$ *are orthonormal. Indeed this is the case because* $\|\boldsymbol{v}_1\| = \|\boldsymbol{v}_2\| = 1$ *and*

$$\boldsymbol{v}_1^T \boldsymbol{v}_2 = \frac{\sqrt{2}}{2} \times \frac{\sqrt{2}}{2} + \frac{\sqrt{2}}{2} \times (-\frac{\sqrt{2}}{2}) = 0.$$

Thus, we have

$$f(\epsilon \boldsymbol{v}_j) = \frac{1}{2}\epsilon^2 \lambda_j, \qquad (j = 1, 2). \tag{13.17}$$

As $\lambda_1 = 1$ *is positive,* f *increases along the direction* $\boldsymbol{v}_1 = \frac{\sqrt{2}}{2}(1\ 1)^T$ *which is indeed along the line* $x = y$*.*

Similarly, for $\lambda_2 = -1$*,* f *will decrease along* $\boldsymbol{v}_2 = \frac{\sqrt{2}}{2}(1\ -1)^T$ *which is exactly along the line* $x = -y$*. As there is no zero eigenvalue, the function will not remain constant in the region around* $(0, 0)$*.*

In essence, the eigenvalues of the Hessian matrix $\boldsymbol{H}$ determine the local behaviour of the function. For example, when $\boldsymbol{H}$ is positive semi-definite, it corresponds to a local minimum.

13.4 Gradient-Based Methods

Gradient-based methods are iterative methods that extensively use the gradient information of the objective function during iterations. For the minimization of a function $f(\boldsymbol{x})$, the essence of this method is

$$\boldsymbol{x}^{(n+1)} = \boldsymbol{x}^{(n)} + \alpha g(\nabla f, \boldsymbol{x}^{(n)}), \tag{13.18}$$

where α is the step size which can vary during iterations. $g(\nabla f, \boldsymbol{x}^{(n)})$ is a function of the gradient ∇f and the current location $\boldsymbol{x}^{(n)}$. Different methods use different forms of $g(\nabla f, \boldsymbol{x}^{(n)})$.

13.4.1 *Newton's Method*

We know that Newton's method is a popular iterative method for finding the zeros of a nonlinear univariate function of $f(x)$ on the interval $[a, b]$. It can be modified for solving optimization problems because it is equivalent to finding the zeros of the first derivative $f'(\boldsymbol{x})$ once the objective function $f(\boldsymbol{x})$ is given.

For a given function $f(\boldsymbol{x})$ which is continuously differentiable, we have the Taylor expansion about a known point $\boldsymbol{x} = \boldsymbol{x}_n$ (with $\Delta\boldsymbol{x} = \boldsymbol{x} - \boldsymbol{x}_n$)

$$f(\boldsymbol{x}) = f(\boldsymbol{x}_n) + (\nabla f(\boldsymbol{x}_n))^T \Delta\boldsymbol{x} + \frac{1}{2}\Delta\boldsymbol{x}^T \nabla^2 f(\boldsymbol{x}_n)\Delta\boldsymbol{x} + ...,$$

which is minimized near a critical point when $\Delta\boldsymbol{x}$ is the solution of the following linear equation

$$\nabla f(\boldsymbol{x}_n) + \nabla^2 f(\boldsymbol{x}_n)\Delta\boldsymbol{x} = 0. \tag{13.19}$$

This leads to

$$\boldsymbol{x} = \boldsymbol{x}_n - \boldsymbol{H}^{-1}\nabla f(\boldsymbol{x}_n), \tag{13.20}$$

where $\boldsymbol{H} = \nabla^2 f(\boldsymbol{x}_n)$ is the Hessian matrix. If the iteration procedure starts from the initial vector $\boldsymbol{x}^{(0)}$ (usually taken to be a guessed point in the domain), then Newton's iteration formula for the n-th iteration is

$$\boldsymbol{x}^{(n+1)} = \boldsymbol{x}^{(n)} - \boldsymbol{H}^{-1}(\boldsymbol{x}^{(n)})f(\boldsymbol{x}^{(n)}). \tag{13.21}$$

It is worth pointing out that if $f(\boldsymbol{x})$ is quadratic, then the solution can be found exactly in a single step. However, this method is not efficient for non-quadratic functions.

In order to speed up the convergence, we can use a smaller step size $\alpha \in (0, 1]$ so that we have modified Newton's method

$$\boldsymbol{x}^{(n+1)} = \boldsymbol{x}^{(n)} - \alpha\boldsymbol{H}^{-1}(\boldsymbol{x}^{(n)})f(\boldsymbol{x}^{(n)}). \tag{13.22}$$

It can usually be time-consuming to calculate the Hessian matrix for second derivatives. A good alternative is to use an identity matrix to approximate the Hessian by using $\boldsymbol{H}^{-1} = \boldsymbol{I}$, and we have the quasi-Newton method

$$\boldsymbol{x}^{(n+1)} = \boldsymbol{x}^{(n)} - \alpha\boldsymbol{I}\nabla f(\boldsymbol{x}^{(n)}), \tag{13.23}$$

which is essentially the steepest descent method.

13.4.2 *Steepest Descent Method*

The essence of this method is to find the lowest possible objective function $f(\boldsymbol{x})$ from the current point $\boldsymbol{x}^{(n)}$. From the Taylor expansion of $f(\boldsymbol{x})$ about $\boldsymbol{x}^{(n)}$, we have

$$f(\boldsymbol{x}^{(n+1)}) = f(\boldsymbol{x}^{(n)} + \Delta \boldsymbol{s}) \approx f(\boldsymbol{x}^{(n)} + (\nabla f(\boldsymbol{x}^{(n)}))^T \Delta \boldsymbol{s}, \tag{13.24}$$

where $\Delta \boldsymbol{s} = \boldsymbol{x}^{(n+1)} - \boldsymbol{x}^{(n)}$ is the increment vector. Since we try to find a lower (better) approximation to the objective function, it requires that the second term on the right hand is negative. That is

$$f(\boldsymbol{x}^{(n)} + \Delta \boldsymbol{s}) - f(\boldsymbol{x}^{(n)}) = (\nabla f)^T \Delta \boldsymbol{s} < 0. \tag{13.25}$$

From vector analysis, we know the inner product $\boldsymbol{u}^T \boldsymbol{v}$ of two vectors $\boldsymbol{u}$ and $\boldsymbol{v}$ is largest when they are parallel but in opposite directions. Therefore, $(\nabla f)^T \Delta \boldsymbol{s}$ becomes the largest when

$$\Delta \boldsymbol{s} = -\alpha \nabla f(\boldsymbol{x}^{(n)}), \tag{13.26}$$

where $\alpha > 0$ is the step size. This is the case when the direction $\Delta \boldsymbol{s}$ is along the steepest descent in the negative gradient direction. As we have seen it earlier, this method is a quasi-Newton method.

The choice of the step size α is very important. A very small step size means slow movement towards the local minimum, while a large step may overshoot and subsequently makes it move far away from the local minimum. Therefore, the step size $\alpha = \alpha^{(n)}$ should be different at each iteration step and should be chosen so that it minimizes the objective function $f(\boldsymbol{x}^{(n+1)}) = f(\boldsymbol{x}^{(n)}, \alpha^{(n)})$. Therefore, the steepest descent method can be written as

$$f(\boldsymbol{x}^{(n+1)}) = f(\boldsymbol{x}^{(n)}) - \alpha^{(n)} (\nabla f(\boldsymbol{x}^{(n)}))^T \nabla f(\boldsymbol{x}^{(n)}). \tag{13.27}$$

In each iteration, the gradient and step size will be calculated. Again, a good initial guess of both the starting point and the step size is useful.

Example 13.4: *Let us minimize the function*

$$f(x_1, x_2) = 10x_1^2 + 5x_1 x_2 + 10(x_2 - 3)^2,$$

where

$$(x_1, x_2) = [-10, 10] \times [-15, 15],$$

using the steepest descent method starting with the initial $\boldsymbol{x}^{(0)} = (10, 15)^T$. *We know that the gradient*

$$\nabla f = (20x_1 + 5x_2, \;\; 5x_1 + 20x_2 - 60)^T,$$

therefore

$$\nabla f(\boldsymbol{x}^{(0)}) = (275, \ \ 290)^T.$$

In the first iteration, we have

$$\boldsymbol{x}^{(1)} = \boldsymbol{x}^{(0)} - \alpha_0 \begin{pmatrix} 275 \\ 290 \end{pmatrix}.$$

The step size α_0 *should be chosen such that* $f(\boldsymbol{x}^{(1)})$ *is at the minimum, which means that*

$$f(\alpha_0) = 10(10 - 275\alpha_0)^2 + 5(10 - 275\alpha_0)(15 - 290\alpha_0) + 10(12 - 290\alpha_0)^2,$$

should be minimized. This becomes an optimization problem for a single independent variable α_0*. All the techniques for univariate optimization problems such as Newton's method can be used to find* α_0*. We can also obtain the solution by setting*

$$\frac{df}{d\alpha_0} = -159725 + 3992000\alpha_0 = 0,$$

whose solution is $\alpha_0 \approx 0.04001$.

At the second step, we have

$$\nabla f(\boldsymbol{x}^{(1)}) = (-3.078, 2.919)^T, \ \ \boldsymbol{x}^{(2)} = \boldsymbol{x}^{(1)} - \alpha_1 \begin{pmatrix} -3.078 \\ 2.919 \end{pmatrix}.$$

The minimization of $f(\alpha_1)$ *gives* $\alpha_1 \approx 0.066$*, and the new location of the steepest descent is*

$$\boldsymbol{x}^{(2)} \approx (-0.797, 3.202)^T.$$

At the third iteration, we have

$$\nabla f(\boldsymbol{x}^{(2)}) = (0.060, 0.064)^T, \ \ \boldsymbol{x}^{(3)} = \boldsymbol{x}^{(2)} - \alpha_2 \begin{pmatrix} 0.060 \\ 0.064 \end{pmatrix}.$$

The minimization of $f(\alpha_2)$ *leads to* $\alpha_2 \approx 0.040$*, and we have*

$$\boldsymbol{x}^{(3)} \approx (-0.8000299, 3.20029)^T.$$

Then, the iterations continue until a prescribed tolerance is met.

From the basic calculus, we know that we can set the first partial derivatives equal to zero

$$\frac{\partial f}{\partial x_1} = 20x_1 + 5x_2 = 0, \ \ \frac{\partial f}{\partial x_2} = 5x_1 + 20x_2 - 60 = 0,$$

we know that the minimum occurs exactly at

$$\boldsymbol{x}_* = (-4/5, 16/5)^T = (-0.8, 3.2)^T.$$

The steepest descent method gives almost the exact solution after only 3 iterations.

In finding the step size α_n in the above steepest descent method, we have used the stationary condition $df(\alpha_n)/d\alpha_n = 0$. Well, you may say that if we use this stationary condition for $f(\alpha_0)$, why not use the same method to get the minimum point of $f(\boldsymbol{x})$ in the first place. There are two reasons here. The first reason is that this is a simple example for demonstrating how the steepest descent method works. The second reason is that even for complicated multiple variables $f(x_1, ..., x_p)$ (say $p = 500$), then $f(\alpha_n)$ at any step n is still a univariate function, and the optimization of such $f(\alpha_n)$ is much simpler compared with the original multivariate problem.

From our example, we know that the convergence from the second iteration to the third iteration is slow. In fact, the steepest descent is typically slow once the local minimization is near. This is because near the local optimality the gradient is nearly zero, and thus the rate of descent is also slow. If high accuracy is needed near the local minimum, other local search methods should be used.

It is worth pointing out that there are many variations of the steepest descent methods. If such optimization aims is to find the maximum, then this method becomes the *hill-climbing* method because the aim is to climb up the hill to the highest peak.

The standard steepest descent method works well for convex functions and near a local peak (valley) of most smooth multimodal functions, though this local peak is not necessarily the global best. However, for some tough functions, it is not a good method. This is better demonstrated by the following example.

Example 13.5: *Let us minimize the so-called banana function introduced by Rosenbrock*

$$f(x_1, x_2) = (1 - x_1)^2 + 100(x_2 - x_1^2)^2,$$

where

$$(x_1, x_2) \in [-5, 5] \times [-5, 5].$$

This function has a global minimum $f_{\min} = 0$ at $(1, 1)$ which can be determined by

$$\frac{\partial f}{\partial x_1} = -2(1 - x_1) - 400x_1(x_2 - x_1^2) = 0,$$

and

$$\frac{\partial f}{\partial x_2} = 200(x_2 - x_1)^2 = 0,$$

whose unique solutions are

$$x_1 = 1, \qquad x_2 = 1.$$

Now we try to find its minimum by using the steepest descent method with the initial guess $\boldsymbol{x}^{(0)} = (5, 5)$. *We know that the gradient is*

$$\nabla f = \big(-2(1 - x_1) - 400x_1(x_2 - x_1^2),\ 200(x_2 - x_1^2) \big)^T.$$

Initially, we have

$$\nabla f(\boldsymbol{x}^{(0)}) = \big(40008,\ -4000 \big)^T.$$

In the first iteration, we have

$$\boldsymbol{x}^{(1)} = \boldsymbol{x}^{(0)} - \alpha_0 \begin{pmatrix} 40008 \\ -4000 \end{pmatrix}.$$

The step size α_0 *should be chosen such that* $f(\boldsymbol{x}^{(1)})$ *reaches its minimum, which means that*

$$f(\alpha_0) = [1 - (5 - 40008\alpha_0)]^2 + 100[(5 + 4000\alpha_0) - (5 - 40008\alpha_0)^2]^2,$$

should be minimized. The stationary condition becomes

$$\frac{df}{d\alpha_0} = 1.0248 \times 10^{21}\alpha_0^3 - 3.8807 \times 10^{17}\alpha_0^2$$

$$+ 0.4546 \times 10^{14}\alpha_0 - 1.6166 \times 10^9 = 0,$$

which has three solutions

$$\alpha_0 \approx 0.00006761,\ \ 0.0001262,\ \ 0.0001848.$$

Whichever these values we use, the new iteration $x_2^{(1)} = x_2^{(0)} + 4000\alpha_0$ *is always greater that* $x_2^{(0)} = 5$, *which moves away from the best solution* $(1, 1)$. *In this case, the simple steepest descent method does not work well. We have to use other more elaborate methods such as the conjugate gradient method.*

The difficulty in the above example arises because of the scaling difference where the factor 100 is associated with the second term. This means that the proper formulation of an optimization problem is important in practice. In fact, Rosenbrock's banana function is a very tough test function for optimization algorithms. Its solution requires more elaborate methods, and we will discuss some of these methods in later chapters.

Chapter 14

Mathematical Programming

Mathematical optimization or mathematical programming consists of two main categories: linear programming and nonlinear programming. In this chapter, we will introduce both briefly.

14.1 Linear Programming

Linear programming is a powerful mathematical programming technique which is widely used in business planning, transport routing and many other optimization applications. The basic idea in linear programming is to find the maximum or minimum of a linear objective under linear constraints.

For example, a small Internet service provider (ISP) can provide two different services x_1 and x_2. The first service is, say, the fixed monthly rate with limited download limits and bandwidth, while the second service is the higher rate with no download limit. The profit of the first service is αx_1 while the second is βx_2, though the profit of the second product is higher $\beta > \alpha > 0$, so the total profit is

$$P(\boldsymbol{x}) = \alpha x_1 + \beta x_2, \qquad \beta/\alpha > 1, \tag{14.1}$$

which is the objective function because the aim of the ISP company is to increase the profit as much as possible. Suppose the provided service is limited by the total bandwidth of the ISP company, thus at most $n_1 = 16$ (in 1000 units) of the first and at most $n_2 = 10$ (in 1000 units) of the second can be provided per unit of time, say, each day. Therefore, we have

$$x_1 \leq n_1, \qquad x_2 \leq n_2. \tag{14.2}$$

If the management of each of the two service packages takes the same staff time, so that a maximum of $n = 20$ (in 1000 units) can be maintained, which means

$$x_1 + x_2 \le n. \tag{14.3}$$

The additional constraints are that both x_1 and x_2 must be non-negative since negative numbers are unrealistic. We now have the following constraints

$$0 \le x_1 \le n_1, \qquad 0 \le x_2 \le n_2. \tag{14.4}$$

The problem now is to find the best x_1 and x_2 so that the profit P is a maximum. Mathematically, we have

$$\underset{(x_1,x_2)\in\mathcal{N}^2}{\text{maximize}} \ P(x_1, x_2) = \alpha x_1 + \beta x_2,$$

$$\text{subject to} \quad x_1 + x_2 \le n,$$

$$0 \le x_1 \le n_1, \ 0 \le x_2 \le n_2. \tag{14.5}$$

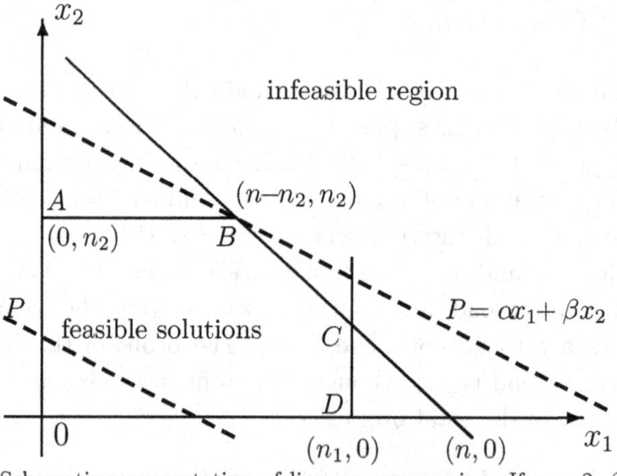

Fig. 14.1 Schematic representation of linear programming. If $\alpha = 2$, $\beta = 3$, $n_1 = 16$, $n_2 = 10$ and $n = 20$, then the optimal solution is at $B(10, 10)$.

Example 14.1: *The feasible solutions to this problem can graphically be represented as the inside region of the polygon OABCD as shown in Fig. 14.1. As the aim is to maximize the profit P, thus the optimal solution is at the extreme point B with $(n - n_2, n_2)$ and $P = \alpha(n - n_2) + \beta n_2$.*

For example, if $\alpha = 2$, $\beta = 3$, $n_1 = 16$, $n_2 = 10$, and $n = 20$, then the optimal solution occurs at $x_1 = n - n_2 = 10$ and $x_2 = n_2 = 10$ with the total profit $P = 2 \times (20 - 10) + 3 \times 10 = 50$ (thousand pounds).

Since the solution (x_1 and x_2) must be integers, an interesting thing is that the solution is independent of β/α if and only if $\beta/\alpha > 1$. However, the profit P does depend on the parameters α and β.

The number of feasible solutions is infinite if x_1 and x_2 are real numbers. Even for $x_1, x_2 \in \mathcal{N}$ are integers, the number of feasible solutions is quite large. Therefore, there is a need to use a systematic method to find the optimal solution. In order to find the best solution, we first plot out all the constraints as straight lines, and all the feasible solutions satisfying all the constraints form the inside region of the polygon $OABCD$. The vertices of the polygon form the set of the extreme points. Then, we plot the objective function P as a family of parallel lines (shown as dashed lines) so as to find the maximum value of P. Obviously, the highest value of P corresponds to the case when the objective line goes through the extreme point B. Therefore, $x_1 = n - n_2$ and $x_2 = n_2$ at the point B are the best solutions. The current example is relatively simple because it has only two decision variables and three constraints, which can be solved easily using a graphic approach. For more complicated problems, we need a formal approach. One of the most widely used methods is the simplex method.

14.2 Simplex Method

The simplex method was introduced by George Dantzig in 1947. The simplex method essentially works in the following way: for a given linear optimization problem such as the example of the ISP service we discussed earlier, it assumes that all the extreme points are known. If the extreme points are not known, the first step is to determine these extreme points or to check whether there are any feasible solutions. With known extreme points, it is easy to test whether or not an extreme point is optimal using the algebraic relationship and the objective function. If the test for optimality is not passed, then move to an adjacent extreme point to do the same test. This process stops until an optimal extreme point is found or the unbounded case occurs.

14.2.1 *Basic Procedure*

Mathematically, the simplex method first transforms the constraint inequalities into equalities by using slack variables.

To convert an inequality such as

$$5x_1 + 6x_2 \leq 20, \tag{14.6}$$

we can use a new variable x_3 or $s_1 = 20 - 5x_1 - 6x_2$ so that the original inequality becomes an equality

$$5x_1 + 6x_2 + s_1 = 20, \tag{14.7}$$

with an auxiliary non-negativeness condition

$$s_1 \geq 0. \tag{14.8}$$

Such a variable is referred to as a slack variable.

Thus, the inequalities in our example

$$x_1 + x_2 \leq n, \qquad 0 \leq x_1 \leq n_1, \qquad 0 \leq x_2 \leq n_2, \tag{14.9}$$

can be written, using three slack variables s_1, s_2, s_3, as the following equalities

$$x_1 + x_2 + s_1 = n, \tag{14.10}$$

$$x_1 + s_2 = n_1, \qquad x_2 + s_3 = n_2, \tag{14.11}$$

and

$$x_i \geq 0 \ (i = 1, 2), \ \ s_j \geq 0 (j = 1, 2, 3). \tag{14.12}$$

The original problem (14.5) becomes

$$\underset{\boldsymbol{x} \in \Re^5}{\text{maximize}} \, P(\boldsymbol{x}) = \alpha x_1 + \beta x_2 + 0 s_1 + 0 s_2 + 0 s_3,$$

$$\text{subject to} \begin{pmatrix} 1\,1\,1\,0\,0 \\ 1\,0\,0\,1\,0 \\ 0\,1\,0\,0\,1 \end{pmatrix} \begin{pmatrix} x_1 \\ x_2 \\ s_1 \\ s_2 \\ s_3 \end{pmatrix} = \begin{pmatrix} n \\ n_1 \\ n_2 \end{pmatrix},$$

$$x_i \geq 0, \qquad (i = 1, 2, ..., 5), \tag{14.13}$$

which has two control variables (x_1, x_2) and three slack variables $x_3 = s_1$, $x_4 = s_2$, $x_5 = s_3$.

In general, a linear programming problem can be written in the following standard form

$$\underset{\boldsymbol{x} \in \Re^n}{\text{maximize}} \, f(\boldsymbol{x}) = Z = \sum_{i=1}^{p} \alpha_i x_i = \boldsymbol{\alpha}^T \boldsymbol{x},$$

$$\text{subject to } \boldsymbol{A}\boldsymbol{x} = \boldsymbol{b},\ x_i \geq 0\ (i = 1, ..., p), \tag{14.14}$$

where $\boldsymbol{A}$ is a $q \times p$ matrix, $\boldsymbol{b} = (b_1, ..., b_q)^T$, and

$$\boldsymbol{x} = [\boldsymbol{x}_p\ \ \boldsymbol{x}_s]^T = (x_1, ..., x_m, s_1, ..., s_{p-m})^T. \tag{14.15}$$

This problem has p variables, and q equalities and all p variables are non-negative. In the standard form, all constraints are expressed as equalities and all variables including slack variables are non-negative.

A basic solution to the linear system $\boldsymbol{A}\boldsymbol{x} = \boldsymbol{b}$ of q linear equations in p variables in the standard form is usually obtained by setting $p - q$ variables equal to zero, and subsequently solving the resulting $q \times q$ linear system to get a unique solution of the remaining q variables. The q variables (that are not bound to zero) are called the basic variables of the basic solution. The $p - q$ variables at zero are called non-basic variables. Any basic solution to this linear system is referred to as a basic feasible solution (BFS) if all its variables are non-negative. The important property of the basic feasible solutions is that there is a unique corner point (extreme point) for each basic feasible solution, and there is at least one basic feasible solution for each corner or extreme point. These corners or extreme points are points on the intersection of two adjacent boundary lines such as A and B in Fig. 14.1. Two basic feasible solutions are said to be adjacent if they have $q - 1$ basic variables in common in the standard form.

Suppose $q = 500$, even the simplest integer equalities $x_i + x_j = 1$ where $i, j = 1, 2, ..., 500$, would give a huge number of combinations 2^{500}. Thus the number of basic feasible solutions will be the order of $2^{500} \approx 3 \times 10^{150}$, which is larger than the number of particles in the whole universe. This huge number of basic feasible solutions and extreme points necessitates a systematic and efficient search method. Simplex method is a powerful method to carry out such a mathematical programming task.

14.2.2 *Augmented Form*

The linear optimization problem is usually converted into the following standard augmented form or the canonical form

$$\begin{pmatrix} 1 & -\boldsymbol{\alpha}^T \\ 0 & \boldsymbol{A} \end{pmatrix} \begin{pmatrix} Z \\ \boldsymbol{x} \end{pmatrix} = \begin{pmatrix} 0 \\ \boldsymbol{b} \end{pmatrix}, \tag{14.16}$$

with the objective to maximize Z. In this canonical form, all the constraints are expressed as equalities for all non-negative variables. All the right-hand sides for all constraints are also non-negative, and each constraint equation has a single basic variable. The intention of writing in this canonical form is to identify basic feasible solutions, and move from one basic feasible solution to another via a so-called pivot operation. Geometrically speaking, this means to find all the corner or extreme points first, then evaluate the objective function by going through the extreme points so as to determine if the current basic feasible solution can be improved or not.

In the framework of the canonical form, the basic steps of the simplex method are: 1) to find a basic feasible solution to start the algorithm. Sometimes, it might be difficult to start, which may either imply there is no feasible solution or that it is necessary to reformulate the problem in a slightly different way by changing the canonical form so that a basic feasible solution can be found; 2) to see if the current basic feasible solution can be improved (even marginally) by increasing the non-basic variables from zero to non-negative values; 3) stop the process if the current feasible solution cannot be improved, which means that it is optimal. If the current feasible solution is not optimal, then move to an adjacent basic feasible solution. This adjacent basic feasible solution can be obtained by changing the canonical form via elementary row operations.

The pivot manipulations are based on the fact that a linear system will remain an equivalent system by multiplying a non-zero constant on a row and adding it to the other row. This procedure continues by going to the second step and repeating the evaluation of the objective function. The optimality of the problem will be reached, or we can stop the iteration if the solution becomes unbounded in the event that you can improve the objective indefinitely.

14.2.3 *A Case Study*

Now we come back to our example, if we use $\alpha = 2$, $\beta = 3$, $n_1 = 16$, $n_2 = 10$ and $n = 20$, we then have

$$\begin{pmatrix} 1 & -2 & -3 & 0 & 0 & 0 \\ 0 & 1 & 1 & 1 & 0 & 0 \\ 0 & 1 & 0 & 0 & 1 & 0 \\ 0 & 0 & 1 & 0 & 0 & 1 \end{pmatrix} \begin{pmatrix} Z \\ x_1 \\ x_2 \\ s_1 \\ s_2 \\ s_3 \end{pmatrix} = \begin{pmatrix} 0 \\ 20 \\ 16 \\ 10 \end{pmatrix}, \tag{14.17}$$

where $x_1, x_2, s_1, ..., s_3 \geq 0$. Now the first step is to identify a corner point or basic feasible solution by setting non-isolated variables $x_1 = 0$ and $x_2 = 0$

(thus the basic variables are s_1, s_2, s_3). We now have

$$s_1 = 20, \; s_2 = 16, \; s_3 = 10. \tag{14.18}$$

The objective function $Z = 0$, which corresponds to the corner point O in Fig. 14.1. In the present canonical form, the corresponding column associated with each basic variable has only one non-zero entry (marked by a box) for each constraint equality, and all other entries in the same column are zero. The non-zero value is usually converted into 1 if it is not unity. This is shown as follows:

$$\begin{array}{cccccc} Z & x_1 & x_2 & s_1 & s_2 & s_3 \end{array} \tag{14.19}$$
$$\begin{pmatrix} 1 & -2 & -3 & 0 & 0 & 0 \\ 0 & 1 & 1 & \boxed{1} & 0 & 0 \\ 0 & 1 & 0 & 0 & \boxed{1} & 0 \\ 0 & 0 & 1 & 0 & 0 & \boxed{1} \end{pmatrix}$$

When we change the set or the bases of basic variables from one set to another, we will aim to convert to a similar form using pivot row operations. There are two ways of numbering this matrix. One way is to call the first row [1 –2 –3 0 0 0] as the zero-th row, so that all other rows correspond to their corresponding constraint equation. The other way is simply to use its order in the matrix, so [1 –2 –3 0 0 0] is simply the first row. We will use this standard notation.

Now the question is whether we can improve the objective by increasing one of the non-basic variables x_1 and x_2? Obviously, if we increase x_1 by a unit, then Z will also increase by 2 units. However, if we increase x_2 by a unit, then Z will increase by 3 units. Since our objective is to increase Z as much as possible, we choose to increase x_2. As the requirement of the non-negativeness of all variables, we cannot increase x_2 without limits. So we increase x_2 while holding $x_1 = 0$, and we have

$$s_1 = 20 - x_2, \; s_2 = 16, \; s_3 = 10 - x_2. \tag{14.20}$$

Thus, the highest possible value of x_2 is $x = 10$ when $s_1 = s_3 = 0$. If x_2 increases further, both s_1 and s_3 will become negative, thus it is no longer a basic feasible solution.

The next step is either to set $x_1 = 0$ and $s_1 = 0$ as non-basic variables or to set $x_1 = 0$ and $s_3 = 0$. Both cases correspond to the point A in our example, so we simply choose $x_1 = 0$ and $s_3 = 0$ as non-basic variables, and the basic variables are thus x_2, s_1 and s_2. Now we have to do some pivot operations so that s_3 will be replaced by x_2 as a new basic variable. Each constraint equation has only a single basic variable in the new canonical form. This means that each column corresponding to each basic variable should have only a single non-zero entry (usually 1). In addition, the right-hand sides of all the constraints are non-negative and increase the value of the objective function at the same time. In order to convert the third column for x_2 to the form with only a single non-zero entry 1 (all other coefficients in the column should be zero), we first multiply the fourth row by 3 and add it to the first row, and the first row becomes

$$Z - 2x_1 + 0x_2 + 0s_1 + 0s_2 + 3s_3 = 30. \tag{14.21}$$

Then, we multiply the fourth row by -1 and add it to the second row, we have

$$0Z + x_1 + 0x_2 + s_1 + 0s_2 - s_3 = 10. \tag{14.22}$$

So the new canonical form becomes

$$\begin{pmatrix} 1 & -2 & 0 & 0 & 0 & 3 \\ 0 & 1 & 0 & 1 & 0 & -1 \\ 0 & 1 & 0 & 0 & 1 & 0 \\ 0 & 0 & 1 & 0 & 0 & 1 \end{pmatrix} \begin{pmatrix} Z \\ x_1 \\ x_2 \\ s_1 \\ s_2 \\ s_3 \end{pmatrix} = \begin{pmatrix} 30 \\ 10 \\ 16 \\ 10 \end{pmatrix}, \tag{14.23}$$

where the third, fourth, and fifth columns (for x_2, s_1 and s_2, respectively) have only one non-zero coefficient. All the values on the right-hand side are non-negative. From this canonical form, we can find the basic feasible solution by setting non-basic variables equal to zero. This is to set $x_1 = 0$ and $s_3 = 0$. We now have the basic feasible solution

$$x_2 = 10, \; s_1 = 10, \; s_2 = 16, \tag{14.24}$$

which corresponds to the corner point A. The objective $Z = 30$.

Now again the question is whether we can improve the objective by increasing the non-basic variables. As the objective function is

$$Z = 30 + 2x_1 - 3s_3, \tag{14.25}$$

Z will increase 2 units if we increase x_1 by 1, but Z will decrease -3 if we increase s_3. Thus, the best way to improve the objective is to increase x_1. The question is what the limit of x_1 is. To answer this question, we hold s_3 at 0, we have

$$s_1 = 10 - x_1, \; s_2 = 16 - x_1, x_2 = 10. \tag{14.26}$$

We can see if x_1 can increase up to $x_1 = 10$, after that s_1 becomes negative, and this occurs when $x_1 = 10$ and $s_1 = 0$. This also suggests that the new adjacent basic feasible solution can be obtained by choosing s_1 and s_3 as the non-basic variables. Therefore, we have to replace s_1 with x_1 so that the new basic variables are x_1, x_2 and s_2.

Using these basic variables, we have to make sure that the second column (for x_1) has only a single non-zero entry. Thus, we multiply the second row by 2 and add it to the first row, and the first row becomes

$$Z + 0x_1 + 0x_2 + 2s_1 + 0s_2 + s_3 = 50. \tag{14.27}$$

We then multiply the second row by -1 and add it to the third row, and we have

$$0Z + 0x_1 + 0x_2 - s_1 + s_2 + s_3 = 6. \tag{14.28}$$

Thus we have the following canonical form

$$\begin{pmatrix} 1 & 0 & 0 & 2 & 0 & 1 \\ 0 & 1 & 0 & 1 & 0 & -1 \\ 0 & 0 & 0 & -1 & 1 & 1 \\ 0 & 0 & 1 & 0 & 0 & 1 \end{pmatrix} \begin{pmatrix} Z \\ x_1 \\ x_2 \\ s_1 \\ s_2 \\ s_3 \end{pmatrix} = \begin{pmatrix} 50 \\ 10 \\ 6 \\ 10 \end{pmatrix}, \tag{14.29}$$

whose basic feasible solution can be obtained by setting non-basic variables $s_1 = s_3 = 0$. We have

$$x_1 = 10, \; x_2 = 10, \; s_2 = 6, \tag{14.30}$$

which corresponds to the extreme point B in Fig. 14.1. The objective value is $Z = 50$ for this basic feasible solution. Let us see if we can improve the objective further. Since the objective becomes

$$Z = 50 - 2s_1 - s_3, \tag{14.31}$$

any increase of s_1 or s_3 from zero will decrease the objective value. Therefore, this basic feasible solution is optimal. Indeed, this is the same solution as that obtained from the graph method. We can see that a major advantage is that we have reached the optimal solution after searching a certain number of extreme points, and there is no need to evaluate other extreme points. This is exactly why the simplex method is so efficient.

The case study we used here is relatively simple, but it is useful to show how the basic procedure works in linear programming. For more practical applications, there are well-established software packages which will do the work for you once you have set up the objective and constraints properly.

14.3 Nonlinear Programming

As most real-world problems are nonlinear, nonlinear mathematical programming forms an important part of mathematical optimization methods. A broad class of nonlinear programming problems is about the minimization or maximization of $f(\boldsymbol{x})$ subject to no constraints, and another important class is the minimization of a quadratic objective function subject to nonlinear constraints. There are many other nonlinear programming problems as well.

Nonlinear programming problems are often classified according to the convexity of the defining functions. An interesting property of a convex function f is that the vanishing of the gradient $\nabla f(\boldsymbol{x}_*) = 0$ guarantees that the point x_* is a global minimum or maximum of f. If a function is not convex or concave, then it is much more difficult to find global minima or maxima.

14.4 Penalty Method

For the simple function optimization with equality and inequality constraints, a common method is the penalty method. For the optimization problem

$$\underset{\boldsymbol{x}\in\Re^n}{\text{minimize}} f(\boldsymbol{x}), \quad \boldsymbol{x} = (x_1, ..., x_n)^T \in \Re^n,$$

$$\text{subject to } \phi_i(\boldsymbol{x}) = 0, \ (i = 1, ..., M),$$

$$\psi_j(\boldsymbol{x}) \le 0, \quad (j = 1, ..., N), \tag{14.32}$$

the idea is to define a penalty function so that the constrained problem is transformed into an unconstrained problem. Now we define $\Pi(\boldsymbol{x}, \mu_i, \nu_j)$

$$\Pi(\boldsymbol{x}, \mu_i, \nu_j) = f(\boldsymbol{x}) + \sum_{i=1}^{M} \mu_i \phi_i^2(\boldsymbol{x}) + \sum_{j=1}^{N} \nu_j \psi_j^2(\boldsymbol{x}), \tag{14.33}$$

where $\mu_i \gg 1$ and $\nu_j \ge 0$.

For example, in order to solve the following problem of Gill-Murray-Wright type

$$\underset{x\in\Re}{\text{minimize}} \ f(x) = 100(x - b)^2 + 1,$$

$$\text{subject to } g(x) = x - a \ge 0, \tag{14.34}$$

where $a > b$ is a given value, we can define a penalty function $\Pi(x)$ using a penalty parameter $\mu \gg 1$. We have

$$\Pi(x, \mu) = f(x) + \frac{\mu}{2} g(x)^T g(x) = 100(x - b)^2 + 1 + \frac{\mu}{2}(x - a)^2, \quad (14.35)$$

where the typical value for μ is $2000 \sim 10000$.

This essentially transforms the original constrained optimization into an unconstrained problem. From the stationary condition $\Pi'(x) = 0$, we have

$$200(x_* - b) - \mu(x_* - a) = 0, \quad (14.36)$$

which gives

$$x_* = \frac{200b + \mu a}{200 + \mu}. \quad (14.37)$$

For $\mu \to \infty$, we have $x_* \to a$. For $\mu = 2000$, $a = 2$ and $b = 1$, we have $x_* \approx 1.9090$. This means the solution depends on the value of μ, and it is very difficult to use extremely large values without causing extra computational difficulties.

14.5 Lagrange Multipliers

Another powerful method without the above limitation of using large μ is the method of Lagrange multipliers. If we want to minimize a function $f(\boldsymbol{x})$

$$\underset{\boldsymbol{x} \in \Re^n}{\text{minimize}} f(\boldsymbol{x}), \qquad \boldsymbol{x} = (x_1, ..., x_n)^T \in \Re^n, \quad (14.38)$$

subject to the following nonlinear equality constraint

$$g(\boldsymbol{x}) = 0, \quad (14.39)$$

then we can combine the objective function $f(\boldsymbol{x})$ with the equality to form a new function, called the Lagrangian

$$\Pi = f(\boldsymbol{x}) + \lambda g(\boldsymbol{x}), \quad (14.40)$$

where λ is the Lagrange multiplier, which is an unknown scalar to be determined.

This again converts the constrained optimization into an unstrained problem for $\Pi(\boldsymbol{x})$, which is the beauty of this method. If we have M equalities,

$$g_j(\boldsymbol{x}) = 0, \qquad (j = 1, ..., M), \quad (14.41)$$

then we need M Lagrange multipliers $\lambda_j (j = 1, ..., M)$. We thus have

$$\Pi(x, \lambda_j) = f(\boldsymbol{x}) + \sum_{j=1}^{M} \lambda_j g_j(\boldsymbol{x}). \tag{14.42}$$

The requirement of stationary conditions leads to

$$\frac{\partial \Pi}{\partial x_i} = \frac{\partial f}{\partial x_i} + \sum_{j=1}^{M} \lambda_j \frac{\partial g_j}{\partial x_i}, \ \ (i = 1, ..., n), \tag{14.43}$$

and

$$\frac{\partial \Pi}{\partial \lambda_j} = g_j = 0, \ \ (j = 1, ..., M). \tag{14.44}$$

These $M + n$ equations will determine the n components of $\boldsymbol{x}$ and M Lagrange multipliers. As $\frac{\partial \Pi}{\partial g_j} = \lambda_j$, we can consider λ_j as the rate of the change of the quantity Π as a functional of g_j.

Example 14.2: *To solve the optimization problem*

$$\underset{(x,y)\in\Re^2}{\text{maximize}} f(x, y) = xy^2,$$

subject to the condition

$$g(x, y) = x^2 + y^2 - 1 = 0,$$

we can now define

$$\Pi = f(x, y) + \lambda g(x, y) = xy^2 + \lambda(x^2 + y^2 - 1).$$

The stationary conditions become

$$\frac{\partial \Pi}{\partial x} = y^2 + 2\lambda x = 0,$$

$$\frac{\partial \Pi}{\partial y} = 2xy + 2\lambda y = 0,$$

and

$$\frac{\partial \Pi}{\partial \lambda} = x^2 + y^2 - 1 = 0.$$

The condition $xy + \lambda y = 0$ implies that $y = 0$ or $\lambda = -x$. The case of $y = 0$ can be eliminated as it leads to $x = 0$ from $y^2 + 2\lambda x = 0$, which does not satisfy the last condition $x^2 + y^2 = 1$. Therefore, the only valid solution is

$$\lambda = -x.$$

From the first stationary condition, we have

$$y^2 - 2x^2 = 0, \quad or \quad y^2 = 2x^2.$$

Substituting this into the third stationary condition, we have

$$x^2 - 2x^2 - 1 = 0,$$

which gives

$$x = \pm 1.$$

So we have four stationary points

$$P_1(1, \sqrt{2}), \ \ P_2(1, -\sqrt{2}), \ \ P_3(-1, \sqrt{2}), \ \ P_4(-1, -\sqrt(2)).$$

The values of the function $f(x, y)$ *at these four points are*

$$f(P_1) = 2, \ \ f(P_2) = 2, \ \ f(P_3) = -2, \ \ f(P_4) = -2.$$

Thus, the function reaches its maxima at $(1, \sqrt{2})$ *and* $(1, -\sqrt{2})$. *The Lagrange multiplier in this case is* $\lambda = -1$.

14.6 Karush-Kuhn-Tucker Conditions

There is a counterpart of the Lagrange multipliers for nonlinear optimization with constraint inequalities. The Karush-Kuhn-Tucker (KKT) conditions concern the requirement for a solution to be optimal in nonlinear programming.

For the nonlinear optimization problem

$$\text{minimize}_{\boldsymbol{x} \in \Re^n} f(\boldsymbol{x}),$$

$$\text{subject to } \phi_i(\boldsymbol{x}) = 0, \ (i = 1, ..., M),$$

$$\psi_j(\boldsymbol{x}) \leq 0, \ \ (j = 1, ..., N). \tag{14.45}$$

If all the functions are continuously differentiable, at a local minimum $\boldsymbol{x}_*$, there exist constants $\lambda_0, \lambda_1, ..., \lambda_q$ and $\mu_1, ..., \mu_p$ such that

$$\lambda_0 \nabla f(\boldsymbol{x}_*) + \sum_{i=1}^{M} \mu_i \nabla \phi_i(\boldsymbol{x}_*) + \sum_{j=1}^{N} \lambda_j \nabla \psi_j(\boldsymbol{x}_*) = 0, \tag{14.46}$$

and

$$\psi_j(\boldsymbol{x}_*) \leq 0, \qquad \lambda_j \psi_j(\boldsymbol{x}_*) = 0, \ \ (j = 1, 2, ..., N), \tag{14.47}$$

where $\lambda_j \geq 0, (i = 0, 1, ..., N)$. The constants satisfy the following condition

$$\sum_{j=0}^{N} \lambda_j + \sum_{i=1}^{M} |\mu_i| \geq 0. \tag{14.48}$$

This is essentially a generalized method of the Lagrange multipliers. However, there is a possibility of degeneracy when $\lambda_0 = 0$ under certain conditions.

14.7 Sequential Quadratic Programming

14.7.1 *Quadratic Programming*

A special type of nonlinear programming is quadratic programming whose objective function is a quadratic form

$$f(\boldsymbol{x}) = \frac{1}{2}\boldsymbol{x}^T \boldsymbol{Q}\boldsymbol{x} + \boldsymbol{b}^T \boldsymbol{x} + \boldsymbol{c}, \tag{14.49}$$

where $\boldsymbol{b}$ and $\boldsymbol{c}$ are constant vectors. $\boldsymbol{Q}$ is a symmetric square matrix. The constraints can be incorporated using Lagrange multipliers and KKT formulations.

14.7.2 *Sequential Quadratic Programming*

Sequential (or successive) quadratic programming (SQP) represents one of the state-of-art and most popular methods for nonlinear constrained optimization. It is also one of the robust methods. For a general nonlinear optimization problem

$$\text{minimize} f(\boldsymbol{x}), \tag{14.50}$$

$$\text{subject to } \ h_i(\boldsymbol{x}) = 0, \ (i = 1, ..., p), \tag{14.51}$$

$$g_j(\boldsymbol{x}) \leq 0, \ (j = 1, ..., q). \tag{14.52}$$

The fundamental idea of sequential quadratic programming is to approximate the computationally extensive full Hessian matrix using a quasi-Newton updating method. Subsequently, this generates a subproblem of quadratic programming (called QP subproblem) at each iteration, and the solution to this subproblem can be used to determine the search direction and next trial solution.

Using the Taylor expansions, the above problem can be approximated, at each iteration, as the following problem

$$\text{minimize } \frac{1}{2}\boldsymbol{s}^T \nabla^2 L(\boldsymbol{x}_k)\boldsymbol{s} + \nabla f(\boldsymbol{x}_k)^T \boldsymbol{s} + f(\boldsymbol{x}_k), \tag{14.53}$$

$$\text{subject to } \ \nabla h_i(\boldsymbol{x}_k)^T \boldsymbol{s} + h_i(\boldsymbol{x}_k) = 0, \ (i = 1, ..., p), \tag{14.54}$$

$$\nabla g_j(\boldsymbol{x}_k)^T \boldsymbol{s} + g_j(\boldsymbol{x}_k) \leq 0, \ (j = 1, ..., q), \tag{14.55}$$

where the Lagrange function, also called the merit function, is defined by

$$L(\boldsymbol{x}) = f(\boldsymbol{x}) + \sum_{i=1}^{p} \lambda_i h_i(\boldsymbol{x}) + \sum_{j=1}^{q} \mu_j g_j(\boldsymbol{x})$$

$$= f(\boldsymbol{x}) + \boldsymbol{\lambda}^T \boldsymbol{h}(\boldsymbol{x}) + \boldsymbol{\mu}^T \boldsymbol{g}(\boldsymbol{x}), \tag{14.56}$$

where $\boldsymbol{\lambda} = (\lambda_1, ..., \lambda_p)^T$ is the vector of Lagrange multipliers, and $\boldsymbol{\mu} = (\mu_1, ..., \mu_q)^T$ is the vector of KKT multipliers. Here we have used the notation $\boldsymbol{h} = (h_1(\boldsymbol{x}), ..., h_p(\boldsymbol{x}))^T$ and $\boldsymbol{g} = (g_1(\boldsymbol{x}), ..., g_q(\boldsymbol{x}))^T$.

To approximate the Hessian $\nabla^2 L(\boldsymbol{x}_k)$ by a positive definite symmetric matrix $\boldsymbol{H}_k$, the standard Broydon-Fletcher-Goldfarbo-Shanno (BFGS) approximation of the Hessian can be used, and we have

$$\boldsymbol{H}_{k+1} = \boldsymbol{H}_k + \frac{\boldsymbol{v}_k \boldsymbol{v}_k^T}{\boldsymbol{v}_k^T \boldsymbol{u}_k} - \frac{\boldsymbol{H}_k \boldsymbol{u}_k \boldsymbol{u}_k^T \boldsymbol{H}_k^T}{\boldsymbol{u}_k^T \boldsymbol{H}_k \boldsymbol{u}_k}, \tag{14.57}$$

where

$$\boldsymbol{u}_k = \boldsymbol{x}_{k+1} - \boldsymbol{x}_k, \tag{14.58}$$

and

$$\boldsymbol{v}_k = \nabla L(\boldsymbol{x}_{k+1}) - \nabla L(\boldsymbol{x}_k). \tag{14.59}$$

The QP subproblem is solved to obtain the search direction

$$\boldsymbol{x}_{k+1} = \boldsymbol{x}_k + \alpha \boldsymbol{s}_k, \tag{14.60}$$

using a line search method by minimizing a penalty function, also commonly called merit function,

$$\Phi(\boldsymbol{x}) = f(\boldsymbol{x}) + \rho \Big[\sum_{i=1}^{p} |h_i(\boldsymbol{x})| + \sum_{j=1}^{q} \max\{0, g_j(\boldsymbol{x})\} \Big], \tag{14.61}$$

where ρ is the penalty parameter.

Sequential Quadratic Programming

begin
Choose a starting point $\boldsymbol{x}_0$ *and approximation* H_0 *to the Hessian*
repeat $k = 1, 2, ...$
 Solve a QP subproblem: QP_k *to get the search direction* $\boldsymbol{s}_k$
 Given $\boldsymbol{s}_k$*, find* α *so as to determine* $\boldsymbol{x}_{k+1}$
 Update the approximate Hessian $\boldsymbol{H}_{k+1}$ *using the BFGS scheme*
 $k = k + 1$
until *(stop criterion)*
end

Fig. 14.2 Procedure of sequential quadratic programming.

It is worth pointing out that any SQP method requires a good choice of $\boldsymbol{H}_k$ as the approximate Hessian of the Lagrangian L. Obviously, if $\boldsymbol{H}_k$

is exactly calculated as $\nabla^2 L$, SQP essentially becomes Newton's method solving the optimality condition. A popular way to approximate the Lagrangian Hessian is to use a quasi-Newton scheme as we used the BFGS formula described earlier.

In this chapter, we have outlined several widely used algorithms without providing any examples. The main reason is that the description of such an example may be lengthy, which also depends on the actual implementation. However, there are both commercial and open source software packages for all these algorithms. For example, the Matlab optimization toolbox implemented all these algorithms.

14.8 No Free Lunch Theorems

The methods used to solve a particular problem depend largely on the type and characteristics of the optimization problem itself. There is no universal method that works for all problems, and there is generally no guarantee of finding the optimal solution in global optimization problems. In fact, there are several so-called Wolpert and Macready's 'No Free Lunch Theorems' (NLF theorems) which state that if any algorithm A outperforms another algorithm B in the search for an extremum of a cost function, then algorithm B will outperform A over other cost functions. NFL theorems apply to the scenario (either deterministic or stochastic) where a set of continuous (or discrete or mixed) parameter θ maps the cost functions into a finite set. Let n_θ be the number of values of θ (either due to discrete values or the finite machine precisions), and n_f be the number of values of the cost function. Then, the number of all possible combinations of cost functions is $N = n_f^{n_\theta}$ which is finite, but usually huge. The NFL theorems prove that the average performance over *all* possible cost functions is the same for all search algorithms.

Mathematically speaking, if $P(s_m^y|f, m, A)$ denotes the performance, based on probability theory, of an algorithm A iterated m times on a cost function f over the sample s_m, then we have the averaged performance for two algorithms

$$\sum_f P(s_m^y|f, m, A) = \sum_f P(s_m^y|f, m, B), \tag{14.62}$$

where $s_m = \{(s_m^x(1), s_m^y(1)), ..., (s_m^x(m), s_m^y(m))\}$ is a time-ordered set of m distinct visited points with a sample of size m. What is interesting is that the performance is independent of algorithm A itself. That is to say, all algorithms for optimization will give the same performance when averaged over *all possible* functions. This means that a universally best method does not exist.

However, it is worth pointing out that the NFL theorems are based on some rigorous assumptions, and two such assumptions are: closed under permutations (CUP) and non-revisiting points (NRP). Recent studies suggested that for continuous problems, CUP is not valid. In addition, for some algorithms, especially for co-evolutionary algorithms, the non-revisiting assumption does not hold either. Therefore, there are potentially 'free lunches' for continuous optimization problems and for co-evolutionary algorithms. All these are under active research at the moment.

Well, you might say, there is no need to formulate new algorithms because all algorithms will perform equally well. The truth is that the performance here is measured in the statistical sense and over *all possible* functions. This does not mean all algorithms perform equally well over some *specific* functions. The reality is that no optimization problems require averaged performance over all possible functions. Even though the NFL theorems are valid mathematically, their influence on parameter search and optimization is limited. For any specific set of functions, some algorithms perform much better than others. In fact, for any specific problem with specific functions, there usually exist some algorithms that are more efficient than others if we do not need to measure their *average* performance. The main problem is probably how to find those algorithms that are better for a particular type of problem.

On the other hand, if the global optimality is not achievable, we have to emphasize the best estimate or sub-optimal solutions under the given conditions. Knowledge of the particular problem concerned is always helpful for the appropriate choice of the best or most efficient methods for the optimization procedure.

The optimization algorithms we discussed so far are conventional methods, and deterministic, as there is no randomness in these algorithms. The main advantage of these algorithms is that they are well-established and benchmarked, but the disadvantages are that they could be trapped in a local optimum, and there is no guarantee that they will find the global optimum even if you run the algorithms *ad infinitum*, because the diversity of the solutions is limited.

Modern optimization algorithms such as particle swarm optimization (PSO), cuckoo search and firefly algorithm often involve a certain degree of randomness so as to increase the diversity of the solutions and also to avoid being trapped in a local optimum. The randomness makes sure that it can 'jump out' of any local optimum. Statistically speaking it can find the global optima as the number of iterations approaches infinite. The employment of randomness is everywhere, especially in the metaheuristic methods such as PSO and cuckoo search. We will study these methods in detail in Part VI.

Part V

Stochastic Methods and Data Modelling

Chapter 15

Stochastic Models

All mathematical models and differential equations we have discussed so far are deterministic systems in the sense that, for given initial and boundary conditions, the solutions of the system can be determined (the only exception is a chaotic system to a certain degree). There is no intrinsic randomness in differential equations. In reality, randomness occurs everywhere, and not all models are deterministic. In fact, it is necessary to use stochastic models and sometimes the only sensible models are stochastic descriptions. In these cases, we have to deal with probability and statistics.

15.1 Random Variables

Randomness such as roulette-rolling and noise arises from the lack of information, or incomplete knowledge of reality. It can also come from the intrinsic complexity, diversity and perturbations of the system. The theory of probability is mainly the studies of random phenomena so as to find non-random regularity. Probability P is a number or an expected frequency assigned to an event A that indicates how likely it is that the event will occur when a random experiment is performed. This probability is often written as $P(A)$ to show that the probability P is associated with event A.

For a discrete random variable X with distinct values such as the number of cars passing through a junction, each value x_i may occur with a certain probability $p(x_i)$. In other words, the probability varies with the random variable. A probability function $p(x_i)$ is a function that defines probabilities to all the discrete values x_i of the random variable X. As an event must occur inside a sample space, the requirement that all the probabilities must be summed to one leads to

$$\sum_{i=1}^{n} p(x_i) = 1. \tag{15.1}$$

The cumulative probability function of X is defined by

$$P(X \leq x) = \sum_{x_i < x} p(x_i). \tag{15.2}$$

For a continuous random variable X that takes a continuous range of values (such as the level of noise), its distribution is continuous and the probability density function $p(x)$ is defined for a range of values $x \in [a, b]$ for given limits a and b [or even over the whole real axis $x \in (-\infty, \infty)$]. In this case, we always use the interval $(x, x+dx]$ so that $p(x)$ is the probability that the random variable X takes the value $x < X \leq x + dx$ is

$$\Phi(x) = P(x < X \leq x + dx) = p(x)dx. \tag{15.3}$$

As all the probabilities of the distribution shall be added to unity, we have

$$\int_a^b p(x)dx = 1. \tag{15.4}$$

The cumulative probability function becomes

$$\Phi(x) = P(X \leq x) = \int_a^x p(x)dx, \tag{15.5}$$

which is the definite integral of the probability density function between the lower limit a up to the present value $X = x$.

Two main measures for a random variable X with a given probability distribution $p(x)$ are its mean and variance. The mean μ or the expectation value of $E[X]$ is defined by

$$\mu \equiv E[X] \equiv <X> = \int x p(x)dx, \tag{15.6}$$

for a continuous distribution and the integration is within the integration limits. If the random variable is discrete, then the integration becomes the weighted sum

$$E[X] = \sum_i x_i p(x_i). \tag{15.7}$$

The variance $\text{var}[X] = \sigma^2$ is the expectation value of the deviation squared $(X - \mu)^2$. That is

$$\sigma^2 \equiv \text{var}[X] = E[(X - \mu)^2] = \int (x - \mu)^2 p(x)dx. \tag{15.8}$$

The square root of the variance $\sigma = \sqrt{\text{var}[X]}$ is called the standard deviation, which is simply σ.

For a discrete distribution, the variance simply becomes the following sum

$$\sigma^2 = \sum_i (x - \mu)^2 p(x_i). \tag{15.9}$$

In addition, any other formulas for a continuous distribution can be converted to their counterpart for a discrete distribution if the integration is replaced by the sum. Therefore, we will mainly focus on the continuous distribution in the rest of the section.

From these definitions, it is straightforward to prove

$$E[\alpha x + \beta] = \alpha E[X] + \beta, \qquad E[X^2] = \mu^2 + \sigma^2, \tag{15.10}$$

and

$$\text{var}[\alpha x + \beta] = \alpha^2 \text{var}[X] \tag{15.11}$$

where α and β are constants.

Other frequently used measures are the mode and median. The mode of a distribution is defined by the value at which the probability density function $p(x)$ is the maximum. For an even number of data sets, the mode may have two values. The median m of a distribution corresponds to the value at which the cumulative probability function $\Phi(m) = 1/2$. The upper and lower quartiles Q_U and Q_L are defined by $\Phi(Q_U) = 3/4$ and $\Phi(Q_L) = 1/4$.

15.2 Binomial and Poisson Distributions

A discrete random variable is said to follow the binomial distribution $B(n, p)$ if its probability distribution is given by

$$B(n, p) = {}^n C_x p^x (1 - p)^{n-x}, \qquad {}^n C_x = \frac{n!}{x!(n - x)!}, \tag{15.12}$$

where $x = 0, 1, 2, ..., n$ are the values that the random variable X may take, and n is the number of trials. There are only two possible outcomes: success or failure. p is the probability of a so-called 'success' of the outcome. Subsequently, the probability of the failure of a trial is $q = 1-p$. Therefore, $B(n, p)$ represents the probability of x successes and $n-x$ failures in n trials. The coefficients come from the coefficients of the binomial expansions

$$(p + q)^n = \sum_{x=0}^{n} {}^n C_x p^x q^{n-x} = 1, \tag{15.13}$$

which is exactly the requirement that all the probabilities should be summed to unity.

Example 15.1: *Tossing a fair coin 20 times, the probability of getting 15 heads is $B(n, 1/2)$. Since $p = 1/2$ and $x = 15$, then we have*

$$^{20}C_{15}(\frac{1}{2})^{15}(\frac{1}{2})^5 = \frac{15504}{1048576} \approx 0.01479.$$

It is straightforward to prove that $\mu = E[X] = np$ and $\sigma^2 = npq = np(1-p)$ for a binomial distribution.

Another related distribution is the geometric distribution whose probability function is defined by

$$P(X = n) = pq^{n-1} = p(1-p)^{n-1}, \tag{15.14}$$

where $n \geq 1$. This distribution is used to calculate the first success, thus the first $n - 1$ trials must be a failure if n trials are needed to observe the first success. The mean and variance of this distribution are $\mu = 1/p$ and $\sigma^2 = (1-p)/p^2$.

The Poisson distribution can be thought of as the limit of the binomial distribution when the number of trial is very large $n \to \infty$ and the probability $p \to 0$ (small probability) with the constraint that $\lambda = np$ is finite. For this reason, it is often called the distribution for small-probability events. Typically, it is concerned with the number of events that occur in a certain time interval (e.g., number of telephone calls in an hour) or spatial area. The Poisson distribution is

$$P(X = x) = \frac{\lambda^x e^{-\lambda}}{x!}, \qquad \lambda > 0, \tag{15.15}$$

where $x = 0, 1, 2, ..., n$ and λ is the mean of the distribution. Using the definition of mean and variance, it is straightforward to prove that $\mu = \lambda$ and $\sigma^2 = \lambda$ for the Poisson distribution. The parameter λ controls the location of the peak as shown in Fig. 15.1.

In probability, an important concept is the moment-generating function which is defined as

$$G_X(\nu) = E[e^{\nu X}], \tag{15.16}$$

where X is the random variable and $G_X(0) = 1$.

The moment-generating function for the Poisson distribution is given by

$$G_X(\nu) = \sum_{x=0}^{\infty} \frac{e^{\nu x} \lambda^x e^{-\lambda}}{x!} = \exp[\lambda(e^\nu - 1)]. \tag{15.17}$$

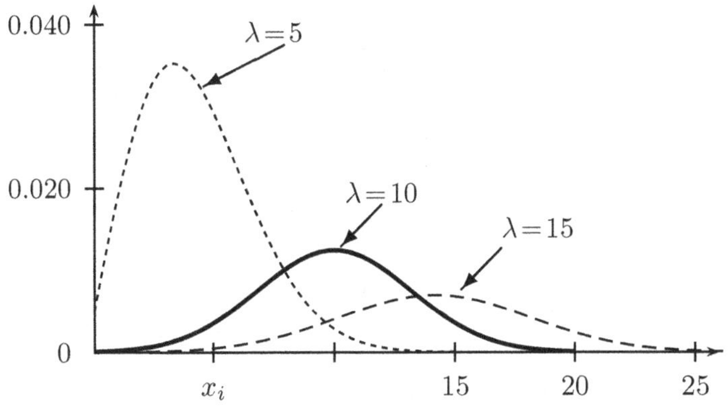

Fig. 15.1 Poisson distributions for different values of $\lambda = 5, 10, 15$.

The n-th moment can be calculated by

$$m_n = E[X^n] = \frac{d^n G_X}{d\nu^n}\Big|_{\nu=0}, \tag{15.18}$$

where $n = 1, 2, ...$ is an integer. Obviously, the mean of a random variable is the first moment, while the variance is the second central moment.

15.3 Gaussian Distribution

The Gaussian distribution or normal distribution is the most important continuous distribution in probability and it has a wide range of applications. For a continuous random variable X, the probability density function (PDF) of a Gaussian distribution is given by

$$p(x) = \frac{1}{\sigma\sqrt{2\pi}} e^{-\frac{(x-\mu)^2}{2\sigma^2}}, \tag{15.19}$$

where $\sigma^2 = \text{var}[X]$ is the variance and $\mu = E[X]$ is the mean of the Gaussian distribution. From the Gaussian integral, it is easy to verify that

$$\int_{-\infty}^{\infty} p(x)dx = 1, \tag{15.20}$$

and this is exactly the reason why the factor $1/\sqrt{2\pi}$ is required in the normalization of all the probabilities. The probability function reaches a peak at $x = \mu$ and the variance σ^2 controls the width of the peak (see Figure 15.2).

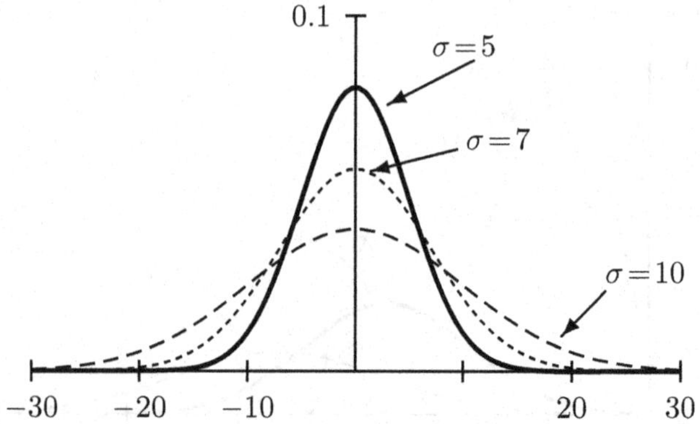

Fig. 15.2 Gaussian distributions for $\sigma = 5, 7, 10$.

The cumulative probability function (CPF) for a normal distribution is the integral of $p(x)$, which is defined by

$$\Phi(x) = P(X < x) = \frac{1}{\sqrt{2\pi\sigma^2}} \int_{-\infty}^{x} e^{-\frac{(\zeta-\mu)^2}{2\sigma^2}} d\zeta. \qquad (15.21)$$

Using the error function defined in early chapters, we can write the above equation as

$$\Phi(x) = \frac{1}{\sqrt{2}}[1 + \mathrm{erf}(\frac{x-\mu}{\sqrt{2}\sigma})], \qquad (15.22)$$

where

$$\mathrm{erf}(x) = \frac{2}{\sqrt{\pi}} \int_0^x e^{-\zeta^2} d\zeta. \qquad (15.23)$$

The moment generating function for the Gaussian distribution is given by

$$G_X(\nu) = e^{\mu\nu + \frac{1}{2}(\sigma\nu)^2}. \qquad (15.24)$$

The Gaussian distribution can be considered as the limit of the Poisson distribution when $\lambda \gg 1$. Using the Sterling's approximation $x! \sim \sqrt{2\pi x}(x/e)^x$ for $x \gg 1$, and setting $\mu = \lambda$ and $\sigma^2 = \lambda$, it can be verified that the Poisson distribution can be written as a Gaussian distribution

$$P(x) \approx \frac{1}{\sqrt{2\pi\lambda}} e^{-\frac{(x-\mu)^2}{2\lambda}}, \qquad (15.25)$$

where $\mu = \lambda$. In statistical applications, the normal distribution is often written as $N(\mu, \sigma)$ to emphasize that the probability density function depends on two parameters μ and σ.

The standard normal distribution is a normal distribution $N(\mu, \sigma)$ with a mean of $\mu = 0$ and standard deviation $\sigma = 1$, that is $N(0, 1)$. This is useful to normalize or standardize data for statistical analysis. If we define a normalized variable

$$\xi = \frac{x - \mu}{\sigma}, \tag{15.26}$$

it is equivalent to give a score so as to place the data above or below the mean in the unit of standard deviation. In terms of the area under the probability density function, ξ sorts where the data falls. It is worth pointing out that some books define $z = \xi = (x - \mu)/\sigma$ in this case, and call the standard normal distribution the Z-distribution.

Table 15.1 Function ϕ defined by equation (15.28).

ξ	$\phi(\xi)$	ξ	$\phi(\xi)$
0.0	0.500	1.0	0.841
0.1	0.540	1.1	0.864
0.2	0.579	1.2	0.885
0.3	0.618	1.3	0.903
0.4	0.655	1.4	0.919
0.5	0.692	1.5	0.933
0.6	0.726	1.6	0.945
0.7	0.758	1.7	0.955
0.8	0.788	1.8	0.964
0.9	0.816	1.9	0.971

Now the probability density function of the standard normal distribution becomes

$$p(x) = \frac{1}{\sqrt{2\pi}} e^{-\frac{\xi^2}{2}}. \tag{15.27}$$

Its cumulative probability function is

$$\phi(\xi) = \frac{1}{\sqrt{2\pi}} \int_{-\infty}^{\xi} e^{-\frac{\xi^2}{2}} d\xi = \frac{1}{2}[1 + \mathrm{erf}(\frac{\xi}{\sqrt{2}})]. \tag{15.28}$$

As the calculations of ϕ and the error function involve the numerical integrations, it is usual practice to tabulate ϕ in a table (see Table 15.1) so that we do not have to calculate their values each time we use it.

15.4 Other Distributions

There are a number of other important distributions such as the exponential distribution, log-normal distribution, uniform distribution and the

χ^2-distribution. The uniform distribution has a probability density function

$$p = \frac{1}{\beta - \alpha}, \qquad x = [\alpha, \beta], \tag{15.29}$$

whose mean is $E[X] = (\alpha + \beta)/2$ and variance is $\sigma^2 = (\beta - \alpha)^2/12$.

The exponential distribution has the following probability density function

$$f(x) = \lambda e^{-\lambda x} \quad (x > 0), \tag{15.30}$$

and $f(x) = 0$ for $x \leq 0$. Its mean and variance are

$$\mu = 1/\lambda, \qquad \sigma^2 = 1/\lambda^2. \tag{15.31}$$

The log-normal distribution has a probability density function

$$f(x) = \frac{1}{x\sqrt{2\pi\sigma^2}} \exp[-\frac{(\ln x - \mu)^2}{2\sigma^2}], \tag{15.32}$$

whose mean and variance are

$$E[X] = e^{\mu + \sigma^2/2}, \qquad \mathrm{var}[X] = e^{\sigma^2 + 2\mu}(e^{\sigma^2} - 1). \tag{15.33}$$

The χ^2-distribution, called chi-square or chi-squared distribution, is very useful in statistical inference and the method of least squares. This distribution is for the quantity

$$\chi_n^2 = \sum_{i=1}^{n} \Big(\frac{X_i - \mu_i}{\sigma_i}\Big)^2, \tag{15.34}$$

where the n independent variables X_i are normally distributed with means μ_i and variances σ_i^2. The probability density function for χ^2-distribution is given by

$$p(x) = \frac{1}{2^{n/2}\Gamma(n/2)} x^{\frac{n}{2}-1} e^{-x/2}, \tag{15.35}$$

where $x \geq 0$, and n is called the degree of freedom. Here the gamma function $\Gamma(n)$ is defined by

$$\Gamma(n) = \int_0^\infty x^{n-1} e^{-x} dx, \tag{15.36}$$

for $n > 0$. In the simplest case, $n = 1$, we have

$$\Gamma(1) = \int_0^\infty x^0 e^{-x} dx = -e^{-x}\Big|_0^\infty = 1. \tag{15.37}$$

Using integration by parts, we have

$$\Gamma(n+1) = \int_0^\infty x^n e^{-x} dx = -xe^{-x}\Big|_0^\infty + \int_0^\infty n x^{n-1} e^{-x} dx$$

$$= n \int_0^\infty x^{n-1} e^{-x} dx = n\Gamma(n), \tag{15.38}$$

which leads to

$$\Gamma(n+1) = n!, \tag{15.39}$$

for any integer $n > 0$.

It can be verified that the mean of the χ^2-distribution is n and its variance is $2n$. For other distributions, readers can refer to any books that are devoted to advanced topics in probability theory and statistical analysis.

15.5 The Central Limit Theorem

The most important theorem in probability is the central limit theorem which concerns a large number of trials and explains why the normal distribution occurs so widely.

Let $X_i (i = 1, 2, ..., n)$ be n independent random variables, each of which is defined by a probability density function $p_i(x)$ with a corresponding mean μ_i and a variance σ_i^2. The sum of all these random variables

$$\Theta = \sum_{i=1}^{n} X_i = X_1 + X_2 + ... + X_n, \tag{15.40}$$

is also a random variable whose distribution approaches the Gaussian distribution as $n \to \infty$. Its mean $E[\Theta]$ and variance $\mathrm{var}[\Theta]$ are given by

$$E[\Theta] = \sum_{i=1}^{n} E[X_i] = \sum_{i=1}^{n} \mu_i, \tag{15.41}$$

and

$$\mathrm{var}[\Theta] = \sum_{i=1}^{n} \mathrm{var}[X_i] = \sum_{i=1}^{n} \sigma_i^2. \tag{15.42}$$

The proof of this theorem is beyond the scope of this book as it involves the moment generating functions, characteristics functions and other techniques. In computational mathematics, we simply use these important results for statistical analysis.

In the special case when all the variables X_i are described by the same probability density function with the same mean μ and variance σ^2, these results become

$$E[\Theta] = n\mu, \qquad \mathrm{var}[\Theta] = n\sigma^2. \tag{15.43}$$

By defining a new variable

$$\xi_n = \frac{\Theta - n\mu}{\sigma\sqrt{n}}, \tag{15.44}$$

then the distribution of ξ_n converges towards the standard normal distribution $N(0,1)$ as $n \to \infty$.

Let us see what the theorem means for a simple experiment of rolling a few dice. For a fair six-sided die, each side will appear equally likely with a probability of $1/6 \approx 0.1667$, thus it obeys a uniform distribution.

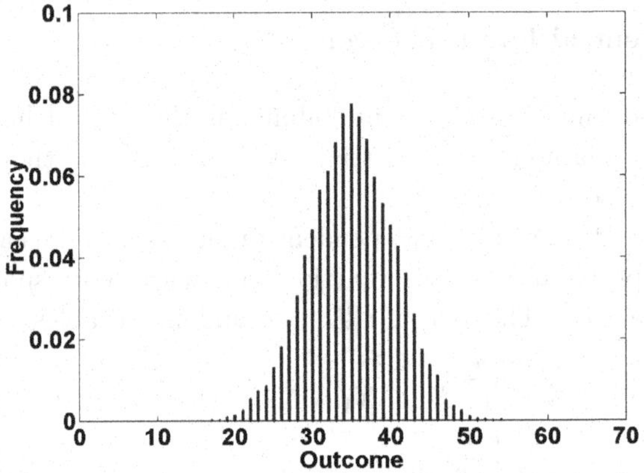

Fig. 15.3 An approximate Gaussian distribution (the outcomes of the sum of face values in rolling 15 dice).

If we roll $n = 15$ independent dice, the sums of the face values vary from 1 to 90. After rolling the 15 dice 10,000 times, the distribution is shown in Figure 15.3 and it approaches to a normal distribution as $n \to \infty$.

15.6 Weibull Distribution

Although the distribution functions of the real-world random processes are dominated by the Gaussian or normal distribution, however, there are some cases where other distributions describe the related phenomena more accurately. Weibull's distribution is such a distribution with many applications in areas such as reliability analysis, engineering design and quality assessment. Therefore, it deserves a special introduction in detail. This distribution was originally developed by Swedish physicist, A. Weibull in

1939, to try to explain the fact, well known but unexplained at that time, that the relative strength of a specimen decreases with its increasing dimension. Since then, it has been applied to study many real-world stochastic processes even including the distributions of wind speeds, rainfalls, energy resources and earthquakes.

Weibull's distribution is a three-parameter distribution given by

$$p(x,\lambda,\beta,n)=\begin{cases}\frac{n}{\lambda}(\frac{x-\beta}{\lambda})^{n-1}\exp[-(\frac{x-\beta}{\lambda})^n] & (x\geq\beta),\\ \\ 0 & (x<\beta),\end{cases} \tag{15.45}$$

where λ is a scaling parameter, and n is the shape parameter, often referred to as the Weibull modulus. The parameter β is the threshold of the distribution. By straightforward integration, we have the cumulative probability density distribution

$$\Phi(x,\lambda,\beta,n)=1-e^{-(\frac{x-\beta}{\lambda})^n}. \tag{15.46}$$

For the fixed values $\lambda=1$ and $\beta=0$, the variation of n will give a broad range of shapes and can be used to approximate various distributions as shown in Fig. 15.4.

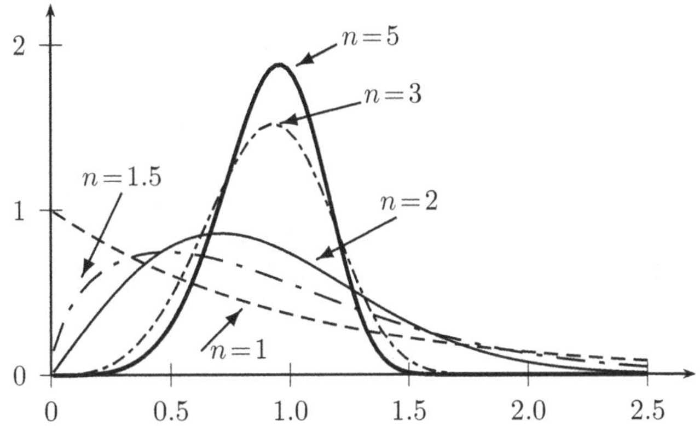

Fig. 15.4 Weibull density function for different values of λ.

In reliability analysis, especially for a large infrastructure such as a dam or a tall building under stress, the survival probability is more conveniently represented as

$$P_s(V)=\exp[\int_V-(\frac{\sigma}{\sigma_0})^n\frac{dV}{V_0}], \tag{15.47}$$

where V is the volume of the system. σ_0 is the failure stress (either tensile or shear) for the reference volume V_0. The failure probability is

$$P_f(V) = 1 - P_s(V). \tag{15.48}$$

For constant stress σ over the whole volume V, we simply have

$$P_s(V) = \exp[-(\frac{\sigma}{\sigma_0})^n \frac{V}{V_0}]. \tag{15.49}$$

At the reference point $\sigma = \sigma_0$ and $V = V_0$ often obtained using laboratory tests, we have $P_s(V_0) = e^{-1} \approx 0.3679$. As the stress becomes extreme, $\sigma \to \infty$, then $P_s \to 0$ and $P_f \to 1$.

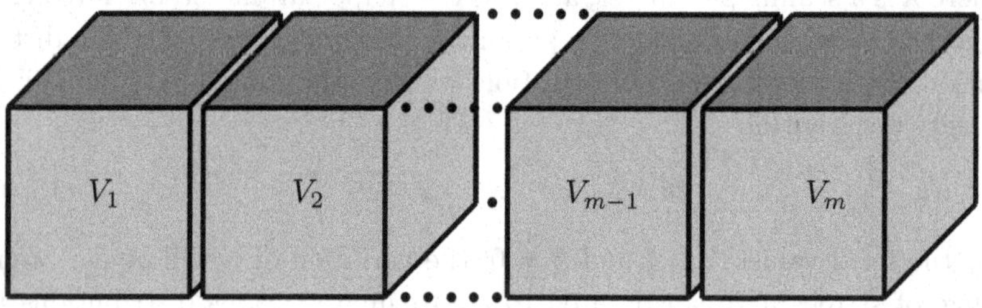

Fig. 15.5 The weakest link analogy.

The fundamental idea of this volume-related probability function is the weakest link theory. The larger the volume of a system, the more likely it is to have critical flaws that could cause potential failure. We can consider that the whole large volume V consists of m small volumes/blocks $V_1, V_2, ..., V_m$ and these small blocks are glued together (see Fig. 15.5), thus the probability of the survival of the whole system is equivalent to the survival of all the subsystem blocks. If any one of these blocks fails, the system is considered flawed and thus failed. In the simplest case, $V_1 = V_2 = ... = V_m = V_0$ and $m = V/V_0$, the survival probability of the whole system under constant stress σ is

$$P_s(V) = P_s(mV_0) = \overbrace{P_s(V_0) \times P_s(V_0) \times ... \times P_s(V_0)}^{m}$$

$$= [P_s(V_0)]^m = [e^{-(\frac{\sigma}{\sigma_0})^n}]^{\frac{V}{V_0}} = \exp[-\frac{V}{V_0}(\frac{\sigma}{\sigma_0})^n]. \tag{15.50}$$

Example 15.2: *A lighting pole vibrates under the pressure of wind. For a given wind speed v, what is the probability of survival of the pole (without seriously damaged under bending)?*

Let us assume that the lighting pole is perfectly cylindrical with a cross-section area of A and height h. For simplicity, we also assume that the failure mechanism is due to the shear bending force (though in reality it is more complicated, we have to consider the bending moment and yield strength).

Let x be the distance along the pole with $x = 0$ at the ground level. The diameter D of the pole is $D = 2\sqrt{\frac{A}{\pi}}$. The shear stress σ at x is

$$\sigma = \frac{v(h-x)D}{A},$$

which means the stress (and thus the moment) is the highest at the root. Suppose a similar system may fail at a critical stress σ_0 with a volume V_0. The probability of survival is given by

$$P_s(V) = \exp[-\int_0^h (\frac{\sigma}{\sigma_0})^n A \frac{dx}{V_0}]$$

$$= \exp[-(\frac{v^n D^n}{A^{n-1} V_0 \sigma_0^n}) \int_0^h (h-x)^n dx] = \exp[-\frac{v^n D^n h^{n+1}}{(n+1)A^{n-1} V_0 \sigma_0^n}].$$

This probability dramatically decreases with the increase of the height and/or the wind speed.

There are a whole range of statistical methods based on solid probability theory. However, we will not discuss any theory further, as our focus in this book is numerical algorithms. Interested readers can refer to more advanced literature.

Chapter 16

Data Modelling

Statistics is the mathematics of data collection and interpretation, and the analysis and characterisation of numerical data by inference from sampling. Statistical methods involve reduction of data, estimates and significance tests, and relationship between two or more variables by analysis of variance, and the test of hypotheses.

16.1 Sample Mean and Variance

If a sample consists of n independent observations $x_1, x_2, ..., x_n$ on a random variable x such as the price of a cup of coffee, two important and commonly used parameters are sample mean and sample variance, which can easily be estimated from the sample. The sample mean is calculated by

$$\bar{x} \equiv <x> = \frac{1}{n}(x_1 + x_2 + ... + x_n) = \frac{1}{n}\sum_{i=1}^{n} x_i, \tag{16.1}$$

which is essentially the arithmetic average of the values x_i.

Generally speaking, if u is a linear combination of n independent random variables $y_1, y_2, ..., y_n$ and each random variable y_i has an individual mean μ_i and a corresponding variance σ_i^2, we have the linear combination

$$u = \sum_{i=1}^{n} \alpha_i y_i = \alpha_1 y_1 + \alpha_2 y_2 + ... + \alpha_n y_n, \tag{16.2}$$

where the parameters $\alpha_i (i = 1, 2, ..., n)$ are the weighting coefficients. From the central limit theorem, we have the mean μ_u of the linear combination

$$\mu_u = E(u) = E(\sum_{i=1}^{n} \alpha_i y_i) = \sum_{i=1}^{n} \alpha_i E(y_i) = \sum \alpha_i \mu_i. \tag{16.3}$$

Then, the variance σ_u^2 of the combination is

$$\sigma_u^2 = E[(u-\mu_u)^2] = E\Big[\sum_{i=1}^{n} \alpha_i (y_i - \mu_i)^2\Big], \tag{16.4}$$

which can be expanded as

$$\sigma_u^2 = \sum_{i=1}^{n} \alpha_i^2 E[(y_i-\mu_i)^2] + \sum_{i,j=1;i\neq j}^{n} \alpha_i \alpha_j E[(y_i-\mu_i)(y_j-\mu_j)], \tag{16.5}$$

where $E[(y_i-\mu_i)^2] = \sigma_i^2$. Since y_i and y_j are independent, we have $E[(y_i - \mu_i)(y_j-\mu_j)] = E[(y_i-\mu_i)]E[(y_j-\mu_j)] = 0$. Therefore, we get

$$\sigma_u^2 = \sum_{i=1}^{n} \alpha_i^2 \sigma_i^2. \tag{16.6}$$

The sample mean defined in equation (16.1) can also be viewed as a linear combination of all the x_i assuming that each of which has the same mean $\mu_i = \mu$ and variance $\sigma_i^2 = \sigma^2$, and the same weighting coefficient $\alpha_i = 1/n$. Hence, the sample mean is an unbiased estimate of the sample due to the fact $\mu_{\bar{x}} = \sum_{i=1}^{n} \mu/n = \mu$. In this case, however, we have the variance

$$\sigma_{\bar{x}}^2 = \sum_{i=1}^{n} \frac{1}{n^2}\sigma^2 = \frac{\sigma^2}{n}, \tag{16.7}$$

which means the variance becomes smaller as the size n of the sample increases by a factor of $1/n$.

The sample variance S^2 is defined by

$$S^2 = \frac{1}{n-1}\sum_{i=1}^{n}(x_i - \bar{x})^2. \tag{16.8}$$

It is worth pointing out that the factor is $1/(n-1)$ not $1/n$ because only $1/(n-1)$ will give the correct and unbiased estimate of the variance. From the probability theory in the previous chapter, we know that $E[x^2] = \mu^2 + \sigma^2$. The mean of the sample variance is

$$\mu_{S^2} = E[\frac{1}{n-1}\sum_{i=1}^{n}(x_i-\bar{x})^2] = \frac{1}{n-1}\sum_{i=1}^{n} E[(x_i^2 - n\bar{x}^2)]. \tag{16.9}$$

Using $E[\bar{x}^2] = \mu^2 + \sigma^2/n$, we get

$$\mu_{S^2} = \frac{1}{n-1}\sum_{i=1}^{n}\{E[x_i^2] - nE[\bar{x}^2]\}$$

$$= \frac{1}{n-1}\{n(\mu^2+\sigma^2) - n(\mu^2+\frac{\sigma^2}{n})\} = \sigma^2. \tag{16.10}$$

Obviously, if we used the factor $1/n$ instead of $1/(n-1)$, we would get $\mu_{S^2} = \frac{n-1}{n}\sigma^2 < \sigma^2$, which would underestimate the sample variance. The other way to think about the factor $1/(n-1)$ is that we need at least one value to estimate the mean, we need at least 2 values to estimate the variance. Thus, for n observations, only $n-1$ different values of variance can be obtained to estimate the total sample variance.

16.2 Method of Least Squares

16.2.1 *Maximum Likelihood*

For a sample of n values $x_1, x_2, ..., x_n$ of a random variable X whose probability density function $p(x)$ depends on a set of k parameters $\beta_1, ..., \beta_k$, the joint probability is then

$$\Phi(\beta_1, ..., \beta_k) = \prod_{i=1}^{n} p(x_i, \beta_1, ..., \beta_k)$$

$$= p(x_1, \beta_1, ..., \beta_k)p(x_2, \beta_1, ..., \beta_k) \cdots p(x_n, \beta_1, ..., \beta_k). \tag{16.11}$$

The essence of the maximum likelihood is to maximize Φ by choosing the parameters β_j. As the sample can be considered as given values, the maximum likelihood requires that

$$\frac{\partial \Phi}{\partial \beta_j} = 0, \qquad (j = 1, 2, ..., k), \tag{16.12}$$

whose solutions for β_j are the maximum likelihood estimates.

16.2.2 *Linear Regression*

For experiments and observations, we usually plot one variable such as pressure or price y against another variable x such as time or spatial coordinates. We try to present the data in such a way that we can see some trend in the data. For a set of n data points (x_i, y_i), the usual practise is to try to draw a straight line $y = a + bx$ so that it represents the major trend. Such a line is often called the regression line or the best fit line as shown in Figure 16.1.

The method of linear least squares is to try to determine the two parameters, a (intercept) and b (slope), for the regression line from n data points,

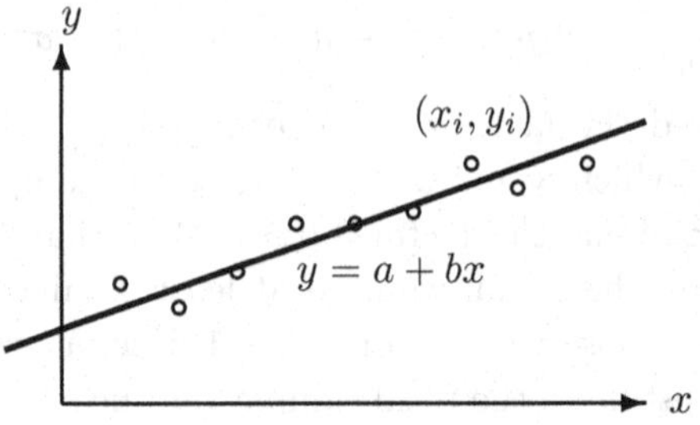

Fig. 16.1 Least square and the best fit line.

assuming that x_i are known more precisely, and the values of y_i obey a normal distribution around the potentially best fit line with a variance σ^2, we have the probability

$$P = \prod_{i=1}^{n} p(y_i) = A \exp\Big\{ -\frac{1}{2\sigma^2} \sum_{i=1}^{n} [y_i - f(x_i)]^2 \Big\}, \tag{16.13}$$

where A is a constant, and $f(x)$ is the function for the regression $[f(x) = a + bx$ for the linear regression$]$. It is worth pointing out that the exponent $\sum_{i=1}^{n} [y_i - f(x_i)]^2 / \sigma^2$ is similar to the quantity χ^2_n defined in the χ^2-distribution.

The essence of the method of least squares is to maximize the probability P by choosing the appropriate a and b. The maximization of P is equivalent to the minimization of the exponent ψ

$$\psi = \sum_{i=1}^{n} [y_i - f(x_i)]^2. \tag{16.14}$$

We see that ψ is the sum of the squares of the deviations $\epsilon_i^2 = (y_i - f(x_i))^2$ where $f(x_i) = a + bx_i$. The minimization means the least sum of the squares, hence the name of the method of least squares.

In order to minimize ψ as a function of a and b, its derivatives should be zero. That is

$$\frac{\partial \psi}{\partial a} = -2 \sum_{i=1}^{n} [y - (a + bx_i)] = 0, \tag{16.15}$$

and

$$\frac{\partial \psi}{\partial b} = -2 \sum_{i=1}^{n} x_i [y_i - (a + bx_i)] = 0. \tag{16.16}$$

By expanding these equations, we have

$$na + b\sum_{i=1}^{n} x_i = \sum_{i=1}^{n} y_i, \tag{16.17}$$

and

$$a\sum_{i=1}^{n} x_i + b\sum_{i=1}^{n} x_i^2 = \sum_{i=1}^{n} x_i y_i, \tag{16.18}$$

which is a system of linear equations for a and b, and it is straightforward to obtain the solutions as

$$a = \frac{1}{n}[\sum_{i=1}^{n} y_i - b\sum_{i=1}^{n} x_i] = \bar{y} - b\bar{x}, \tag{16.19}$$

$$b = \frac{n\sum_{i=1}^{n} x_i y_i - (\sum_{i=1}^{n} x_i)(\sum_{i=1}^{n} y_i)}{n\sum_{i=1}^{n} x_i^2 - (\sum_{i=1}^{n} x_i)^2}, \tag{16.20}$$

where

$$\bar{x} = \frac{1}{n}\sum_{i=1}^{n} x_i, \qquad \bar{y} = \frac{1}{n}\sum_{i=1}^{n} y_i. \tag{16.21}$$

If we use the following notations

$$K_x = \sum_{i=1}^{n} x_i, \qquad K_y = \sum_{i=1}^{n} y_i, \tag{16.22}$$

and

$$K_{xx} = \sum_{i=1}^{n} x_i^2, \qquad K_{xy} = \sum_{i=1}^{n} x_i y_i, \tag{16.23}$$

then the above equations for a and b become

$$a = \frac{K_{xx}K_y - K_x K_{xy}}{nK_{xx} - (K_x)^2}, \tag{16.24}$$

and

$$b = \frac{nK_{xy} - K_x K_y}{nK_{xx} - (K_x)^2}. \tag{16.25}$$

The residual error is defined by

$$\epsilon_i = y_i - (a + bx_i), \tag{16.26}$$

whose sample mean is given by

$$\mu_\epsilon = \frac{1}{n}\sum_{i=1}^{n} \epsilon_i = \frac{1}{n}y_i - a - b\frac{1}{n}\sum_{i=1}^{n} x_i = \bar{y} - a - b\bar{x} = 0. \tag{16.27}$$

The sample variance S^2 is

$$S^2 = \frac{1}{n-2}\sum_{i=1}^{n}[y_i - (a + bx_i)]^2, \tag{16.28}$$

where the factor $1/(n-2)$ comes from the fact that two constraints are needed for the best fit, and therefore the residuals have $n-2$ degrees of freedom.

16.3 Correlation Coefficient

The correlation coefficient $r_{x,y}$ is a very useful parameter for finding any potential relationship between two sets of data x_i and y_i for two random variables x and y, respectively. If x has a mean μ_x and a sample variance S_x^2, and y has a mean μ_y and a sample variance S_y^2, the correlation coefficient is defined by

$$r_{x,y} = \frac{\mathrm{cov}(x,y)}{S_x S_y} = \frac{E[xy] - \mu_x \mu_y}{S_x S_y}, \tag{16.29}$$

where

$$\mathrm{cov}(x,y) = E[(x-\mu_x)(y-\mu_y)], \tag{16.30}$$

is the covariance. If the two variables are independent or $\mathrm{cov}(x,y) = 0$, there is no correlation between them ($r_{x,y} = 0$). If $r_{x,y}^2 = 1$, then there is a linear relationship between these two variables. $r_{x,y} = 1$ is an increasing linear relationship where the increase of one variable will lead to the increase of another. On the other hand, $r_{x,y} = -1$ is a decreasing relationship when one increases while the other decreases.

For a set of n data points (x_i, y_i), the correlation coefficient can be calculated by

$$r_{x,y} = \frac{n\sum_{i=1}^n x_i y_i - \sum_{i=1}^n x_i \sum_{i=1}^n y_i}{\sqrt{[n\sum x_i^2 - (\sum_{i=1}^n x_i)^2][n\sum_{i=1}^n y_i^2 - (\sum_{i=1}^n y_i)^2]}},$$

or

$$r_{x,y} = \frac{nK_{xy} - K_x K_y}{\sqrt{(nK_{xx} - K_x^2)(nK_{yy} - K_y^2)}}, \tag{16.31}$$

where $K_{yy} = \sum_{i=1}^n y_i^2$.

Example 16.1: *Is there any relationship between shoe size and height among the general population? By collecting data randomly, say, we have the following data:*

Height (h): 162, 167, 168, 171, 174, 176, 183, 179 (cm);

Shoe size (s): 5.5, 6, 7.5, 7.5, 8.5, 10, 11, 12.

Let us try the linear regression. We have

$$s = a + bh.$$

Since

$$K_h = \sum_{i=1}^{8} h_i = 1380, \quad K_s = \sum_{i=1}^{8} s_i = 68,$$

$$K_{hs} = \sum_{i=1}^{8} h_i s_i = 11835.5, \quad K_{hh} = \sum_{i=1}^{8} h_i^2 = 238380,$$

we get

$$a = \frac{K_{hh}K_s - K_h K_{xy}}{nK_{hh} - K_h^2}$$

$$= \frac{238380 \times 68 - 1380 \times 11835.5}{8 \times 238380 - 1380^2} \approx -46.6477,$$

and

$$b = \frac{nK_{hs} - K_h K_s}{nK_{hh} - K_h^2}$$

$$= \frac{8 \times 11835.5 - 1380 \times 68}{8 \times 238380 - 1380^2} \approx 0.3197.$$

So the regression line becomes

$$s = -46.6477 + 0.3197h,$$

which is shown in Fig. 16.2.

From the data set, we know that the sample means are $\mu_h = 172.5$, *and* $\mu_s = 8.5$. *The covariance* $\mathrm{cov}(h, s) = E[(h - \mu_h)(s - \mu_s)] = 13.2$. *We also have the standard deviation of height* $S_h = 6.422$ *and the standard deviation of shoe size* $S_s = 2.179$. *Therefore, their correlation coefficient* r *is given by*

$$r = \frac{\mathrm{cov}(h, s)}{S_h S_s} \approx \frac{13.2}{6.422 * 2.179} \approx 0.94.$$

This is a relatively strong correlation indeed. It is worth pointing out this conclusion is based on a small set of samples, which may not extend to the general population.

16.4 Linearization

Sometimes, some obviously nonlinear functions can be transformed into linear forms so as to carry out linear regression, instead of more complicated nonlinear regression.

For example, the following nonlinear function

$$f(x) = \alpha e^{-\beta x}, \tag{16.32}$$

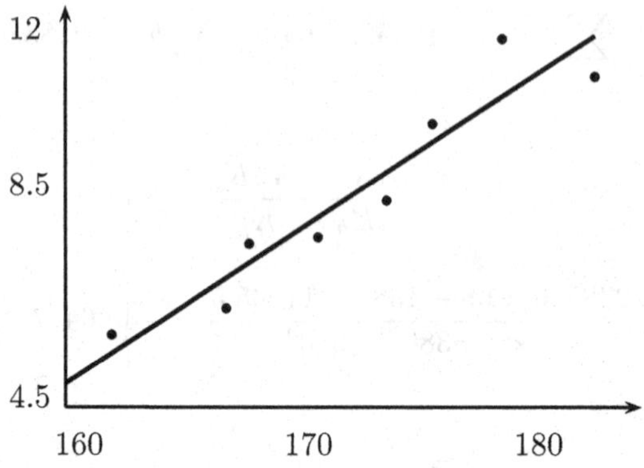

Fig. 16.2 A best fit line by linear least square.

can be transformed into a linear form by taking logarithms of both sides. We have

$$\ln f(x) = \ln(\alpha) - \beta x, \tag{16.33}$$

which is equivalent to $y = a + bx$ if we let $y = \ln f(x)$, $a = \ln(\alpha)$ and $b = -\beta$. In addition,

$$f(x) = \alpha e^{-\beta x + \gamma} = A e^{-\beta x},$$

where $A = \alpha e^{\gamma}$ is essentially the same as the above function.

Similarly, function

$$f(x) = \alpha x^{\beta}, \tag{16.34}$$

can also be transformed into

$$\ln[f(x)] = \ln(\alpha) + \beta \ln(x), \tag{16.35}$$

which is a linear regression $y = a + b\zeta$ between $y = \ln[f(x)]$ and $\zeta = \ln(x)$ where $a = \ln(\alpha)$ and $b = \beta$.

Furthermore, function

$$f(x) = \alpha \beta^{x}, \tag{16.36}$$

can also be converted into the standard linear form

$$\ln f(x) = \ln \alpha + x \ln \beta, \tag{16.37}$$

by letting $y = \ln[f(x)]$, $a = \ln \alpha$, and $b = \ln \beta$.

It is worth pointing out that the data points involving zeros should be taken out due to the potential singularity of the logarithm. Fortunately, these points are rarely in the regression for the functions in the above form.

Example 16.2: *If a set of data can fit to a nonlinear function*

$$y = ax\exp(-x/b),$$

in the range of $(0, \infty)$, *it is then possible to convert it to a linear regression.*

As $x = 0$ *is just a single point, so we can leave this out. For* $x \neq 0$, *we can divide both sides by* x, *we have*

$$\frac{y}{x} = a\exp(-x/b).$$

Taking the logarithm of both sides, we have

$$\ln\frac{y}{x} = \ln a - \frac{1}{b}x,$$

which is a linear regression of y/x *versus* x.

16.5 Generalized Linear Regression

The most widely used linear regression is the so-called generalized least square as the linear combination of the basic functions. Fitting to a polynomial of degree p

$$y(x) = \alpha_0 + \alpha_1 x + \alpha_2 x^2 + .. + \alpha_p x^p, \tag{16.38}$$

is probably the most widely used. This is equivalent to the regression to the linear combination of the basis functions 1, x, x, ..., and x^p. However, there is no particular reason why we have to use these basis functions. In fact, the basis functions can be any arbitrary known functions such as $\sin(x)$, $\cos(x)$ and even $\exp(x)$, and the main requirement is that they can be explicitly expressed as basic functions. In this sense, the generalized least square can be written as

$$y(x) = \sum_{j=0}^{p} \alpha_j f_j(x), \tag{16.39}$$

where the basis functions f_j are known functions of x. Now the least square is defined as

$$\psi = \sum_{i=1}^{n} \frac{[y_i - \sum_{j=0}^{p} \alpha_j f_j(x_i)]^2}{\sigma_i^2}, \tag{16.40}$$

where $\sigma_i (i = 1, 2, ..., n)$ are the standard deviations of the i-th data point at (x_i, y_i). There are n data points in total. In order to determine the coefficients uniquely, it requires that

$$n \geq p + 1. \tag{16.41}$$

In the case of unknown standard deviations σ_i, we can always set all the values σ_i as the same constant $\sigma_i = \sigma = 1$.

Let $\boldsymbol{D}$ be the design matrix which is given by

$$D_{ij} = \frac{f_j(x_i)}{\sigma_i}. \tag{16.42}$$

The minimum of ψ is determined by

$$\frac{\partial \psi}{\partial \alpha_j} = 0, \qquad (j = 0, 1, ..., p). \tag{16.43}$$

That is

$$\sum_{i=1}^{n} \frac{f_k(x_i)}{\sigma_i}[y_i - \sum_{j=0}^{p} \alpha_j f_j(x_i)] = 0, \qquad k = 0, ..., p. \tag{16.44}$$

Rearranging the terms and interchanging the order of summations, we have

$$\sum_{j=0}^{p} \sum_{i=1}^{n} \frac{f_j(x_i) f_k(x_i)}{\sigma_i^2} = \sum_{i=1}^{n} \frac{y_i f_k(x_i)}{\sigma_i^2}, \tag{16.45}$$

which can be written compactly as the following matrix equation

$$\sum_{j=0}^{p} A_{kj} \alpha_j = b_k, \tag{16.46}$$

or

$$\boldsymbol{A}\boldsymbol{\alpha} = \boldsymbol{b}, \tag{16.47}$$

where

$$\boldsymbol{A} = \boldsymbol{D}^{\boldsymbol{T}} \cdot \boldsymbol{D},$$

is a $(p+1) \times (p+1)$ matrix. That is

$$A_{kj} = \sum_{i=1}^{n} \frac{f_k(x_i) f_j(x_i)}{\sigma_i^2}. \tag{16.48}$$

Here b_k is a column vector given by

$$b_k = \sum_{i=1}^{n} \frac{y_i f_k(x_i)}{\sigma_i^2}, \tag{16.49}$$

where $(k = 0, ..., p)$. Equation (16.46) is a linear system of the so-called normal equations which can be solved using the standard methods for solving linear systems. The solution of the coefficients is $\boldsymbol{\alpha} = \boldsymbol{A}^{-1}\boldsymbol{b}$ or

$$\alpha_k = \sum_{j=0}^{p} [A]_{kj}^{-1} b_j, \qquad (k = 0, ..., p), \tag{16.50}$$

where $\boldsymbol{A}^{-1} = [A]_{ij}^{-1}$.

Example 16.3: *If we have a data set which can fit to a sine function, we can assume that*

$$y = a + b\sin x.$$

For simplicity, we also assume that $\sigma_i = 1$. *Thus, we have*

$$\psi = \sum_{i=1}^{n}[y_i - (a + b\sin x_i)]^2.$$

The requirement of least squares leads to

$$\frac{\partial \psi}{\partial a} = 2na + \sum_{i=1}^{n}(-2y_i + 2b\sin x_i) = 0,$$

and

$$\frac{\partial \psi}{\partial b} = \sum_{i=1}^{n}(-2y_i \sin x_i + 2a\sin x_i + 2b\sin^2 x_i) = 0.$$

Solving the first equation to get a *in terms of* b, *we have*

$$a = \frac{1}{n}\sum_{i=1}^{n}(y_i - b\sin x_i).$$

Substituting it into the other equation, we can obtain b. *Therefore, the solutions for* a *and* b *are*

$$a = \frac{1}{\Delta}[(\sum_{i=1}^{n} y_i)(\sum_{i=1}^{n}\sin^2 x_i) - (\sum_{i=1}^{n}\sin x_i)(\sum_{i=1}^{n} y_i \sin x_i)],$$

$$b = \frac{1}{\Delta}[n(\sum_{i=1}^{n} y_i \sin x_i) - (\sum_{i=1}^{n}\sin x_i)(\sum_{i=1}^{n} y_i)],$$

where

$$\Delta = n\sum_{i=1}^{n}\sin^2 x_i - (\sum_{i=1}^{n}\sin x_i)^2.$$

A special case of the generalized linear least squares is the so-called polynomial least squares when the basis functions are simple power functions $f_i(x) = x^i, (i = 0, 1, ..., q)$. For simplicity, we assume that $\sigma_i = \sigma = 1$.

The matrix equation (16.46) simply becomes

$$\begin{pmatrix} \sum_{i=1}^n 1 & \sum_{i=1}^n x_i & \dots & \sum_{i=1}^n x_i^p \\ \sum_{i=1}^n x_i & \sum_{i=1}^n x_i^2 & \dots & \sum_{i=1}^n x_i^{p+1} \\ \vdots & & \ddots & \\ \sum_{i=1}^n x_i^p & \sum_{i=1}^n x_i^{p+1} & \dots & \sum_{i=1}^n x_i^{2p} \end{pmatrix} \begin{pmatrix} \alpha_0 \\ \alpha_1 \\ \vdots \\ \alpha_p \end{pmatrix} = \begin{pmatrix} \sum_{i=1}^n y_i \\ \sum_{i=1}^n x_i y_i \\ \vdots \\ \sum_{i=1}^n x_i^p y_i \end{pmatrix}.$$

In the simplest case when $p = 1$, it becomes the standard linear regression $y = \alpha_0 + \alpha_1 x = a + bx$. Now we have

$$\begin{pmatrix} n & \sum_{i=1}^n x_i \\ \sum_{i=1}^n x_i & \sum_{i=1}^n x_i^2 \end{pmatrix} \begin{pmatrix} \alpha_0 \\ \alpha_1 \end{pmatrix} = \begin{pmatrix} \sum_{i-1}^n y_i \\ \sum_{i=1}^n x_i y_i \end{pmatrix}. \tag{16.51}$$

Its solution is

$$\begin{pmatrix} \alpha_0 \\ \alpha_1 \end{pmatrix} = \frac{1}{\Delta} \begin{pmatrix} \sum_{i=1}^n x_i^2 & -\sum_{i=1}^n x_i \\ -\sum_{i=1}^n x_i & n \end{pmatrix} \begin{pmatrix} \sum_{i=1}^n y_i \\ \sum_{i=1}^n x_i y_i \end{pmatrix}$$

$$= \frac{1}{\Delta} \begin{pmatrix} (\sum_{i=1}^n x_i^2)(\sum_{i=1}^n y_i) - (\sum_{i=1}^n x_i)(\sum_{i=1}^n x_i y_i) \\ n \sum_{i=1}^n x_i y_i - (\sum_{i=1}^n x_i)(\sum_{i=1}^n y_i) \end{pmatrix}, \tag{16.52}$$

where

$$\Delta = n \sum_{i=1}^n x_i^2 - (\sum_{i=1}^n x_i)^2. \tag{16.53}$$

These are exactly the same coefficients as those in Eq. (16.25).

Example 16.4: *We now use a quadratic function to best fit the following data:*

$$x: -0.98, \;\; 1.00, \;\; 2.02, \;\; 3.03, \;\; 4.00$$

$$y: \;\; 2.44, \;\; -1.51, \;\; -0.47, \;\; 2.54, \;\; 7.52.$$

For the formula $y = \alpha_0 + \alpha_1 x + \alpha_2 x^2$, *we have*

$$\begin{pmatrix} n & \sum_{i=1}^n x_i & \sum_{i=1}^n x_i^2 \\ \sum_{i=1}^n x_i & \sum_{i=1}^n x_i^2 & \sum_{i=1}^n x_i^3 \\ \sum_{i=1}^n x_i^2 & \sum_{i=1}^n x_i^3 & \sum_{i=1}^n x_i^4 \end{pmatrix} \begin{pmatrix} \alpha_0 \\ \alpha_1 \\ \alpha_2 \end{pmatrix} \begin{pmatrix} \sum_{i=1}^n y_i \\ \sum_{i=1}^n x_i y_i \\ \sum_{i=1}^n x_i^2 y_i \end{pmatrix}.$$

Using the data set, we have $n = 5$, $\sum_{i=1}^{n} x_i = 9.07$ *and* $\sum_{i=1}^{n} y_i = 10.52$. *Other quantities can be calculated in a similar way. Therefore, we have*

$$\begin{pmatrix} 5.0000 & 9.0700 & 31.2217 \\ 9.0700 & 31.2217 & 100.119 \\ 31.2217 & 100.119 & 358.861 \end{pmatrix} \begin{pmatrix} \alpha_0 \\ \alpha_0 \\ \alpha_2 \end{pmatrix} = \begin{pmatrix} 10.52 \\ 32.9256 \\ 142.5551 \end{pmatrix}.$$

By direct inversion, we have

$$\begin{pmatrix} \alpha_0 \\ \alpha_1 \\ \alpha_2 \end{pmatrix} = \begin{pmatrix} -0.5055 \\ -2.0262 \\ 1.0065 \end{pmatrix}.$$

Finally, the best fit equation is

$$y(x) = -0.5055 - 2.0262x + 1.0065x^2,$$

which is quite close to the formula $y = x^2 - 2x - 1/2$ *used to generate the original data with a random component of about* 2.5%.

16.6 Nonlinear Regression

There are many other functions that cannot be transformed into any linear forms. For example, $f(x) = a + \sin(bx)$ where a and b are the parameters to be determined by the best fit. In this case, we have to deal with generalized nonlinear regression.

For any nonlinear function

$$y = f(x, \boldsymbol{\alpha}), \tag{16.54}$$

where $\boldsymbol{\alpha} = (\alpha_1, \alpha_2, ..., \alpha_p)^T$ is a vector of parameters. The least square requires the minimization of ψ

$$\psi = \sum_{i=1}^{n} \frac{[y_i - f(x_i, \boldsymbol{\alpha})]^2}{\sigma_i^2}, \tag{16.55}$$

where the data points are $(x_i, y_i)(i = 1, 2, ..., N)$.

The stationary conditions lead to

$$\frac{\partial \psi}{\partial \alpha_j} = -2 \sum_{i=1}^{n} \frac{[y_i - f(x_i, \boldsymbol{\alpha})]}{\sigma_i^2} \frac{\partial f(x_i, \boldsymbol{\alpha})}{\partial \alpha_j}, \tag{16.56}$$

for $j = 1, 2, ..., p$. In general, ψ is a function of $\boldsymbol{\alpha}$ which can be approximated, about a fixed $\boldsymbol{\alpha_0}$ (so that $\delta\boldsymbol{\alpha} = \boldsymbol{\alpha} - \boldsymbol{\alpha_0}$), as the following quadratic form

$$\psi(\boldsymbol{\alpha}) = \psi_0(\boldsymbol{\alpha_0}) + \boldsymbol{G}^T \delta\boldsymbol{\alpha} + \frac{1}{2} \delta\boldsymbol{\alpha}^T \boldsymbol{H} \delta\boldsymbol{\alpha} + ... \tag{16.57}$$

where $\boldsymbol{G}$ is the gradient vector and $\boldsymbol{H}$ is the Hessian matrix. The minimization of ψ leads

$$\frac{\delta\psi}{\delta\boldsymbol{\alpha}} = \boldsymbol{G} + \boldsymbol{H}\delta\boldsymbol{\alpha} = 0, \tag{16.58}$$

where we have used $\delta\boldsymbol{\alpha}^T\boldsymbol{H}\boldsymbol{\alpha} \equiv \boldsymbol{\alpha}^T\boldsymbol{H}\delta\boldsymbol{\alpha}$. This is in fact the steepest descent method that was discussed earlier. The descent increment $\delta\boldsymbol{\alpha}$ from the previous value to the next value is determined by solving

$$\boldsymbol{H}\delta\boldsymbol{\alpha} = -\boldsymbol{G}. \tag{16.59}$$

As there is a factor of 2, for simplicity and convenience, we can define a vector (half of the negative gradient) $\boldsymbol{b} = [b_j] = -\boldsymbol{G}/2$ as

$$b_j = -\frac{1}{2}\frac{\partial\psi}{\partial\alpha_j}. \tag{16.60}$$

Similarly, we can also define the half Hessian matrix $\boldsymbol{A} = [A_{ij}] = \frac{1}{2}\boldsymbol{H}$, called the curvature matrix, as

$$A_{jk} \equiv \frac{1}{2}\frac{\partial^2\psi}{\partial\alpha_j\partial\alpha_k} = \sum_{i=1}^{n}\frac{1}{\sigma_i^2}\{\frac{\partial f(x_i,\boldsymbol{\alpha})}{\partial\alpha_j}\frac{\partial f(x_i,\boldsymbol{\alpha})}{\partial\alpha_k}$$

$$- [y_i - f(x_i,\boldsymbol{\alpha})]\frac{\partial^2 f(x_i,\boldsymbol{\alpha})}{\partial\alpha_k\partial\alpha_j}\}. \tag{16.61}$$

Using the above notations, Eq. (16.59) becomes

$$\sum_{j=1}^{p} A_{ij}\delta\alpha_j = b_i, \qquad i = 1, 2, ..., p. \tag{16.62}$$

The increment $\delta\boldsymbol{\alpha}$ is thus obtained by

$$\delta\boldsymbol{\alpha} = \boldsymbol{A}^{-1}\boldsymbol{b}. \tag{16.63}$$

Starting with an initial guess $\alpha^{(0)}$, say, $(1, 1, ..., 1)$, we can solve the above equation to obtain $\delta\boldsymbol{\alpha}$, the final optimal $\boldsymbol{\alpha}_{\min}$ when ψ reaches the minimum can be determined by iterations. However, the inverse of the Hessian could cause potential convergence problems if the model is not well fitted, especially in the case of many odd data points.

The convergence of the solution of this nonlinear regression can easily be achieved by using the Levenberg-Marquardt method which combines the method of inverting the Hessian matrix and the steepest descent method. The iterations vary between these two extremes. The detailed discussion can be found in more advanced literature.

In most applications, the number of parameters is limited to a few, and some parameters can be solved or expressed in terms of others, which means only a limited number of parameters need to be estimated by iterations.

For example, the nonlinear function

$$y(x) = \alpha + \beta e^{-\gamma x}, \tag{16.64}$$

has three parameters to be fitted. For a set of n data points (x_i, y_i), we try to minimize ψ.

$$\psi = \sum_{i=1}^{n} [y_i - (\alpha + \beta e^{-\gamma x_i})]^2. \tag{16.65}$$

The stationary conditions become

$$\frac{\partial \psi}{\partial \alpha} = 2n\alpha + 2\sum_{i=1}^{n} [\beta e^{-\gamma x_i} - y_i] = 0, \tag{16.66}$$

$$\frac{\partial \psi}{\partial \beta} = 2\sum_{i=1}^{n} [\alpha e^{-\gamma x_i} + \beta e^{-2\gamma x_i} - y_i e^{-\gamma x_i}] = 0, \tag{16.67}$$

$$\frac{\partial \psi}{\partial \gamma} = 2\sum_{i=1}^{n} [\beta y_i x_i e^{-\gamma x_i} - \beta \alpha x_i e^{-\gamma x_i} - \beta^2 x_i e^{-2\gamma x_i}] = 0. \tag{16.68}$$

The first two equations give

$$\alpha = \frac{(\sum_{i=1}^{n} y_i)(\sum_{i=1}^{n} e^{-2\gamma x_i}) - (\sum_{i=1}^{n} e^{-\gamma x_i})(\sum_{i=1}^{n} y_i e^{-\gamma x_i})}{n\sum_{i=1}^{n} e^{-2\gamma x_i} - (\sum_{i=1}^{n} e^{-\gamma x_i})^2},$$

$$\beta = \frac{n\sum_{i=1}^{n} y_i e^{-\gamma x_i} - (\sum_{i=1}^{n} y_i)(\sum_{i=1}^{n} e^{-\gamma x_i})}{n\sum_{i=1}^{n} e^{-2\gamma x_i} - (\sum_{i=1}^{n} e^{-\gamma x_i})^2}. \tag{16.69}$$

Substituting them into (16.68), we get a single equation for the only unknown γ, which can be solved using iteration methods such as the Newton-Raphson method. Let us look at an example.

Example 16.5: *The reaction rate of the photosynthesis in green plants can be approximated by the following expression*

$$v = \frac{\alpha}{1 + \beta / I},$$

where I is the light intensity, α is the maximum rate, and β is the affinity. The observed data are:

$$I : 10.27, 20.12, 30.20, 40.17, 50.24$$

$$v : 15.23, 20.07, 22.54, 24.06, 25.04$$

The nonlinear least-square best fit requires the minimization of ψ

$$\psi = \sum_{i=1}^{n}[v_i - \frac{\alpha}{1+\beta/I_i}]^2,$$

which leads to

$$\frac{\partial\psi}{\partial\alpha} = -2\sum_{i=1}^{n}[\frac{v_i}{1+\beta/I_i} - \frac{\alpha}{(1+\beta/I_i)^2}] = 0,$$

$$\frac{\partial\psi}{\partial\beta} = 2\alpha\sum_{i=1}^{n}[\frac{v_i}{I_i(1+\beta/I_i)^2} - \frac{\alpha}{I_i(1+\beta/I_i)^3}] = 0.$$

The first equation gives

$$\alpha = \frac{\sum_{i=1}^{n}\frac{v_i}{1+\beta/I_i}}{\sum_{i=1}^{n}\frac{1}{(1+\beta/I_i)^2}}.$$

Substituting this into the second equation, we have

$$\sum_{i=1}^{n}\frac{v_i}{I_i(1+\beta/I_i)^2} - \frac{\sum_{i=1}^{n}\frac{v_i}{1+\beta/I_i}}{\sum_{i=1}^{n}\frac{1}{(1+\beta/I_i)^2}}\sum_{i=1}^{n}\frac{1}{I_i(1+\beta/I_i)^3} = 0.$$

Using the $n = 5$ *data points and evaluating the above expression iteratively we have* $\beta \approx 9.97$ *after about 25 iterations starting from an initial guess* $\beta^{(0)} = 0$. *This will give* $\alpha \approx 30.03$.

This iterative procedure is very time-consuming. It is always a good idea to try to transform the function to the linear form if possible.

As I_i *and* v_i *are non-zero, so we can rearrange the equation as*

$$\frac{1}{v_i} = \frac{1+\beta/I_i}{\alpha} = \frac{1}{\alpha} + \frac{\beta/\alpha}{I_i},$$

which can be written as a linear regression of $y_i = 1/v_i$ *versus* $p_i = 1/I_i$

$$y_i = a + bp_i,$$

where

$$a = \frac{1}{\alpha}, \qquad b = \frac{\beta}{\alpha}.$$

Now we can convert the data points into

$$p_i : 0.0974, 0.497, 0.0331, 0.0249, 0.0199$$

$$y_i : 0.0657, 0.0498, 0.0444, 0.0416, 0.0399.$$

From the standard linear regression discussed in the previous sections, we have

$$a = \frac{(\sum_{i=1}^n p_i^2)(\sum_{i=1}^n y_i) - (\sum_{i=1}^n p_i)(\sum_{i=1}^n y_i)}{n \sum_{i=1}^n p_i^2 - (\sum_{i=1}^n p_i)^2} \approx 0.0333$$

and

$$b = \frac{n \sum_{i=1}^n p_i y_i - (\sum_{i=1}^n p_i)(\sum_{i=1}^n y_i)}{n \sum_{i=1}^n p_i^2 - (\sum_{i=1}^n p_i)^2} \approx 0.3321.$$

Finally, we have

$$\alpha = \frac{1}{a} \approx 30.03, \qquad \beta = \frac{b}{a} \approx 9.97.$$

The results are the same, however, the second method using linearization is much more efficient as it uses fewer calculations.

16.7 Hypothesis Testing

16.7.1 *Confidence Interval*

The confidence interval is defined as the interval $\theta_1 \le X \le \theta_2$ so that the probabilities at these two limits θ_1 and θ_2 are equal to a given probability $\gamma = 1 - \alpha$ (say, 95% or 99%). That is

$$P(\theta_1 \le X \le \theta_2) = \gamma = 1 - \alpha. \tag{16.70}$$

The predetermined parameter γ is always near 1 so that it can be expressed as a small deviation $\alpha \ll 1$ from 1 (see Figure 16.3). If we choose $\gamma = 95\%$, it means that we can expect that about 95% of the sample will fall within the confidence interval while 5% of the data will not.

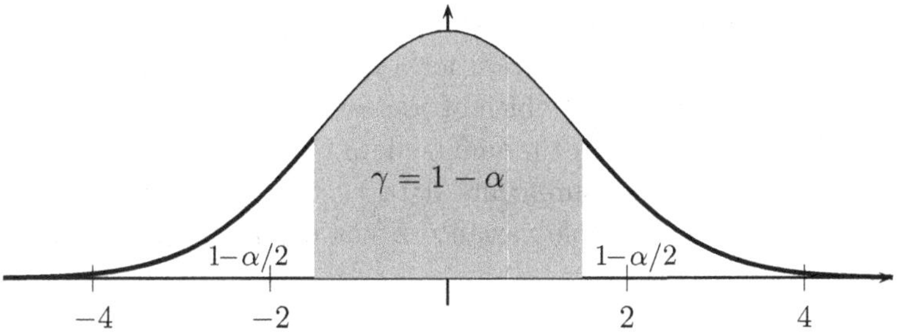

Fig. 16.3 Confidence interval $\gamma = 1 - \alpha$.

For the standard normal distribution, this means $P(-\theta \leq \xi \leq \theta) = 1 - \alpha$, so that

$$\phi(\xi \leq \theta) = 1 - \frac{\alpha}{2}. \tag{16.71}$$

If $\alpha = 0.05$, we have $\phi(\xi \leq \theta) = 0.975$ or $\theta = 1.960$. That is to say, $-\theta \leq \xi \leq \theta$ or $\mu - \theta\sigma \leq x \leq \mu + \theta\sigma$. We also know that if you repeat an experiment n times, the variance will decrease from σ^2 to σ^2/n, which is equivalent to saying that the standard deviation becomes $\sigma/\sqrt{n}$ for a sample size n. If $\alpha = 0.01$, then $\theta = 2.579$, we have

$$\mu - 2.579\frac{\sigma}{\sqrt{n}} \leq x \leq \mu + 2.579\frac{\sigma}{\sqrt{n}}. \tag{16.72}$$

On the other hand, for $\theta = 1$, we get $\mu - \sigma \leq x \leq \mu + \sigma$ and $\gamma = 0.682$. In other words, only 68.2% of the sample data will fall within the interval $[\mu - \sigma, \mu + \sigma]$ or

$$x = \mu \pm \sigma, \tag{16.73}$$

with a 68.2% confidence level.

It is conventional to use $\gamma = 0.95$ for probably significant, 0.99 for significant, and 0.999 for highly significant.

16.7.2 *Student's t-Distribution*

The Student's t-test is a very powerful method for testing the null hypothesis to see if the means of two normally distributed samples are equal. This method was designed by W. S. Gosset in 1908 and he had to use a pen name 'Student' because of his employer's policy in publishing research results at that time. This is a powerful method for hypothesis testing using small-size samples. This test can also be used to test if the slope of the regression line is significantly different from 0. It has become one of the most popular methods for hypothesis testing. The theoretical basis of the t-test is the Student's t-distribution for a sample population with the unknown standard deviation σ, which of course can be estimated in terms of the sample variance S^2 from the sample data.

For n independent measurements/data $x_1, x_2, ..., x_n$ with an estimated sample mean $\bar{x}$ and a sample variance S^2 as defined by Eq. (16.8), the t-variable is defined by

$$t = \frac{\bar{x} - \mu}{(S/\sqrt{n})}. \tag{16.74}$$

The Student's t-distribution with $k = n - 1$ degrees of freedom is the distribution for the random variable t, and the probability density function is

$$p(t) = \frac{\Gamma(\frac{k+1}{2})}{\sqrt{k\pi}\Gamma(k/2)}[1 + \frac{t^2}{k}]^{-\frac{k+1}{2}}. \qquad (16.75)$$

It can be verified that the mean is $E[t] = 0$. The variance is $\sigma^2 = k/(k-2)$ for $k > 2$ and infinite for $0 < k \leq 2$.

The corresponding cumulative probability function is

$$F(t) = \frac{\Gamma(\frac{k+1}{2})}{\sqrt{k\pi}\Gamma(k/2)} \int_{-\infty}^{t} [1 + \frac{\zeta^2}{k}]^{-\frac{k+1}{2}} d\zeta. \qquad (16.76)$$

This integral leads to a hypergeometric function, which is not straightforward to calculate, which is why they are tabulated in many statistical tables. For a confidence level of $\gamma = 1 - \alpha$, the confidence interval is given by

$$F(\theta) = 1 - \frac{\alpha}{2}, \qquad (16.77)$$

which is usually tabulated. For $\alpha = 0.05$ and 0.01 (or $1 - \alpha/2 = 0.975$ and 0.995), the values are tabulated in Table 16.1.

Table 16.1 Limits defined by $F(\theta) = 1 - \alpha/2$ in Eq. (16.77).

k	$F(\theta)_{0.975}$	$F(\theta)_{0.995}$
1	12.7	63.7
2	4.30	9.93
3	3.18	5.84
4	2.78	4.60
5	2.57	4.03
6	2.45	3.71
7	2.37	3.50
8	2.31	3.36
9	2.26	3.25
10	2.23	3.17
20	2.09	2.85
50	2.01	2.68
100	1.98	2.63
∞	1.96	2.58

Suppose we are dealing with the 95% confidence interval, we have $p(-\theta \leq t \leq \theta) = 1 - \alpha = 0.95$ or $p(t \leq \theta) = 1 - \alpha/2 = 0.975$, we have $\theta = t_{\alpha,k} = 12.70$ $(k = 1)$, 4.30 $(k = 2)$, 3.18 $(k = 3)$, ..., 2.228 $(k = 10)$, ..., 1.959 for $k \to \infty$. Hence,

$$\mu - \theta \frac{S}{\sqrt{n}} \leq t \leq \mu + \theta \frac{S}{\sqrt{n}}. \tag{16.78}$$

This is much more complicated than its counterpart, the standard normal distribution.

16.7.3 *Student's t-Test*

There are quite a few variations of the Student's t-test, and the most common t-tests are the one-sample t-test and the two-sample t-test. The one sample t-test is used for measurements that are randomly drawn from a population to compare the sample mean with a known number.

In order to do statistical testing, we first have to pose precise questions or form a hypothesis, which is conventionally called the null hypothesis. The basic steps of a t-test are as follows:

1. The null hypothesis: H_0: $\mu = \mu_0$ (often known value) for one sample, or H_0: $\mu_1 = \mu_2$ for two samples;
2. Calculate the t-test statistic t and find the critical value θ for a given confidence level $\gamma = 1 - \alpha$ by using $F(t \leq \theta) = 1 - \alpha/2$;
3. If $|t| > \theta$, reject the hypothesis. Otherwise, accept the hypothesis.

Another important t-test is the two-sample paired test. Assuming that two pairs of n sample data sets U_i and V_i are independent and drawn from the same normal distribution, the paired t-test is used to determine whether they are significantly different from each other. The t-variable is defined by

$$t = \frac{(\bar{U} - \bar{V})}{S_d/\sqrt{n}} = (\bar{U} - \bar{V}) \sqrt{\frac{n(n-1)}{\sum_{i=1}^n (\tilde{U}_i - \tilde{V}_i)^2}}, \tag{16.79}$$

where $\tilde{U}_i = U_i - \bar{U}$ and $\tilde{V}_i = V_i - \bar{V}$. In addition,

$$S_d^2 = \frac{1}{n-1} \sum_{i=1}^{n} (\tilde{U}_i - \tilde{V}_i)^2. \tag{16.80}$$

This is equivalent to applying the one-sample test to the difference $U_i - V_i$ data sequence.

Let us look at a simple example.

Example 16.6: *A novel teaching method of teaching children sciences was tried in a class (say class B), while a standard method was used in another class (say class A). At the end of the assessment, 8 students are randomly drawn from each class, and their science scores are as follows:*
Class A: $U_i =$ *76, 77, 76, 81, 77, 76, 75, 82;*
Class B: $V_i =$ *79, 81, 77, 86, 82, 81, 82, 80.*
At a 95% *confidence level, can you say the new method is really better than the standard method?*

If we suppose that the two methods do not produce any difference in results, that is to say, their means are the same. Thus the null hypothesis is:

$$H_0 : \mu_A = \mu_B.$$

We know that $\bar{U} = 77.5$, $\bar{V} = 81$. *The combined sample variance* $S_d = 2.828$. *We now have*

$$t = \frac{\bar{U} - \bar{V}}{S_d/\sqrt{n}} = \frac{77.5 - 81}{2.828/\sqrt{8}} = -3.5.$$

We know from the statistical table that the critical value $\theta = 2.37$ *for* $F(\theta) = 1 - \alpha/2$ *and* $k = n - 1 = 7$. *As* $t < -\theta$ *or* $t > \theta$, *we can reject the null hypothesis. That is to say, the new method does produce better results in teaching sciences.*

The variance analysis and hypothesis testing are important topics in applied statistics, and there are many excellent books on these topics. Readers can refer to the relevant books listed at the end of this book. It is worth pointing out that other important methods for hypothesis testing are Fisher's F-test, χ^2-test, and non-parametric tests. What we have discussed in this chapter is just the tip of the iceberg; however, it forms the solid basis for further studies.

Chapter 17

Data Mining, Neural Networks and Support Vector Machine

The evolution of the Internet and social media has resulted in the huge increase of data in terms of both volumes and complexity. In fact, 'big data' has become a buzzword nowadays, and the so-called big data science is becoming an important area. Data mining has expanded beyond the traditional data modelling techniques such as statistical models and regression methods. Data mining now also includes clustering and classifications, feature selection and feature extraction, and machine learning techniques such as decision tree methods, hidden Markov models, artificial neural networks, and support vector machine. To introduce these methods systematically can take a whole book, and it is not possible to cover even a good fraction of these methods in a book chapter. Therefore, we will here introduce only clustering and classification methods.

Clustering and classification methods are rather rich with a wide spectrum of methods. We introduce the basic k-mean method for clustering and support vector machine for classification. Artificial neural networks (ANN) are a class of methods with different variations and variants, and ANN can have many applications in a diverse range of areas, including clustering, classification, machine learning, computational intelligence, feature extraction and selection and others. In the rest of this chapter, we will briefly introduce the essence of these methods.

17.1 Clustering Methods

17.1.1 *Hierarchy Clustering*

For a given set of n observations, the aim is to divide them into some clusters (say, k different clusters) so as to minimize certain clustering measures or objectives. There are many key issues here. Firstly, we usually do not know how many clusters the data may intrinsically have. Secondly, the data sets can be massive ($n \gg 1$, for example, $n = O(10^9)$ or even $n = O(10^{18})$). Thirdly, the data may not be clean enough with useless information and/or noisy data. Finally, the data can be incomplete, and thus may lack sufficient information needed for correct clustering. Obviously, there are other issues, too, such as time factors, unstructured data and distance metrics.

Hierarchy clustering usually works well for small datasets. It starts with every point in its own cluster (that is, $k = n$ for n data points), followed by a simple iterative procedure

(1) Each point belongs to its own cluster $k = n$.
(2) While (stopping criterion),
(3) Choose two nearest clusters to merge into one cluster;
(4) Update $k \leftarrow k - 1$;
(5) Repeat until the metric measure goes up;
(6) End.

This iterative procedure can lead to one big cluster $k = 1$ in the end. But it does result in a complex decision tree, which provides an informative summary and some insight into the structures and relationship within the data. However, if a distance metric such as the Euclidean metric is defined properly, the metric will start to decrease at the initial stage when two clusters are merged. In the final stage, this metric usually starts to increase, which is an indication to stop and the number of the clusters can be the true number of clusters. However, this is not straightforward in practice, and there may not exist any unique k value at all.

In the case when Euclidean distance measures are used, the distance between centroids is defined as the cluster distance. The complexity of this algorithm is $O(n^3)$ where n is the number of points. For $n = 10^9$, this can lead to $O(10^{27})$ floating-point operations, which is quite computationally expensive.

17.1.2 *k-Means Clustering Method*

The main aim of the k-means clustering method is to divide a set of n observations into k different clusters in such a way that each cluster belongs to the nearest cluster with the shortest distance to its corresponding cluster mean or centroid.

Suppose we have n observation points $\boldsymbol{x}_1, \boldsymbol{x}_2, ..., \boldsymbol{x}_n$ in a d-dimensional vector space, our aim is to partition these observations into k clusters $S_1, S_2, ..., S_k$) with centroid means $(\xi_1, \xi_1, ..., \xi_k)$ so that the cluster-wise sum of squares, also called within-cluster sum of squares, can be minimized. That is,

$$\text{minimize} \sum_{j=1, \boldsymbol{x}_i \in S_i}^{k} ||\boldsymbol{x}_i - \xi_j||^2, \tag{17.1}$$

where $1 < k \leq n$ and typically $k \ll n$.

This k-means method for dividing n points into k clusters can be summarized schematically as follows:

(1) Choose randomly k points as the initial centroids of the k clusters.
(2) For each remaining point i,
(3) Assign i to the cluster with the closest centroid;
(4) Update the centroid of that cluster (containing i);
(5) End.

There are some key issues concerning this method. The choice of k points as the initial centroids is not efficient. In the worst case, the k points selected randomly can belong to the same cluster. One possible remedy is to choose k points with the largest distances from each other. This is often carried out by starting from a random point, and then try to find the second point that is as far as possible to the first point, and then try to find the third point that is as far as possible from the previous two points. This continues until the first k points are initialized. This method is an improvement over the previous random selection method, but there is still no guarantee that the choice of these initial points will lead to the best clustering solutions. Therefore, some sort of random restart and multiple runs are needed.

On the other hand, the algorithm complexity of this method is typically $O(n^{kd+1} \log(n))$ where d is the dimension of the data. Even for $n = 10^6$, $k = 3$ and $d = 2$, this becomes $O(10^{43})$, which is extremely computationally expensive. However, it is worth pointing out that such complexity is just theoretical, and these methods can sometime work surprisingly well in practice (at least for small datasets). In the worst cases, such algorithm complexity can become NP-hard complexity.

Recent developments show that other methods may be more suitable for large datasets. For example, the Bradley-Fayyad-Reina (BFR) algorithm and Clustering Using REpresentative (CURE) algorithm have shown good results.

In addition, recent trends tend to combine traditional algorithms with optimization algorithms that are based on swarm intelligence. The basic idea is to use optimization techniques to optimize the centroids and then use clustering methods such as k-means to carry out clustering. Recent studies suggest such hybrid methods can produce very promising results. For example, the firefly algorithm can be used to do clustering and classifications with superior performance.

For any methods to be efficient and useful, large matrices should be avoided and there is no need to try every possible combination. The methods used to solve a large-scale problem should be efficient enough to produce good results in a practically acceptable time scale. However, in general, there is no guarantee that the global optimality can be found.

It is worth pointing out that distance metrics are also very important. Even with the most efficient methods, if the metric measure is not defined properly, the results may be incorrect or meaningless. Most clustering methods use the Euclidean distance and Jaccard similarity, though other distances such as the edit distance and Hamming distance are also widely used. Briefly speaking, the Euclidean distance $d(x, y)$ between two data points x and y is the L_p-norm given by

$$d(x,y) = \Big(\sum_i |x_i - y_i|^p\Big)^{1/p}. \tag{17.2}$$

In most cases, $p = 2$ is used. Jaccard's similarity index of two sets U and V is defined as

$$J(U,V) = |U \cap V|/|U \cup V|, \tag{17.3}$$

which leads to $0 \leq J(U,V) \leq 1$. and the Jaccard distance is defined as

$$d_J(U,V) = 1 - J(U,V). \tag{17.4}$$

The edit distance between two strings U and V is the smallest number of insertions and deletions of single characters that will convert U to V. For example, $U=$ '*abcde*' and $V=$ '*ackdeg*', the edit distance is $d(U,V) = 3$. By deleting b, inserting k after c, and then inserting g after e, string U can be converted to V.

On the other hand, the Hamming distance is the number of components which two vectors/strings differ. For example, the Hamming distance between 10101 and 11110 is 3. Obviously, other distance metrics are also being used in the literature. Interested readers can refer to more advance literature about data mining.

17.2 Artificial Neural Networks

Many applications use artificial neural networks (ANN), which is especially true in machine learning, artificial intelligence and pattern recognition. The fundamental idea of ANNs is to learn from data and make predictions. For a given set of input data, a neural network maps the input data into some outputs. The relationships between the inputs and the outputs are quite complicated, and it is usually impossible to express such relationships in any exact or analytical forms. By comparing the outputs with the true outputs, the network system can adjust its weights so as to better match its outputs. If there is a sufficient number of data, the network can be 'trained' well, and thus the trained network can even make predictions for new data, which is the essence of artificial neural networks.

17.2.1 *Artificial Neuron*

The basic mathematical model of an artificial neuron was first proposed by W. McCulloch and W. Pitts in 1943, and this fundamental model is referred to as the McCulloch-Pitts model. Other models and neural networks are based on it.

An artificial neuron with n inputs or impulses and an output y_k will be activated if the signal strength reaches a certain threshold θ. Each input has a corresponding weight w_i (see Fig. 17.1). The output of this neuron is given by

$$y_l = \Phi\Big(\sum_{i=1}^{n} w_i u_i\Big), \tag{17.5}$$

where the weighted sum $\xi = \sum_{i=1}^{n} w_i u_i$ is the total signal strength, and Φ is the so-called activation function, which can be taken as a step function. That is, we have

$$\Phi(\xi) = \begin{cases} 1 & \text{if } \xi \geq \theta, \\ 0 & \text{if } \xi < \theta. \end{cases} \tag{17.6}$$

We can see that the output is only activated to a non-zero value if the overall signal strength is greater than the threshold θ.

The step function has discontinuity, sometimes, it is easier to use a nonlinear, smooth function, called a Sigmoid function

$$S(\xi) = \frac{1}{1+e^{-\xi}}, \tag{17.7}$$

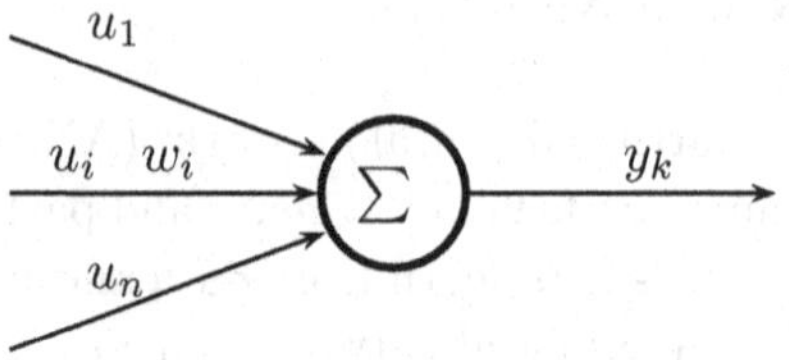

Fig. 17.1 A simple neuron.

which approaches 1 as $U \to \infty$, and becomes 0 as $U \to -\infty$. An interesting property of this function is

$$S'(\xi) = S(\xi)[1 - S(\xi)]. \tag{17.8}$$

17.2.2 *Artificial Neural Networks*

A single neuron can only perform a simple task – on or off. Complex functions can be designed and performed using a network of interconnecting neurons or perceptrons. The structure of a network can be complicated, and one of the most widely used is to arrange them in a layered structure, with an input layer, an output layer, and one or more hidden layer (see Fig. 17.2). The connection strength between two neurons is represented by its corresponding weight. Some artificial neural networks (ANNs) can perform complex tasks and simulate complex mathematical models, even if there is no explicit functional form mathematically. Neural networks have developed over last few decades and have been applied in almost all areas of science and engineering.

The construction of a neural network involves the estimation of the suitable weights of a network system with some training/known data sets. The task of training is to find the suitable weights ω_{ij} so that the neural networks not only can best-fit the known data, but also can predict outputs for new inputs. A good artificial neural network should be able to minimize both errors simultaneously – the fitting/learning errors and the prediction errors.

The errors can be defined as the difference between the calculated (or predicated) output o_k and real output y_k for all n_o output neurons in the least-square sense

$$E = \frac{1}{2} \sum_{k=1}^{n_o} (o_k - y_k)^2. \tag{17.9}$$

Here the output o_k is a function of inputs/activations and weights. In order to minimize this error, we can use the standard minimization techniques to find the solutions of the weights.

A simple and yet efficient technique is the steepest descent method. For any initial random weights, the weight increment for w_{hk} is

$$\Delta w_{hk} = -\eta \frac{\partial E}{\partial w_{hk}} = -\eta \frac{\partial E}{\partial o_k} \frac{\partial o_k}{\partial w_{hk}}, \tag{17.10}$$

where η is the learning rate. Typically, we can choose $\eta = 1$.

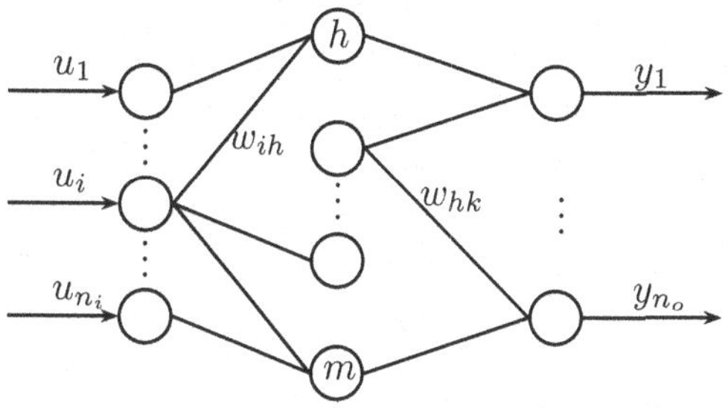

Fig. 17.2 Schematic representation of a three-layer neural network with n_i inputs, m hidden nodes and n_o outputs.

From

$$S_k = \sum_{h=1}^{m} w_{hk} o_h, \qquad (k = 1, 2, ..., n_o), \tag{17.11}$$

and

$$o_k = f(S_k) = \frac{1}{1 + e^{-S_k}}, \tag{17.12}$$

we have

$$f' = f(1 - f), \tag{17.13}$$

$$\frac{\partial o_k}{\partial w_{hk}} = \frac{\partial o_k}{\partial S_k} \frac{\partial S_k}{\partial w_{hk}} = o_k(1 - o_k) o_h, \tag{17.14}$$

and

$$\frac{\partial E}{\partial o_k} = (o_k - y_k). \tag{17.15}$$

Therefore, we have

$$\Delta w_{hk} = -\eta \delta_k o_h, \tag{17.16}$$

where

$$\delta_k = o_k(1 - o_k)(o_k - y_k). \tag{17.17}$$

For a given structure of the neural network, there are many ways of calculating weights by supervised learning. One of the simplest and widely used methods is to use the back propagation algorithm for training neural networks, often called back propagation neural networks (BPNNs).

17.2.3 *Back Propagation Algorithm*

The basic idea of a BPNN is to start from the output layer and propagate backwards so as to estimate and update the weights (see Fig. 17.3).

From any initial random weighting matrices w_{ih} (for connecting the input nodes to the hidden layer) and w_{hk} (for connecting the hidden layer to the output nodes), we can calculate the outputs of the hidden layer o_h

$$o_h = \frac{1}{1 + \exp[-\sum_{i=1}^{n_i} w_{ih} u_i]}, \qquad (h = 1, 2, ..., m), \tag{17.18}$$

and the outputs for the output nodes

$$o_k = \frac{1}{1 + \exp[-\sum_{h=1}^{m} w_{hk} o_h]}, \qquad (k = 1, 2, ..., n_o). \tag{17.19}$$

BPNN

begin
Initialize weight matrices W_{ih} *and* W_{hk} *randomly*
for *all training data points*
  while *(residual errors are not zero)*
    Calculate the output for the hidden layer o_h *using (17.18)*
    Calculate the output for the output layer o_k *using (17.19)*
    Compute errors δ_k *and* δ_h *using (17.20) and (17.21)*
    Update weights w_{ih} *and* w_{hk} *via (17.22) and (17.23)*
  end while
end for
end

Fig. 17.3 Pseudocode of back propagation neural networks.

The errors for the output nodes are given by

$$\delta_k = o_k(1 - o_k)(y_k - o_k), \qquad (k = 1, 2, ..., n_o), \tag{17.20}$$

where $y_k(k = 1, 2, ..., n_o)$ are the data (real outputs) for the inputs $u_i(i = 1, 2, ..., n_i)$. Similarly, the errors for the hidden nodes can be written as

$$\delta_h = o_h(1 - o_h) \sum_{k=1}^{n_o} w_{hk}\delta_k, \qquad (h = 1, 2, ..., m). \tag{17.21}$$

The updating formulae for weights at iteration t are

$$w_{hk}^{t+1} = w_{hk}^{t} + \eta\delta_k o_h, \tag{17.22}$$

and

$$w_{ih}^{t+1} = w_{ih}^{t} + \eta\delta_h u_i, \tag{17.23}$$

where $0 < \eta \leq 1$ is the learning rate.

Here we can see that the weight increments are

$$\Delta w_{ih} = \eta\delta_h u_i, \tag{17.24}$$

with similar updating formulae for w_{hk}. An improved version is to use the so-called weight momentum α to increase the learning efficiency

$$\Delta w_{ih} = \eta\delta_h u_i + \alpha w_{ih}(\tau - 1), \tag{17.25}$$

where τ is an extra parameter. For more details, readers can refer to more advanced literature.

There are many good software packages for artificial neural networks, and there are dozens of good books fully dedicated to theory and implementations. Therefore, we will not provide any code here.

17.3 Support Vector Machine

Support vector machines are a class of powerful tools which become increasingly popular in classifications, data mining, pattern recognition, artificial intelligence, and optimization.

17.3.1 *Classifications*

In many applications, the aim is to separate some complex data into different categories. For example, in pattern recognition, we may need to simply separate circles from squares. That is to label them into two different classes. In other applications, we have to answer a yes-no question, which is a binary classification.

If there are k different classes, we can in principle first classify them into two classes: class, say 1, and non-class 1. We then focus on the non-class 1 and divide them into two different classes, and so on and so forth.

Mathematically speaking, for a given set but scattered data, the objective is to separate them into different regions/domains or types. In the simplest case, the outputs are just class either A or B; In other words, that is, either +1 or −1.

17.3.2 *Statistical Learning Theory*

For the case of two-class classifications, we have the learning examples or data as $(\boldsymbol{x}_i, y_i)$ where $i = 1, 2, ..., n$ and $y_i \in \{-1, +1\}$. The aim of such learning is to find a function $f_\alpha(\boldsymbol{x})$ from allowable functions $\{f_\alpha : \alpha \in \Omega\}$ such that

$$f_\alpha(\boldsymbol{x}_i) \mapsto y_i, \qquad (i = 1, 2, ..., n), \tag{17.26}$$

and that the expected risk $E(\alpha)$ is minimal. That is the minimization of the risk

$$E(\alpha) = \frac{1}{2} \int |f_\alpha(x) - y| dP(\boldsymbol{x}, y), \tag{17.27}$$

where $P(\boldsymbol{x}, y)$ is an unknown probability distribution, which makes it impossible to calculate $E(\alpha)$ directly. A simple approach is to use the so-called empirical risk

$$E_p(\alpha) \approx \frac{1}{n} \sum_{i=1}^{n} \frac{1}{2} \big| f_\alpha(\boldsymbol{x}_i) - y_i \big|. \tag{17.28}$$

A main drawback of this approach is that a small risk or error on the training set does not necessarily guarantee a small error on prediction if the number n of training data is small.

In the framework of structural risk minimization and statistical learning theory, there exists an upper bound for such errors. For a given probability of at least $1 - p$, the Vapnik bound for the errors can be written as

$$E(\alpha) \le R_p(\alpha) + \phi\Big(\frac{h}{n}, \frac{\log(p)}{n}\Big), \tag{17.29}$$

where

$$\phi\Big(\frac{h}{n}, \frac{\log(p)}{n}\Big) = \sqrt{\frac{1}{n}\Big[h(\log\frac{2n}{h} + 1) - \log(\frac{p}{4})\Big]}. \tag{17.30}$$

Here h is a parameter, often referred to as the Vapnik-Chervonenskis dimension (or simply VC-dimension). This dimension describes the capacity for prediction of the function set f_α. In the simplest binary classification with only two values of +1 and −1, h is essentially the maximum number of points which can be classified into two distinct classes in all possible 2^h combinations.

17.3.3 *Linear Support Vector Machine*

The basic idea of classification is to try to separate different samples into different classes. For binary classifications such as the triangles and spheres (or solid dots) as shown in Fig. 17.4, we intend to construct a hyperplane

$$\boldsymbol{w} \cdot \boldsymbol{x} + \boldsymbol{b} = 0, \tag{17.31}$$

so that these samples can be divided into classes with triangles on one side and the spheres on the other side. Here the normal vector $\boldsymbol{w}$ and $\boldsymbol{b}$ have the same size as $\boldsymbol{x}$, and they can be determined using the data, though the method of determining them is not straightforward. This requires the existence of a hyperplane; otherwise, this approach will not work. In this case, we have to use other methods.

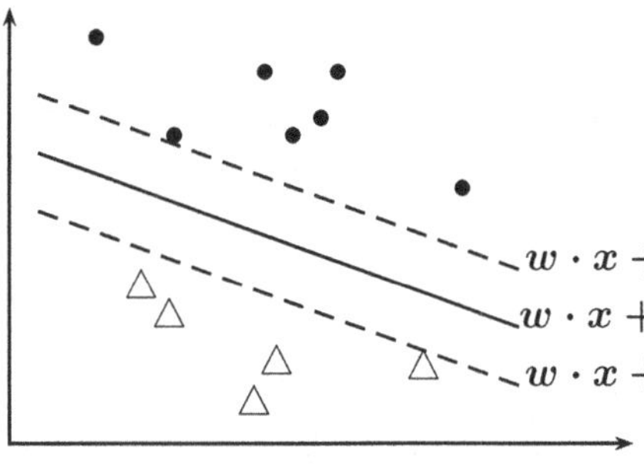

Fig. 17.4 Hyperplane, maximum margins and a linear support vector machine (SVM).

In essence, if we can construct such a hyperplane, we should construct two hyperplanes (shown as dashed lines) so that the two hyperplanes should be as far away as possible and no samples should be between these two planes. Mathematically, this is equivalent to two equations

$$\boldsymbol{w} \cdot \boldsymbol{x} + \boldsymbol{b} = +1, \tag{17.32}$$

and

$$\boldsymbol{w} \cdot \boldsymbol{x} + \boldsymbol{b} = -1. \tag{17.33}$$

From these two equations, it is straightforward to verify that the normal (perpendicular) distance between these two hyperplanes is related to the norm $||\boldsymbol{w}||$ via

$$d = \frac{2}{||\boldsymbol{w}||}. \tag{17.34}$$

A main objective of constructing these two hyperplanes is to maximize the distance or the margin between the two planes. The maximization of d is equivalent to the minimization of $||w||$ or more conveniently $||w||^2/2$. Here, $||w||$ is the standard 2-norm defined in earlier chapters. From the optimization point of view, the maximization of margins can be written as

$$\text{minimize } \frac{1}{2}||\boldsymbol{w}||^2 = \frac{1}{2}(\boldsymbol{w} \cdot \boldsymbol{w}). \tag{17.35}$$

If we can classify all the samples completely, for any sample $(\boldsymbol{x}_i, y_i)$ where $i = 1, 2, ..., n$, we have

$$\boldsymbol{w} \cdot \boldsymbol{x}_i + \boldsymbol{b} \geq +1, \qquad \text{if } (\boldsymbol{x}_i, y_i) \in \text{one class}, \tag{17.36}$$

and

$$\boldsymbol{w} \cdot \boldsymbol{x}_i + \boldsymbol{b} \leq -1, \qquad \text{if } (\boldsymbol{x}_i, y_i) \in \text{the other class}. \tag{17.37}$$

As $y_i \in \{+1, -1\}$, the above two equations can be combined as

$$y_i(\boldsymbol{w} \cdot \boldsymbol{x}_i + \boldsymbol{b}) \geq 1, \qquad (i = 1, 2, ..., n). \tag{17.38}$$

However, in reality, it is not always possible to construct such a separating hyperplane. A very useful approach is to use non-negative slack variables

$$\eta_i \geq 0, \qquad (i = 1, 2, ..., n), \tag{17.39}$$

so that

$$y_i(\boldsymbol{w} \cdot \boldsymbol{x}_i + \boldsymbol{b}) \geq 1 - \eta_i, \qquad (i = 1, 2, ..., n). \tag{17.40}$$

Now the optimization problem for the support vector machine becomes

$$\text{minimize } \Psi = \frac{1}{2}||\boldsymbol{w}||^2 + \lambda \sum_{i=1}^{n} \eta_i, \tag{17.41}$$

subject to

$$y_i(\boldsymbol{w} \cdot \boldsymbol{x}_i + \boldsymbol{b}) \geq 1 - \eta_i, \tag{17.42}$$

$$\eta_i \geq 0, \qquad (i = 1, 2, ..., n), \tag{17.43}$$

where $\lambda > 0$ is a parameter to be chosen appropriately. Here, the term $\sum_{i=1}^{n} \eta_i$ is essentially a measure of the upper bound of the number of misclassifications on the training data.

By using Lagrange multipliers $\alpha_i \geq 0$, we can rewrite the above constrained optimization into an unconstrained version, and we have

$$L = \frac{1}{2}||\boldsymbol{w}||^2 + \lambda \sum_{i=1}^{n} \eta_i - \sum_{i=1}^{n} \alpha_i [y_i(\boldsymbol{w} \cdot \boldsymbol{x}_i + \boldsymbol{b}) - (1 - \eta_i)]. \quad (17.44)$$

From this, we can write the Karush-Kuhn-Tucker conditions as

$$\frac{\partial L}{\partial \boldsymbol{w}} = \boldsymbol{w} - \sum_{i=1}^{n} \alpha_i y_i \boldsymbol{x}_i = 0, \quad (17.45)$$

$$\frac{\partial L}{\partial \boldsymbol{b}} = -\sum_{i=1}^{n} \alpha_i y_i = 0, \qquad y_i(\boldsymbol{w} \cdot \boldsymbol{x}_i + \boldsymbol{b}) - (1 - \eta_i) \geq 0, \quad (17.46)$$

$$\alpha_i [y_i(\boldsymbol{w} \cdot \boldsymbol{x}_i + \boldsymbol{b}) - (1 - \eta_i)] = 0, \qquad (i = 1, 2, ..., n), \quad (17.47)$$

$$\alpha_i \geq 0, \qquad \eta_i \geq 0, \qquad (i = 1, 2, ..., n). \quad (17.48)$$

From the first KKT condition, we get

$$\boldsymbol{w} = \sum_{i=1}^{n} y_i \alpha_i \boldsymbol{x}_i. \quad (17.49)$$

It is worth pointing out here that only the nonzero coefficients α_i contribute to the overall solution. This comes from the KKT condition (17.47), which implies that when $\alpha_i \neq 0$, the inequality (17.42) must be satisfied exactly, while $\alpha_0 = 0$ means the inequality is automatically met. In this latter case, $\eta_i = 0$. Therefore, only the corresponding training data $(\boldsymbol{x}_i, y_i)$ with $\alpha_i > 0$ can contribute to the solution, and thus such $\boldsymbol{x}_i$ form the support vectors (hence, the name support vector machine). All the other data with $\alpha_i = 0$ become irrelevant. It can be shown that the solution for α_i can be found by solving the following quadratic programming

$$\text{maximize} \sum_{i=1}^{n} \alpha_i - \frac{1}{2} \sum_{i,j=1}^{n} \alpha_i \alpha_j y_i y_j (\boldsymbol{x}_i \cdot \boldsymbol{x}_j), \quad (17.50)$$

subject to

$$\sum_{i=1}^{n} \alpha_i y_i = 0, \qquad 0 \leq \alpha_i \leq \lambda, \qquad (i = 1, 2, ..., n). \quad (17.51)$$

From the coefficients α_i, we can write the final classification or decision function as

$$f(\boldsymbol{x}) = \text{sgn}\Big[\sum_{i=1}^{n} \alpha_i y_i (\boldsymbol{x} \cdot \boldsymbol{x}_i) + \boldsymbol{b}\Big] \quad (17.52)$$

where sgn is the classic sign function.

17.3.4 *Kernel Functions and Nonlinear SVM*

In reality, most problems are nonlinear, and the above linear SVM cannot be used. Ideally, we should find some nonlinear transformation ϕ so that the data can be mapped onto a high-dimensional space where the classification becomes linear (see Fig. 17.5). The transformation should be chosen in a certain way so that their dot product leads to a kernel-style function $K(\boldsymbol{x}, \boldsymbol{x}_i) = \phi(\boldsymbol{x}) \cdot \phi(\boldsymbol{x}_i)$, which enables us to write our decision function as

$$f(\boldsymbol{x}) = \mathrm{sgn}\Big[\sum_{i=1}^{n} \alpha_i y_i K(\boldsymbol{x}, \boldsymbol{x}_i) + \boldsymbol{b}\Big]. \tag{17.53}$$

From the theory of eigenfunctions, we know that it is possible to expand functions in terms of eigenfunctions. In fact, we do not need to know such transformations, we can directly use kernel functions $K(\boldsymbol{x}, \boldsymbol{x}_i)$ to complete this task. This is the so-called kernel function trick. Now the main task is to chose a suitable kernel function for a given problem.

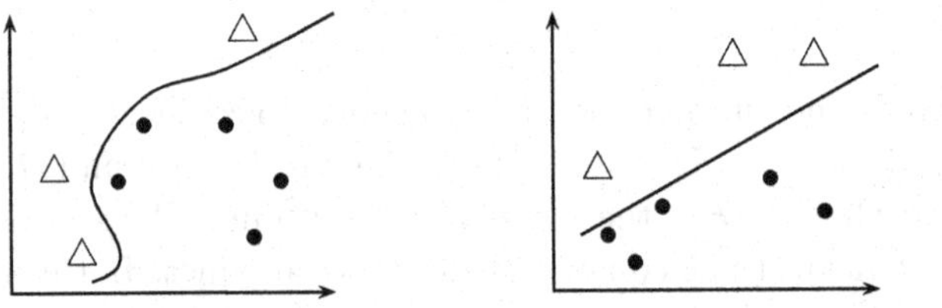

Fig. 17.5 Kernel functions by nonlinear transformation.

For most problems concerning a nonlinear support vector machine, we can use $K(\boldsymbol{x}, \boldsymbol{x}_i) = (\boldsymbol{x} \cdot x_i)^d$ for polynomial classifiers, $K(\boldsymbol{x}, \boldsymbol{x}_i) = \tanh[k(\boldsymbol{x} \cdot \boldsymbol{x}_i) + \Theta)]$ for neural networks. The most widely used kernel is the Gaussian radial basis function (RBF)

$$K(\boldsymbol{x}, \boldsymbol{x}_i) = \exp\Big[- ||\boldsymbol{x} - \boldsymbol{x}_i||^2/(2\sigma^2)\Big] = \exp\Big[- \gamma ||\boldsymbol{x} - \boldsymbol{x}_i||^2\Big], \tag{17.54}$$

for nonlinear classifiers. This kernel can easily be extended to any high dimensions. Here σ^2 is the variance and $\gamma = 1/2\sigma^2$ is a constant.

Following a similar procedure as discussed earlier for linear SVM, we can obtain the coefficients α_i by solving the following optimization problem

$$\text{maximize} \sum_{i=1}^{n} \alpha_i - \frac{1}{2} \alpha_i \alpha_j y_i y_j K(\boldsymbol{x}_i, \boldsymbol{x}_j). \tag{17.55}$$

It is worth pointing out under Mercer's conditions for kernel functions, the matrix $\boldsymbol{A} = y_i y_j K(\boldsymbol{x}_i, \boldsymbol{x}_j)$ is a symmetric positive definite matrix, which implies that the above maximization problem is a quadratic programming problem, and can thus be solved efficiently by standard quadratic programming techniques. There are many software packages (commercial or open source) which are easily available, so we will not provide any discussion of the implementation. In addition, some methods and their variants are still an area of active research. Interested readers can refer to more advanced literature.

Chapter 18

Random Number Generators and Monte Carlo Method

Random numbers commonly used in computer programs are not really random as they are generated via algorithms which are deterministic in some way. Truly random numbers should be generated using physical methods. Strictly speaking, all random numbers generated by algorithmic generators are pseudo-random numbers. Therefore, such generators are called pseudo-random number generators (PRNG) in the literature. However, modern pseudo-random generators can generate numbers which are almost indistinguishable from true random numbers.

18.1 Linear Congruential Algorithms

There are several classes of random number generators, including linear congruential generators, Fibonacci generators, inverse transform generators and twisters.

A linear congruential generator, first formulated by D. H. Lehmer, uses an iterative procedure

$$d_i = (ad_{i-1} + c) \bmod m, \tag{18.1}$$

where a, c and m are all integers; a and m are relatively prime. Here the 'mod' means to take the reminder after division. The starting or initial value d_0 is often called the seed of the generator. In most algorithms, the modulus m takes the form

$$m = 2^k,\ 2^k \pm 1. \tag{18.2}$$

This generator has a possible maximum period m which is attainable only if $c > 0$ and m are relatively prime, and $(a - 1)$ is a multiple of 4. The classic example is the IBM generator with $a = 65539 = 2^{16} + 3$, $c = 0$ and

$m = 2^{31}$ or

$$d_i = (2^{16} + 3)d_{i-1} \bmod m. \tag{18.3}$$

However, this generator has many problems because

$$d_i = (2^{16} + 3)d_{i-1} \bmod m = (2^{16} + 3) \cdot [(2^{16} + 3)^2 d_{i-2}] \bmod m$$

$$= (2^{16} + 3)^2 d_{i-2} \bmod m = (2^{32} + 6 \cdot 2^{16} + 9)d_{i-2} \bmod m$$

$$= 6 \cdot (2^{16} + 3)d_{i-2} - 9d_{i-2} = 6d_{i-1} - 9d_{i-2}, \tag{18.4}$$

where we have used $(2^{32} \bmod m) = 0$ due to $m = 2^{31}$. This is a recursive relationship, and there is a strong correlation among any consecutive three numbers d_{i-2}, d_{i-1} and d_i. Consequently, the numbers generated are not so 'random' from the statistical point of view.

Nowadays, modern computers widely use $a = 1,103,515,245$, $c = 12345$ and $m = 2^{32}$.

If the multiplier a and modulus m are carefully chosen, a seemingly simple generator could produce pseudo-random numbers of high quality. Park and Miller's generator use $c = 0$ so that

$$d_i = a d_{i-1} \bmod m, \tag{18.5}$$

where $a = 16807 = 7^5$ and $m = 2^{31} - 1 = 2,147,483,647$.

In the above generators, we have used multiplications and divisions extensively. If we can use additions or subtractions in the formulation, we may increase the efficiency of a generator. In fact, the Lagged Fibonacci pseudo-random generator uses the following formula

$$d_i = (d_{i-k} + d_{i-n}) \bmod m, \tag{18.6}$$

where k and n are integers. In the popular Mitchell-Moore generator, $k = 24$, $n = 55$ and $m = 2^{32}$.

18.2 Uniform Distribution

All the generators we have discussed so far in this chapter generate uniformly-distributed integers from 0 to $m - 1$. In many applications, we are concerned with the uniform distribution in a range, say, $[0, 1]$. In order to obtain scaled floating-point numbers, we have to divide by the period m, so we have d_i/m in the range of $[0, 1]$.

In linear congruential generators and lagged Fibonacci generators, we have to use divisions via mod and multiplications which are not computationally efficient. A better approach is to use additions and subtractions only, and this leads to Marsaglia's generator.

Linear congruential generators produce pseudo-random integers. On the other hand, Marsaglia generators, first introduced by G. Marsaglia, can produce floating-point numbers. For example, in the popular Matlab program, the following 'substract-with-borrow' generator is used. We have

$$d_i = d_{i+20} - d_{i+5} - k, \tag{18.7}$$

where the indices $(i, i+5, i+20)$ only take values mod 32, which means that only their last five bits are used since $2^5 = 32$. Here k is the residual from the previous step so that k is non-negative. That is to say, $k = 0$ if d_i is positive; however, $k = 2^{-53}$ (limited by the machine precision) if d_i is negative and $d_i = d_i + 1$ so that a modified $d_i > 0$ is stored. The period of this generator is about 2^{1492} in theory. However, because of the $k = \epsilon = 2^{-53}$ or $k = 0$, then all the numbers generated are the multiple of $\epsilon = 2^{-53}$ which is a little defect, though it is not noticeable for most applications. Consequently, the generated floating-points are in the range of $[\epsilon, 1-\epsilon]$ or $[2^{-53}, 1-2^{-53}]$. Surprisingly, 0 and 1 cannot be generated by this generator.

This so-called machine epsilon (ϵ) is determined by the way that the floating-point numbers are represented. Most modern computers use double precision real numbers that are represented by 64-bit strings or words in the standard IEEE arithmetic format specified by IEEE standards in 1977 or later 1985. In general, a 64-bit word consists of 53 bits for a signed fraction in base 2, and 11 bits for a signed exponent. The sign is represented by a single bit. Therefore, the accuracy of the computer is measured by the small quantity

$$\epsilon = 2^{-53} \approx 1.11 \times 10^{-16}.$$

Since the exponent of a real number has 11 bits with 1 bit for sign, so the real numbers in this IEEE format are in the range of $2^{\pm 2^{10}} \approx 10^{\pm 308}$. For almost all computations, these representations are sufficient.

Modern algorithms can be exceptionally elaborate and more generators are being invented, with the intention to produce distributions of better quality with higher efficiency. For example, the recent Mersenne twister algorithm could have a period of $2^{19937} - 1$, which is sufficient by any statistical standards.

18.3 Generation of Other Distributions

We have seen that most generators just produce pseudo-random numbers for the uniform distribution, often in the range $[0,1]$; however, we have to deal with other distributions such as normal distributions in many applications.

The simplest method is probably to use the central limit theorem. Let u_i $(i = 1, 2, ..., n)$ be n independent uniformly-distributed numbers in $[0,1]$. Then, the central limit theorem implies that the random variable

$$v = \sum_{i=1}^{n} u_i - \frac{n}{2}, \tag{18.8}$$

obeys the standard normal distribution $N(0,1)$ with a zero mean and a unitary variance. The probability density function of the standard normal distribution is given by

$$p(x) = \frac{1}{\sqrt{2\pi}} e^{-x^2/2}. \tag{18.9}$$

However, this method is very inefficient and with poor tails. So it is rarely used, and more rigorous methods should be used.

The generation of other distributions can be obtained in two major ways: the inverse transform method and the acceptance-rejection method. The former is more computationally extensive but easier to implement, while the latter is not easy to implement though it is usually more efficient. As the speed of modern computers increases, it seems that it does not really matter much which method we choose in most cases.

The basic idea of an acceptance-rejection method is better illustrated in terms of a specific distribution. In order to generate a normal distribution with a bell-shaped probability density function (pdf), we consider the shape of the pdf curve

$$f(x) = \frac{1}{\sqrt{2\pi}} e^{-\frac{x^2}{2}}, \tag{18.10}$$

in a two-dimensional plane, often a truncated box. We first generate random points $P(x, y)$ which distribute uniformly in the plane. If the point P falls under this curve $f(x)$, it is accepted as a valid value; otherwise, it is rejected. The resulting distribution for the accepted points is the normal distribution. This is the basic idea, however, its implementation requires more elaborate algorithms such as the Ziggurat algorithm, which divides the area under the curve into n equal sections, and the generated points are sorted and scaled.

On the other hand, the inverse transform method or sampling intends to generate random numbers by transforming uniformly-distributed numbers. Such transforms use the cumulative distribution function (cdf). The basic procedure of this method of generating a non-uniform distribution $f(x)$ with a given cumulative distribution function $\Phi(v)$ is as follows: First, we can generate a uniformly-distributed number u, then we try to find the value v so that

$$\Phi(v) = u, \tag{18.11}$$

which is equivalent to finding the inverse

$$v = \Phi^{-1}(u). \tag{18.12}$$

The sample set of v obtained by this inversion obeys the original distribution $f(x)$. Let us look at an example.

Example 18.1: *We know that the probability density of the exponential distribution is*

$$f(x) = \lambda e^{-\lambda x}, \qquad x \geq 0,$$

and its cumulative distribution function is

$$\Phi(v) = \int_0^v f(x)dx = 1 - e^{-\lambda v}, \qquad v \geq 0.$$

For a generated uniform distribution $u(0,1)$ in $(0,1)$, excluding 0 and 1, we can produce the numbers of the exponential distribution by the inverse transform

$$\Phi(v) = u(0,1), \qquad or \qquad v = \Phi^{-1}(u) = -\frac{1}{\lambda}\ln(1-u).$$

In fact, $v = -\frac{1}{\lambda}\ln u$ also obeys the same exponential distribution as both u and $1-u$ are uniformly distributed any way.

For the standard normal distribution $N(0,1)$

$$p(x) = \frac{1}{\sqrt{2\pi}} e^{-x^2/2},$$

its cumulative distribution function is given by

$$\Phi(v) = \frac{1}{\sqrt{2\pi}} \int_{-\infty}^{v} e^{-x^2/2} dx = \frac{1}{2}[1 + \text{erf}(\frac{v}{\sqrt{2}})],$$

where the error function erf is $\text{erf}(x) = \frac{2}{\sqrt{\pi}}\int_0^x e^{-\zeta^2} d\zeta$. From a uniform distribution $u(0,1)$, we have $\Phi(v) = u$. Thus, the inverse becomes

$$v = \Phi^{-1}(u) = \sqrt{2}\,\text{erf}^{-1}(2u-1),$$

where erf^{-1} is the inverse error function which is defined by

$$\mathrm{erf}(y) = x, \qquad \text{or} \qquad y = \mathrm{erf}^{-1}(x). \tag{18.13}$$

As the error function and its inverse involve integration, it is not straightforward to estimate their values. Apart from numerical integration, the power series expansions often give good approximations by expanding $e^{-\zeta^2}$ as

$$e^{-\zeta^2} = \sum_{n=0}^{\infty} \frac{(-1)^n \zeta^{2n}}{n!} = 1 - \zeta^2 + \frac{\zeta^4}{2!} - \cdots. \tag{18.14}$$

Therefore, we have

$$\mathrm{erf}(x) = \frac{2}{\sqrt{\pi}} \sum_{n=0}^{\infty} \frac{(-1)^n x^{2n+1}}{n!(2n+1)} = \frac{2}{\sqrt{\pi}}[x - \frac{x^3}{3} + \frac{x^5}{10} - \frac{x^7}{42} + ...].$$

In a similar way, the inverse error function can be estimated by

$$\mathrm{erf}^{-1}(x) = \frac{\sqrt{\pi}}{2}\Big[x + \frac{\pi x^3}{12} + \frac{7\pi^2 x^5}{480} + \frac{127\pi^3 x^7}{40320} + ...\Big]. \tag{18.15}$$

Let us look at a specific example.

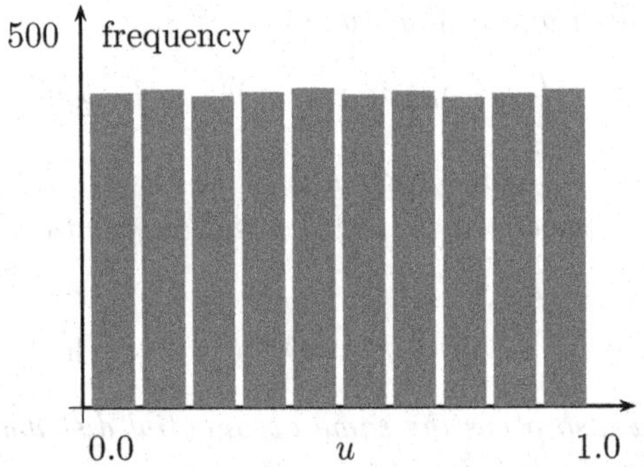

Fig. 18.1 Histogram of 5000 random numbers generated from a uniform distribution in the range $(0, 1)$.

Example 18.2: *Suppose we generate 5000 uniformly-distributed numbers:* $u = 0.9022, 0.4985, ..., 0.1019, ..., 0.4990.$

In order to get a normal distribution, we use

$$v = \sqrt{2}\mathrm{erf}^{-1}(2u - 1),$$

to transform the above numbers. For example, the first number 0.9022 *will be transformed into*

$$v = \sqrt{2}\mathrm{erf}^{-1}(2 \times 0.9022 - 1) = 1.2941.$$

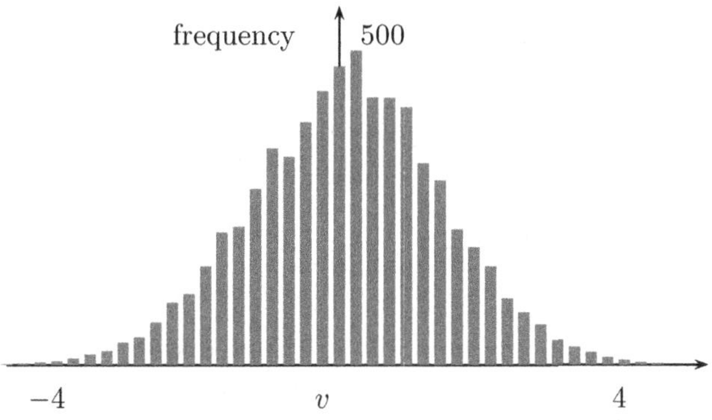

Fig. 18.2 Histogram of the normally-distributed numbers generated by the simple inverse transform method.

We finally have $v = 1.2941, \ -0.0037, \ ..., \ -1.2705, \ ..., \ 0.0025$. *The histograms of the two distributions are shown in Figs. 18.1 and 18.2.*

We have seen that the standard inverse transform method involves the calculations of erf^{-1} which are difficult to evaluate. The Box-Müller method provides a more efficient way to generate a normal distribution.

Let u_1, u_2 be two series of random numbers which are drawn from a uniform distribution in the standard range of $(0, 1)$. The Box-Müller method uses the following transform pair

$$v_1 = \sqrt{-2\ln u_1}\cos(2\pi u_2), \quad v_2 = \sqrt{-2\ln u_1}\sin(2\pi u_2). \tag{18.16}$$

Then, v_1 and v_2 obeys the standard normal distribution $N(0, 1)$

$$dv_1 dv_2 = \frac{1}{2\pi} e^{-(u_1^2+u_2^2)/2} du_1 du_2. \tag{18.17}$$

Here u_1 serves as the radius (with a uniform distribution) of a unit circle, while the angle $2\pi u_2$ distributes uniformly over the range $(0, 2\pi)$.

The disadvantage of this method is that it involves the trigonometric functions $\sin(u)$ and $\cos(u)$, which may slow down the evaluations. A further improvement is to use the polar form. From two uniformly-distributed numbers u_1 and u_2 in $(0, 1)$, we can obtain two uniformly-distributed numbers s_1 and s_2 in the range $(-1, 1)$. We have

$$s_1 = 2u_1 - 1, \qquad s_2 = 2u_2 - 1. \tag{18.18}$$

The condition $r^2 = s_1^2 + s_2^2 \leq 1$ corresponds to the case that the two numbers (s_1, s_2) fall within the unit circle in the region $(-1, 1) \times (-1, 1)$. Inside this unit circle, the angle $\theta = \tan^{-1} s_2/s_1 \in (0, 2\pi)$ and the quantity

$p = r^2 = s_1^2 + s_2^2 \in (0, 1)$ are independent, and they both are uniformly distributed. Straightforward trigonometric manipulations lead to

$$v_1 = \sqrt{-2 \ln p} \cos(\theta) = s_1 \sqrt{\frac{-2 \ln p}{p}}, \tag{18.19}$$

$$v_2 = \sqrt{-2 \ln p} \sin(\theta) = s_2 \sqrt{\frac{-2 \ln p}{p}}. \tag{18.20}$$

Again, both v_1 and v_2 are normally distributed with a zero mean and a unitary variance. Other more complicated distributions can be generated in a similar manner.

18.4 Metropolis Algorithms

Inverse transform methods can be used to generate random samples with various distributions. However, they are not efficient enough for many applications, especially for multivariate distributions in higher dimensions. In this case, random samples can be generated more efficiently using the Metropolis algorithms first developed in 1953, and the generated samples obey the prescribed probability density function $p(x, y, ..., z)$.

In Metropolis algorithms, the transition probability ϕ from state S_i to state S_j obeys the transition rule

$$p_i \phi(S_i \to S_j) = p_j \phi(S_j \to S_i), \tag{18.21}$$

where $p_i = p(S_i)$ is the probability density function of the random variable concerned. Using the notation $\phi_{i \to j}$ for the transition probability $\phi(S_i \to S_j)$, we have

$$\frac{\phi_{i \to j}}{\phi_{j \to i}} = \frac{p_j}{p_i}. \tag{18.22}$$

This provides an iterative procedure to generate sampling points or states from an initial ϕ_0, and the new sampling is accepted or not by an accepting rate. The new states are accepted if the quantity (or energy difference)

$$\Delta E = - \ln \frac{p_j}{p_i}, \tag{18.23}$$

is negative. Otherwise, the new states are only accepted with a probability of p_j/p_i. A major advantage of this method is that we never have to evaluate p_i itself, only the relative probability p_j/p_i matters, which makes many evaluations much simpler. Another advantage is that new sampling points are selected such that more important regions are sampled more intensively. That is equivalent to, say, measuring the depths of a lake. The sampling points are chosen more heavily inside the lake, and the region on the land is rarely measured or not measured at all.

For example, in many applications, we have to evaluate the average

$$<u> = \frac{\sum_i u_i e^{-\gamma E_i}}{\sum_i e^{-\gamma E_i}}, \tag{18.24}$$

where $\gamma = 1/\kappa T$ and κ is the Boltzmann constant and T is the temperature. The states are the energy level E_i and E_j. In this case, we have

$$\frac{p_j}{p_i} = e^{-\gamma(E_j - E_i)}. \tag{18.25}$$

The transition probability $\phi_{i \to j} = 1$ if $p_j > p_i$ or $E_j < E_i$. Otherwise, $\phi_{i \to j} = e^{-\gamma(E_j - E_i)}$ or a random number between 0 and 1. That is

$$\phi_{i \to j} = \begin{cases} 1 & \text{if } p_j > p_i \ \text{(or } E_j < E_i), \\ e^{-\gamma(E_j - E_i)} & \text{if } E_j \geq E_i. \end{cases} \tag{18.26}$$

In other words, all low-energy states are accepted, while high-energy states are accepted with a small probability. This idea forms the basis of the powerful method of simulated annealing.

18.5 Monte Carlo Methods

In many applications, the shear number of possible combinations and states is so astronomical that it makes it impossible to carry out evaluations over all possible combinations systematically. In this case, the Monte Carlo method is one of best alternatives. Monte Carlo is in fact a class of methods now widely used in computer simulations. Since the pioneer studies in the 1940s and 1950s, especially the work by Ulam, von Newmann, and Metropolis, it has now applied in almost all area of simulations, from Ising model to financial markets, from molecular dynamics to engineering, and from the routing of the Internet to climate simulations.

The basic procedure of Monte Carlo is to generate a (large but finite) number of random points so that they distribute uniformly inside the domain Ω of interest. The evaluations of functions (or any quantities of interest) are then carried out over these discrete locations. Then the system characteristics can be expected to be derived or represented by certain quantities averaged over these finite number evaluations.

A classic example is the geometrical probability problem which provides a method for estimating π using Buffon's needle-throwing technique first described by Compte de Buffon in 1777. The basic idea is to consider a simulation of tossing a needle with a length L (or many needles of the same kind) onto a floor (or a table) with equally-spaced parallel lines with a distance of w apart. You can either drop a needle many times, or drop many needles one by one. What is the probability of the needle(s) crossing these lines (see Figure 18.3)?

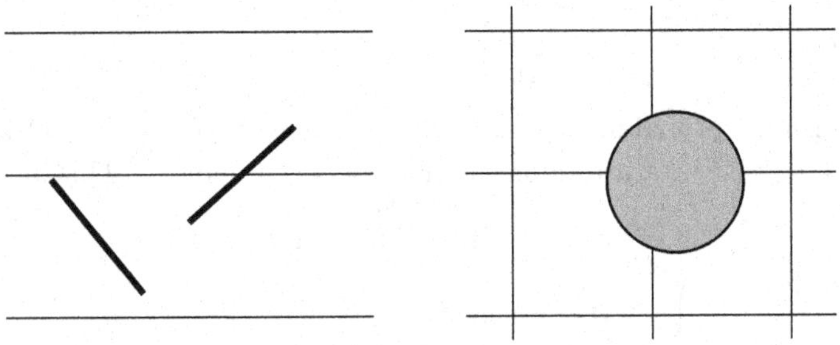

Fig. 18.3 Estimation of π by dropping needles or tossing coins.

Let u be the distance or location of the centre of the needle to the nearest line, and θ be the acute angle of the needle formed with the line. For simplicity, we can also assume $L \leq w$ so that a needle can only cross one line.

Since the random nature of this process, both the location u and the angle θ are uniformly distributed. The probability density function of u in the range of 0 and $w/2$ is simply $2du/w$, while the probability density function of θ in the range of 0 and $\pi/2$ is $2d\theta/\pi$. Therefore, the joint probability function is

$$p(u, \theta)dud\theta = \frac{4}{w\pi}dud\theta. \tag{18.27}$$

Since the condition for a needle crossing a line is $u \leq L/2\sin(\theta)$, we have

$$P = \frac{4}{\pi w}\int_0^{\pi/2}\int_0^{L\sin(\theta)/2} dud\theta = \frac{2L}{\pi w}, \tag{18.28}$$

which is the probability of a needle crossing a line. The interesting feature is that when $L = w$, we have

$$P = \frac{2}{\pi}, \tag{18.29}$$

which is independent of the length of the needle. Suppose we drop N needles with n needles crossing lines, we have

$$\lim_{N\to\infty}\frac{n}{N}=\frac{2}{\pi}, \qquad \text{or} \qquad \pi=\lim_{N\to\infty}\frac{2N}{n}. \tag{18.30}$$

For example, in a simulation, we dropped $N = 4000$, and found $n = 2547$ needles crossing lines. The estimate of π is

$$\pi=\frac{2\times 4000}{2547}\approx 3.1410. \tag{18.31}$$

However, the error in such kind of simulation decreases with $1/\sqrt{N}$, and is thus very sensitive to the change of even the last digit. Suppose, in a similar experiment, we get $n = 2546$ instead of 2547, we get

$$\pi=\frac{2\times 4000}{2546}\approx 3.1422. \tag{18.32}$$

For better accuracy, we have to use large N, otherwise, the results might be suspiciously lucky. For example, if we ran a hypothetical simulation with $N = 22$, we got (luckily) 14 needles crossing the lines, which leads to an estimate $\pi \approx 2\times 22/14 = 22/7 \approx 3.142857$. However, this estimate is biased because $22/7$ is a good approximation of π anyway. Suppose, you get 15 instead of 14, then the new estimate becomes $\pi = 2 \times 22/15 \approx 2.933$. An infamous example is the Lazzarini's estimate in 1901 with $N = 3408$ to get $\pi \approx 355/113$, which seems unlikely to be true.

Furthermore, π can also be estimated by tossing coins of a diameter D onto a chequerboard with the uniform grid spacing of w (see Fig. 18.3). For simplicity, we assume that $D \leq w$. For a coin to cross a corner, its centre has to be within a quarter of a circle from the corner (or the distance $s \leq D/2$), which means that, for a unit grid, it is the area of a whole circle. So the probability P of a coin crossing a corner is the area of a coin $(\pi D^2/4)$ dividing by the area of a unit grid (w^2). That is

$$P=\frac{\pi D^2}{4w^2}. \tag{18.33}$$

In the simple case of $w = D$, we have $P = \pi/4$.

Interestingly, the probability P_n of a coin not crossing any line is $P_n = (1 - D/w)^2$ because a coin has a probability of $(w - D)/w = (1 - D/w)$ of not crossing a line in one direction. The two events of not crossing any line in two directions are independent. Therefore, the probability P_a of a coin crossing any line (not necessarily a corner) is $P_a = 1 - P_n = 1 - (1 - D/w)^2$. In the case of $w = 2D$, we have $P_n = 1/4$ and $P_a = 3/4$.

There are other ways of estimating π using randomised evaluations. The random sampling using the Monte Carlo method is another classic example.

Example 18.3: *For a unit square with an inscribed circle, we know that the area of the circle is* $A_o = \pi/4$ *and the area of the unit square is* $A_s = 1$.

If we generate N *random points uniformly distributed inside the unit square, there are* n *points fall within the inscribed circle. Then, the probability of a point falling within the circle is*

$$\frac{A_o}{A_s} = \frac{\pi}{4} = \frac{n}{N},$$

when $N \to \infty$. *So* π *can be estimated by* $\pi \approx \frac{4n}{N}$. *Suppose in a simulation, we have* $n = 15707$ *points inside the circle among* $N = 20000$, *we obtain*

$$\pi \approx \frac{4n}{N} \approx \frac{4 \times 15707}{20000} \approx 3.1414.$$

We know this method of estimating π *is inefficient; however, it does demonstrate simply how the Monte Carlo method works.*

18.6 Monte Carlo Integration

Monte Carlo integration is a very efficient method for estimating multi-dimensional integrals. However, for 1D and 2D integrals, it is less efficient than Gaussian integration. Therefore, Monte Carlo integration is recommended mainly for complicated integrals in higher dimensions.

The fundamental idea of Monte Carlo integration is to randomly sample the domain of integration inside a control volume (often a regular region), and the integral of interest is estimated using the fraction of the random sampling points and the volume of the control region. Mathematical speaking, that is to say,

$$I = \int_\Omega f dV \approx V[\frac{1}{N} \sum_{i=1}^{N} f_i] + O(\epsilon), \tag{18.34}$$

where V is the volume of the domain Ω, and f_i is the evaluation of $f(x, y, z, ...)$ at the sampling point $(x_i, y_i, z_i, ...)$. The error estimate ϵ is given by

$$\epsilon = VS = V\sqrt{\frac{\mu_2 - \mu^2}{N}}, \tag{18.35}$$

where

$$\mu = \frac{1}{N}\sum_{i=1}^{N} f_i, \qquad \mu_2 = \frac{1}{N}\sum_{i=1}^{n} f_i^2. \tag{18.36}$$

Here the sample variance S^2 can be estimated by

$$S^2 = \frac{1}{N-1}\sum_{i=1}^{N}(f_i - \mu)^2 \approx \frac{1}{N}\sum_{i=1}^{N} f_i^2 - \mu^2, \tag{18.37}$$

which is the approximation of the variance σ_f^2

$$\sigma_f^2 = \frac{1}{V}\int (f - \mu)^2 dV. \tag{18.38}$$

The law of large number asymptotics implies that $\epsilon \to 0$ as $N \to \infty$. That is to say

$$V\mu = \lim_{N\to\infty} \frac{V}{N}\sum_{i=1}^{N} \to I. \tag{18.39}$$

In the simplest case in the domain $[0, 1]$, we have

$$\int_0^1 f(x)dx = \lim_{N\to\infty}[\frac{1}{N}\sum_{i=1}^{N} f(x_i)]. \tag{18.40}$$

We can see that the error of Monte Carlo integration decreases with N in a manner of $1/\sqrt{N}$, which is independent of the number (p) of dimensions. This becomes advantageous over other conventional methods for multiple integrals in higher dimensions.

The basic procedure is to generate the random points so that they distribute uniformly inside the domain. In order to calculate the volume in higher dimensions, it is better to use a regular control domain to enclose the domain Ω of the integration.

For simplicity of discussion, we now use the integral of a univariate function $f(x)$ over the interval $[a, b]$ (see Figure 18.4). Let us estimate the integral

$$I = \int_a^b f(x)dx = (b-a)\Big[\frac{1}{N}\sum_{i=1}^{N} f(x_i)\Big], \tag{18.41}$$

where a and b are finite integration limits. Now we first use a regular control volume or a bounding box so that it encloses the interval $[a, b]$ and the curve $f(x)$ itself. As the length of the domain is $(b - a)$ and the height of the

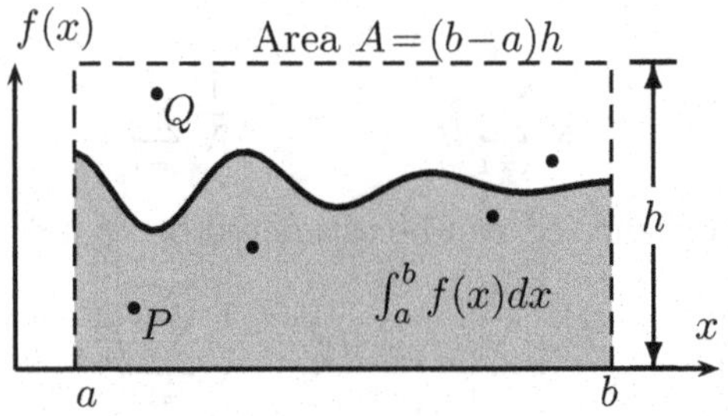

Fig. 18.4 Representation of Monte Carlo integration.

box is h, the area A (or more generally the volume in higher dimension) is simply

$$A = (b - a)h. \tag{18.42}$$

We know the integral I is the shaded area under the curve inside the bounding box, then the fraction or ratio I/A of the integral (area) to the total area of the bounding box is statistically equivalent to the fraction or probability of uniformly-distributed random sampling points falling in the shaded area inside the box. Suppose we generate N sampling points which are uniformly distributed inside the box. If there are K points that are under the curve inside the box (shaded region), then the integral I can be estimated by

$$I \approx A\frac{K}{N} = \frac{(b-a)hK}{N}. \tag{18.43}$$

For the estimation of a multi-dimensional integral in the domain Ω

$$I = \int_\Omega \beta f d\Omega = \int dx \int dy ... \int \beta f dz; \tag{18.44}$$

where $\beta \in \Re$ is a real constant, then the basic steps of Monte Carlo integration can be summarized as the pseudo code shown in Figure 18.5. For simplicity, we can use $\beta = 1$ for most integrals, though it may become a convenient scaling factor for some integrals. Now let us look at a simple example.

Example 18.4: *We know that the integral*

$$I = \text{erf}(2) = \frac{2}{\sqrt{\pi}}\int_0^2 e^{-x^2} dx \approx 0.995322.$$

Monte Carlo Integration

begin
  Define a bounding box/volume $\mathcal{V}$ *for* $I = \int_\Omega \beta f d\Omega$;
  Calculate the area/volume V *of* $\mathcal{V}$;
  Set the counter $S = 0$ *and max number* N *of sampling points;*
  for $i = 1$ *to* N;
    Generate a point P *randomly inside the volume* $\mathcal{V}$;
    if P *is inside the domain* $\Omega \in \mathcal{V}$;
      $S = S + \beta$;
    end if
  end for
  Estimate the integral $I = \frac{VS}{N}$;
end

Fig. 18.5 Pseudo code for Monte Carlo integration.

Suppose we try to use the Monte Carlo integration to estimate this, and we have $a = 0$, $b = 2$ *and* $\beta = 1$. *The function is simply*

$$f(x) = \frac{2}{\sqrt{\pi}} e^{-x^2}.$$

Since the maximum of $f_{\max} = 2/\sqrt{\pi} \approx 1.128$ *occurs at* $x = 0$, *we can use* $h = 1.5$. *Therefore, the area of the bounding box is* $A = (b - a)h = 3$. *For* $N = 5000$, *our simulation suggests that there are* $K = 1654$ *sampling points falling within the shaded region (under the curve inside the box in Figure 18.4). The estimate of the integral is given by*

$$I \approx A\frac{K}{N} \approx 3 \times \frac{1654}{5000} \approx 0.9924,$$

which is within 0.3% *of the true value* 0.995322.

Well, this is a relatively simple example, and it is not efficient to use Monte Carlo integration. The control region here is regular (a rectangular box). In many applications, the domain of the interest is irregular and it is impossible to calculate the integral analytically. In this case, Monte Carlo integration becomes a powerful alternative. Therefore, Monte Carlo integration is recommended for estimating the integrals with irregular domain and for evaluating multi-dimensional integrals.

18.7 Importance of Sampling

In the simple Monte Carlo integration, we have used the uniform sampling. For an integrand which varies rapidly in a narrow region such as a sharp peak (e.g., $f(x) = e^{-(100x)^2}$), the only sampling points that are important are near the peaks, the sampling points far outside will contribute less. Thus, it seems that a lot of unnecessary sampling points are wasted. There are two main ways to use the sampling points more effectively, and they are change of variables and importance sampling.

The change of variables uses the integrand itself so that it can be transformed to a more uniform (flat) function. For example, the integral

$$I = \int_a^b f(u)du, \tag{18.45}$$

can be transformed using a known function $u = g(v)$

$$I = \int_{a_v}^{b_v} f[g(v)]\frac{dg}{dv}dv. \tag{18.46}$$

The idea is to make sure that the new integrand is or close to a constant A

$$\phi(v) = f[g(v)]\frac{dg(v)}{dv} = A, \tag{18.47}$$

where $v = g^{-1}(u)$. This means that a uniform sampling can be used for ϕ. The new integration limits are $a_v = g^{-1}(a)$ and $b_v = g^{-1}(b)$.

For example, for an exponential function $f(u) = e^{-\alpha u}$, we can use $u = g(u) = -\ln(\alpha v)/\alpha$ so that $f(u)dg/dv = -1$. We then have

$$I = \int_a^b e^{-\alpha u}du = \int_{e^{-\alpha a}/\alpha}^{e^{-\alpha b}/\alpha} (-1)dv. \tag{18.48}$$

Then, we can use a uniform sampling set to estimate the integral. A slightly different method is to use the stratified sampling. That is to decompose the domain into subregions. For example, we can use

$$I = \int_a^b f(x)dx = \int_a^c f(x)dx + \int_c^b f(x)dx, \tag{18.49}$$

where c is a known limit which divides the original region using the characteristics of the integrand. In each of the subregions, the function $f(x)$ should be relatively smooth and flat.

Both these methods are limited either to the case when g^{-1} exists uniquely and has explicit expressions in terms of basic functions, or to the case when $f(x)$ are flat in subregions. Otherwise, it is not easy, or even impossible, to make such transformations or domain-decomposition. A far more efficient method is to use the importance sampling.

As the integration is the area (or volume) under a curve (or surface), the region(s) with higher values of integrand will contribute more, thus, we should put more weight on these important points. Importance sampling is just the method for doing such weighted sampling. The integral of interest is often rewritten as the weighted form or a product such that

$$I = \int_a^b f(x)dx = \int_a^b h(x)p(x)dx = \int_a^b \frac{f(x)}{p(x)} p(x)dx, \tag{18.50}$$

where $h(x) = f(x)/p(x)$. Obviously, it is required that $p(x) \neq 0$ in $[a, b]$. Here the function $p(x)$ acts as the probability density function whose integration over the region $[a, b]$ should be always equal to 1. That is

$$\int_a^b p(x)dx = 1. \tag{18.51}$$

The evaluation of the integral becomes the estimation of the expected value of $E = < h(x) > = < f(x)/p(x) >$. The idea is to choose a function $p(x)$ such that the sampling points become more important when $f(x)$ is in the region with higher values. That is equivalent to a weighted sum with $p(x)$ as the weighting coefficients.

The choice of $p(x)$ should make $h(x) = f(x)/p(x)$ as close to constant as possible. In a special case $p(x) = 1/(b - a)$, equation (18.50) becomes (18.41).

In addition, the error in Monte Carlo integration decreases in the form of $1/\sqrt{N}$ as N increases. As the true randomness of sampling points is not essential as long as the sampling points can be distributed as uniformly as possible. In fact, studies show that it is possible to sample the points in a certain deterministic way so as to minimize the error of the Monte Carlo integration. In this case, the error may decrease in terms of $(\ln N)^p/N$ where p is the dimension if appropriate methods such as Halton sequences are used. Readers interested in such topics can refer to more advanced literature.

18.8 Quasi-Monte Carlo Methods

Standard Monte Carlo methods extensively use pseudo-random numbers and have typically the order of convergence of $O(1/\sqrt{N})$ where N is the sample size as discussed earlier. Such a rate of convergence is limited by the so-called law of the iterated logarithms. However, there are ways to improve the rate of convergence by carefully distributing the sampling points over the domain of interest. Quasi-Monte Carlo methods are a class of such methods.

In essence, quasi-Monte Carlo methods operate in the same way as standard Monte Carlo methods, but use low-discrepancy sequences in terms of quasi-random or sub-random numbers instead of pseudo-random numbers. The sampling points using quasi-random numbers look more 'regular' without clustering together comparing with their pseudonumber counterparts. The idea of low-discrepancy intends to generate random points as away as possible from other numbers so as to avoid clustering.

Numerical analysis suggests that Quasi-Monte Carlo methods have a typical rate of convergence $O(1/N)$. More specifically, for d-dimensional integration, the rate of convergence is about $O((\log N)^d/N$. If regular grids or lattice (rather than quasi-random numbers) are used, the rate of convergence is estimated to be $O(N^{-2/d})$, which means that regular grids have higher convergence rates for higher dimenions $d > 4$. Therefore, for higher dimensions, the best methods are quasi-Monte Carlo as the errors decrease is approximately $O(1/N)$, which is bounded by the Koksma-Hlawka inequality. Now the question is how to generate quasi-random numbers?

18.9 Quasi-Random Numbers

Quasi-random numbers are designed to have a high level of uniformity in multi-dimensional space; however, they are not statistically independent, which may counter-intuitively lead to some advantages over standard Monte Carlo methods. There are many ways of generating quasi-random numbers using deterministic sequences, including radical inverse methods. Probably the most widely used is the van der Corput sequence, developed by the Dutch mathematician J. G. van der Corput in 1935. In this sequence, an arbitrary (decimal) integer n is expressed by a unique expansion in terms of a prime base b in the form

$$n = \sum_{j=0}^{m} a_j(n) b^j, \tag{18.52}$$

where the coefficients $a_j(n)$ can only take values $\{0, 1, 2, ..., b-1\}$. Here m is the smallest integer which leads to $a_j(n) = 0$ for all $j > m$. Then the expression in base b is reversed or reflected,

$$\phi_b(n) = \sum_{j=0}^{m} a_j(n) \frac{1}{b^{j+1}}, \tag{18.53}$$

and the reflected decimal number is a low-discrepancy number whose distribution is uniform in the interval $[0, 1)$.

Example 18.5: *For example, the integer* $n = 6$ *in base* 2 *can be expressed as* 110 *because* $6 = 2 \times 2^2 + 1 \times 2^1 + 0 \times 2^0$, *and* $m = 2$. *The expression* 110 *in base* 2 *is reversed or reflected as* 011. *When expressed in terms of a decimal number, it becomes the van der Corput number* $\phi_2(6) = 0 \times 2^{-1} + 1 \times 2^{-2} + 1 \times 2^{-3} = 3/8$. *This number is in the unit interval* $[0, 1]$. *For example, for the integers* $0, 1, 2, ..., 15$, *we have* $0, \frac{1}{2}, \frac{1}{4}, \frac{3}{4}, \frac{1}{8}, \frac{5}{8}, \frac{3}{8}, ..., \frac{15}{16}$. *If we plot these points in the unit interval, we can see that these points seems to 'fill the gaps'.*

Similarly, if we use base 3, *the integer* 5 *can be expressed as* 12 *since* $5 = 1 \times 3^1 + 2 \times 3^0$. *Now the reflection of the coefficients becomes* 21. *The number generated by van der Corput sequence becomes* $\phi_3(5) = 2 \times 3^{-1} + 1 \times 3^{-2} = \frac{7}{9}$.

Low-discrepancy sequences have been extended to higher dimensions, including Halton sequence (1960), Sobol sequence (1967), Faure sequence (1982), and Niederreiter sequence (1987). For example, Sobol's quasi-random sequence developed in 1967 is among the most widely used in quasi-Monte Carlo simulations.

Part VI

Computational Intelligence

Chapter 19

Evolutionary Computation

The optimization methods we discussed in Part IV are deterministic in the sense that solutions can uniquely be determined by an iterative procedure starting from an initial guess. The only randomness is the starting point which is guessed randomly. For almost all analytical methods and search algorithms, we seem to try to avoid randomness, and the only exception is in probability and statistics. On the other hand, modern trends in solving tough optimization problems tend to use evolutionary algorithms and nature-inspired metaheuristic algorithms, especially those based on swarm intelligence intelligence.

19.1 Introduction to Evolutionary Computation

In reality, randomness is everywhere and there is no strong reason for not using randomness in developing algorithms. In fact, many modern search algorithms use randomness very effectively. For example, heuristic methods use a trial-and-error approach to find the solutions to many difficult problems. Modern stochastic search methods are often evolutionary algorithms, or more recently called 'meta-heuristic'. Here *meta* means 'beyond' or 'higher level', while *heuristic* means 'to find' or 'to discover by trial and error'.

All these algorithms form part of the evolutionary computation. Loosely speaking, evolutionary computation is part of computational intelligence in the wide context of computer science and artificial intelligence. However, the definitions and boundary between different disciplines can be rather arbitrary.

Two major characteristics of modern metaheuristic methods are nature-inspired, and a balance between randomness and regularity. Almost all modern heuristic methods such as genetic algorithms (GA), particle swarm optimization (PSO), cuckoo search (CS) and firefly algorithm (FA) are nature-inspired as they have been developed based on the study of natural phenomena, learning from the beauty and effectiveness of nature.

In addition, a balanced use of randomness with a proper combination with certain deterministic components is in fact the essence of making such algorithms so powerful and effective. If the randomness in an algorithm is too high, then the solutions generated by the algorithm do not converge easily as they could continue to 'jump around' in the search space. If there is no randomness at all, then they can suffer the same disadvantages as those of deterministic methods (such as the gradient-based search). Therefore, a certain tradeoff is needed. Many NP-hard problems such as the travelling salesman problem can be solved using metaheuristic methods. In this chapter and later chapters, we will introduce some of these metaheuristic methods.

19.2 Simulated Annealing

Simulated annealing (SA) is a random search technique for global optimization problems, and it mimics the annealing process in materials processing when a metal cools and freezes into a crystalline state with the minimum energy and larger crystal size so as to reduce the defects in metallic structures. The annealing process involves the careful control of temperature and cooling rate (often called annealing schedule).

The application of simulated annealing into optimization problems was pioneered by Kirkpatrick, Gelatt and Vecchi in 1983. Since then, there have been extensive studies. Unlike the gradient-based methods and other deterministic search methods which have the disadvantage of becoming trapped in local minima, the main advantage of simulated annealing is its ability to avoid being trapped in local minima. In fact, it has been proved that simulated annealing will converge to its global optimality if enough randomness is used in combination with very slow cooling.

Metaphorically speaking, the iterations in SA are equivalent to dropping some bouncing balls over a landscape. As the balls bounce and lose energy, they will settle down to some local minima. If the balls are allowed to bounce enough times and lose energy slowly enough, some of the balls will eventually fall into the lowest global locations, hence the global minimum will be reached.

The basic idea of the simulated annealing algorithm is to use random search which not only accepts changes that improve the objective function, but also keeps some changes that are not ideal. In a minimization problem, for example, any better moves or changes that decrease the cost (or the value) of the objective function f will be accepted, however, some changes that increase f will also be accepted with a probability p. This probability p, also called the transition probability, is determined by

$$p = \exp[-\frac{\delta E}{k_B T}], \tag{19.1}$$

where k_B is the Boltzmann's constant, and T is the temperature for controlling the annealing process. δE is the change of the energy level. This transition probability is based on the Boltzmann distribution in physics. The simplest way to link δE with the change of the objective function δf is to use

$$\delta E = \gamma \delta f, \tag{19.2}$$

where γ is a real constant. For simplicity without losing generality, we can use $k_B = 1$ and $\gamma = 1$. Thus, the probability p simply becomes

$$p(\delta f, T) = e^{-\frac{\delta f}{T}}. \tag{19.3}$$

Whether or not to accept a change, we usually use a random number r (drawn from a uniform distribution in [0,1]) as a threshold. Thus, if $p > r$ or

$$p = e^{-\frac{\delta f}{T}} > r, \tag{19.4}$$

it is accepted.

Here the choice of the right temperature is crucially important. For a given change δf, if T is too high ($T \to \infty$), then $p \to 1$, which means almost all changes will be accepted. If T is too low ($T \to 0$), then any $\delta f > 0$ (worse solution) will rarely be accepted as $p \to 0$ and thus the diversity of the solution is limited, but any improvement δf will almost always be accepted. In fact, the special case $T \to 0$ corresponds to the gradient-based method because only better solutions are accepted, and the system is essentially climbing up or descending a hill. Therefore, if T is too high, the system is at a high energy state on the topological landscape, and the minima are not easily reached. If T is too low, the system may be trapped in a local minimum (not necessarily the global minimum), and there is not enough energy for the system to jump out of the local minimum to explore other potential global minima. So a proper, initial temperature should be calculated.

Another important issue is how to control the cooling process so that the system cools down gradually from a higher temperature to ultimately freeze to a global minimum state. There are many ways to control the cooling rate or the decrease in temperature.

Two commonly used cooling schedules are: linear and geometric cooling. For a linear cooling process, we have

$$T = T_0 - \beta t,$$

(or $T \to T - \delta T$ with a temperature increment δT). Here, T_0 is the initial temperature, and t is the pseudo time for iterations. β is the cooling rate, and it should be chosen in such a way that $T \to 0$ when $t \to t_f$ (maximum number of iterations), which usually gives $\beta = T_0/t_f$.

The geometric cooling essentially decreases the temperature by a cooling factor $0 < \alpha < 1$ so that T is replaced by αT or

$$T(t) = T_0 \alpha^t, \qquad t = 1, 2, ..., t_f. \tag{19.5}$$

The advantage of the second method is that $T \to 0$ when $t \to \infty$, and thus there is no need to specify the maximum number of iterations t_f. For this reason, we will use this geometric cooling schedule. The cooling process should be slow enough to allow the system to stabilize easily. In practise, $\alpha = 0.7 \sim 0.99$ is commonly used.

In addition, for a given temperature, multiple evaluations of the objective function are needed. If there are too few evaluations, there is a danger that the system will not stabilize and subsequently will not converge to its global optimality. If there are too many evaluations, it is time-consuming, and the system will usually converge too slowly as the number of iterations to achieve stability may be exponential to the problem size.

Therefore, there is a balance between the number of evaluations and solution quality. We can either do many evaluations at a few temperature levels or do few evaluations at many temperature levels. There are two major ways to set the number of iterations: fixed or varied. The first uses a fixed number of iterations at each temperature, while the second is designed to increase the number of iterations at lower temperatures so that the local minima can be fully explored.

The basic procedure of the simulated annealing algorithm can be summarized as the pseudocode shown in Fig. 19.1.

In order to find a suitable starting temperature T_0, we can use any information about the objective function. If we know the maximum change $\max(\delta f)$ of the objective function, we can use it to estimate an initial

Simulated Annealing Algorithm

begin
    Objective function $f(\boldsymbol{x})$, $\boldsymbol{x} = (x_1, ..., x_d)^T$
    Initialize initial temperature T_0 *and initial guess* $\boldsymbol{x}^{(0)}$
    Set final temperature T_f *and max number of iterations* N
    Define cooling schedule $T \mapsto \alpha T$, $(0 < \alpha < 1)$
        while ($T > T_f$ *and* $n < N$)
            Move randomly to new locations:
                $\boldsymbol{x}_{n+1} = \boldsymbol{x}_n + randn$
            Calculate $\delta f = f_{n+1}(\boldsymbol{x}_{n+1}) - f_n(\boldsymbol{x}_n)$
            Accept the new solution if better
            if *not improved*
            Generate a random number r
            Accept if $p = \exp[-\delta f/T] > r$
            end if
            Update the best $\boldsymbol{x}_*$ *and* f_*
        end while
end

Fig. 19.1 Simulated annealing algorithm.

temperature T_0 for a given probability p_0. That is

$$T_0 \approx -\frac{\max(\delta f)}{\ln p_0}. \tag{19.6}$$

If we do not know the possible maximum change of the objective function, we can use a heuristic approach. We can start evaluations from a very high temperature (so that almost all changes are accepted) and reduce the temperature quickly until about 50% or 60% of the worse moves are accepted, and then use this temperature as the new initial temperature T_0 for proper and relatively slow cooling.

For the final temperature, in theory it should be zero so that no worse move can be accepted. However, if $T_f \to 0$, more unnecessary evaluations are needed. In practise, we simply choose a very small value, say, $T_f = 10^{-10} \sim 10^{-5}$, depending on the required quality of the solutions and time constraints.

Example 19.1: *For Rosenbrock's banana function*

$$f(x, y) = (1 - x)^2 + 100(y - x^2)^2,$$

we know that its global minimum $f_* = 0$ *occurs at* $(1, 1)$. *This is a standard test function and quite tough for most algorithms.*

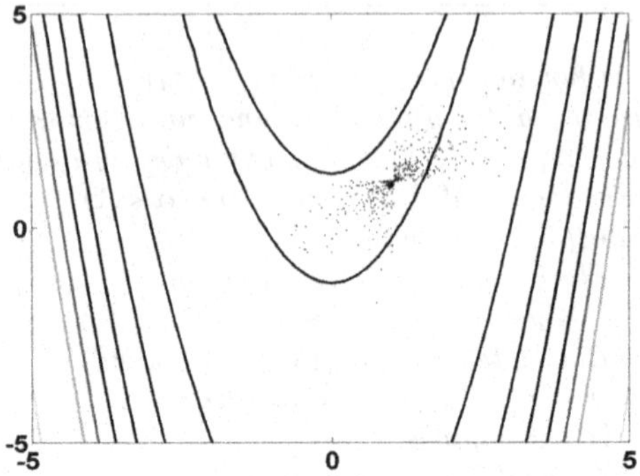

Fig. 19.2 Last 500 evaluations during a typical run in the simulated annealing (the global best is marked with •).

However, it is relatively easy to find this global minimum if we use the simulated annealing algorithm. The typical results, say the last 500 evaluations, during the simulated annealing are shown in Fig. 19.2.

The implementation of simulated annealing to optimize the banana function can be demonstrated using the following Matlab/Octave program.

Here we have used the initial temperature $T_0 = 1.0$, the final temperature $T_f = 10^{-10}$, and a geometric cooling schedule with $\alpha = 0.9$. A demo code in Matlab can be found at the Mathworks[1].

19.3 Genetic Algorithms

Strictly speaking, the genetic algorithm (GA) is not so modern, but it is an evolutionary algorithm and probably the most widely used. It is becoming a conventional and classic method. However, it does have fundamental genetic operators that inspired many later algorithms so we will introduce it in detail.

The genetic algorithm (GA), developed by John Holland and his collaborators in the 1960s and 1970s, is a model or abstraction of biological evolution based on Charles Darwin's theory of natural selection. Holland was the first to use crossover and recombination, mutation, and selection in the study of adaptive and artificial systems. These genetic operators form the essential part of the genetic algorithm as a problem-solving strategy.

[1]http://www.mathworks.co.uk/matlabcentral/fileexchange/29739-simulated-annealing-for-constrained-optimization

Since then, many variants of genetic algorithms have been developed and applied to a wide range of optimization problems, from graph colouring to pattern recognition, from discrete systems (such as the travelling salesman problem) to continuous systems (e.g., the efficient design of airfoils in aerospace engineering), and from the financial market to multiobjective engineering optimization.

There are many advantages of genetic algorithms over traditional optimization algorithms, the two most noticeable advantages are the ability to deal with complex problems, and parallelism. Genetic algorithms can deal with various types of optimization whether the objective (fitness) function is stationary or non-stationary (change with time), linear or nonlinear, continuous or discontinuous, or with random noise. As multiple offsprings in a population act like independent agents, the population (or any subgroup) can explore the search space in many directions simultaneously. This feature makes it ideal to parallelize the algorithms for implementation. Different parameters and even different groups of strings can be manipulated at the same time. Such advantages also map onto the algorithms based on swarm intelligence (SI) and thus SI-based algorithms such as particle swarm optimization and firefly algorithm to be introduced later also possess such good advantages.

However, genetic algorithms also have some disadvantages. The formulation of the fitness function, the population size, the choice of the important parameters such as the rate of mutation and crossover, and the selection criteria of new populations should be carried out carefully. Any inappropriate choice will make it difficult for the algorithm to converge, or it simply produces meaningless results.

19.3.1 *Basic Procedure*

The essence of genetic algorithms involves the encoding of an optimization function as arrays of bits or character strings to represent the chromosomes, the manipulation operations of strings by genetic operators, and the selection according to their fitness in the aim of finding a solution to the problem concerned. This is often done by the following procedure: 1) encoding of the objectives or optimization functions; 2) defining a fitness function or selection criterion; 3) creating a population of individuals; 4) evolution cycle or iterations by evaluating the fitness of all the individuals in the population, creating a new population by performing crossover, and mutation, fitness-proportionate reproduction, etc., and replacing the old population and iterating again using the new population; 5) decoding the results to obtain the solution to the problem.

Genetic Algorithm

begin
  Objective function $f(\boldsymbol{x}),\ \ \boldsymbol{x} = (x_1, ..., x_d)^T$
  Encode the solution into chromosomes (binary strings)
  Define fitness F *(eg,* $F \propto f(\boldsymbol{x})$ *for maximization)*
  Generate the initial population
  Initial probabilities of crossover (p_c) *and mutation* (p_m)
  while ($t <$ *Max number of generations*)
    Generate new solution by crossover and mutation
    if $p_c >$*rand, Crossover;* **end if**
    if $p_m >$*rand, Mutate;* **end if**
    Accept the new solutions if their fitness increase
    Select the current best for new generation (elitism)
  end while
  Decode the results and visualization
end

Fig. 19.3 Pseudocode of genetic algorithms.

These steps can be represented schematically as the pseudocode of genetic algorithms shown in Fig. 19.3. One iteration of creating a new population is called a generation. The fixed-length character strings are used in most genetic algorithms during each generation although there is substantial research on the variable-length strings and coding structures. The coding of the objective function is usually in the form of binary arrays or real-valued arrays in the adaptive genetic algorithms. For simplicity, we use binary strings for encoding and decoding. The genetic operators include crossover, mutation, and selection from the population.

The crossover of two parent strings is the main operator with a higher probability p_c and is carried out by swapping one segment of one chromosome with the corresponding segment on another chromosome at a random position (see Fig. 19.4). The crossover carried out in this way is a single-point crossover. Crossover at multiple points is also used in many genetic algorithms to increase the efficiency of the algorithms.

The mutation operation is achieved by flopping the randomly selected bits (see Fig. 19.5), and the mutation probability p_m is usually small. The selection of an individual in a population is carried out by the evaluation of its fitness, and it can remain in the new generation if a certain threshold

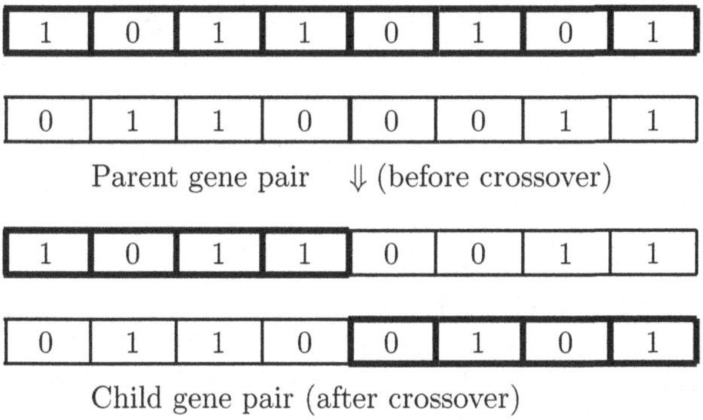

Fig. 19.4 Diagram of crossover at a random crossover point in genetic algorithms.

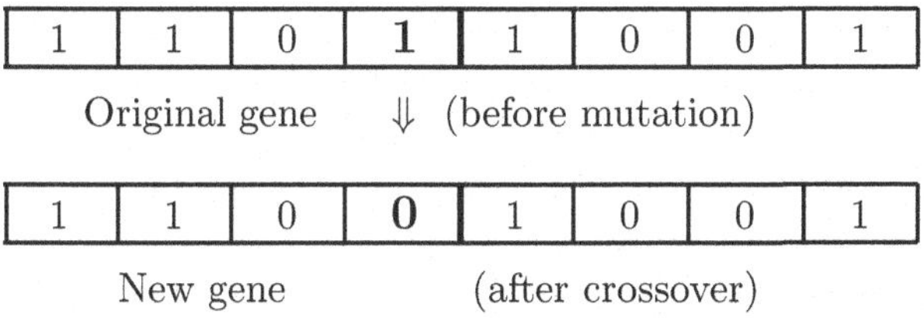

Fig. 19.5 Schematic representation of mutation at a single site by flipping a randomly selected bit ($1 \to 0$).

of the fitness is reached or the reproduction of a population is fitness-proportionate. That is to say, the individuals with higher fitness are more likely to reproduce.

19.3.2 *Choice of Parameters*

An important issue is the formulation or choice of an appropriate fitness function that determines the selection criterion in a particular problem. For the minimization of a function using genetic algorithms, one simple way of constructing a fitness function is to use the simplest form $F = A - y$ with A being a large constant (though $A = 0$ will do) and $y = f(\mathbf{x})$, thus the objective is to maximize the fitness function and subsequently minimize the objective function $f(\mathbf{x})$. However, there are many different ways of defining a fitness function. For example, we can use the individual fitness assignment relative to the whole population

$$F(x_i) = \frac{f(\xi_i))}{\sum_{i=1}^{N} f(\xi_i)}, \tag{19.7}$$

where ξ_i is the phenotypic value of individual i, and N is the population size. The appropriate form of the fitness function will ensure that the solutions with higher fitness should be selected efficiently. Poorly defined fitness functions may result in incorrect or meaningless solutions.

Another important issue is the choice of various parameters. The crossover probability p_c is usually very high, typically in the range of $0.7 \sim 1.0$. On the other hand, the mutation probability p_m is usually small (usually $0.001 \sim 0.05$). If p_c is too small, then the crossover occurs sparsely, which is not efficient for evolution. If the mutation probability is too high, the solutions could still 'jump around' even if the optimal solution is approaching.

The selection criterion is also important; how to select the current population so that the best individuals with higher fitness are preserved and passed on to the next generation. That is often carried out in association with a certain elitism. The basic elitism is to select the most fit individual (in each generation) which will be carried over to the new generation without being modified by genetic operators. This ensures that the best solution is achieved more quickly.

Other issues include multiple sites for mutation and the population size. Mutation at a single site is not very efficient; mutation at multiple sites will increase the evolution efficiency. However, too many mutants will make it difficult for the system to converge, or even make the system go astray to the wrong solutions. In reality, if the mutation is too high under high selection pressure, then the whole population might go extinct.

In addition, the choice of the right population size is also very important. If the population size is too small, there is not enough evolution going on, and there is a risk that the whole population may go extinct. In the real world, for a species with a small population, ecological theory suggests that there is a real danger of extinction. Even though the system carries on, there is still a danger of premature convergence. In a small population, if a significantly more fit individual appears too early, it may reproduce enough offspring to overwhelm the whole (small) population. This will eventually drive the system to a local optimum (not the global optimum). On the other hand, if the population is too large, more evaluations of the objective function are needed, which will require an extensive computing time.

Furthermore, more complex and adaptive genetic algorithms are under active research and the literature about these topics is vast.

Using the basic procedure described here, we can implement the genetic algorithms in any programming language. In fact, there is no need to do any programming (if you prefer) because there are many software packages (either freeware or commercial) about genetic algorithms. For example, Matlab itself has an optimization toolbox.

19.4 Differential Evolution

Differential evolution (DE) was developed by R. Storn and K. Price in 1996 and 1997. It is a vector-based algorithm, which has some similarity to pattern search and genetic algorithms due to its use of crossover and mutation. In fact, DE can be considered as a further development to genetic algorithms with explicit updating equations, which makes it possible to do some theoretical analysis. DE is a stochastic search algorithm with self-organizing tendency and does not use the information of derivatives. Thus, it is a population-based, derivative-free method. In addition, DE uses real-number as solution strings, thus no encoding and decoding is needed.

For a d-dimensional optimization problem with d parameters, a population of n solution vectors are initially generated, we have $\boldsymbol{x}_i$ where $i = 1, 2, ..., n$. For each solution $\boldsymbol{x}_i$ at any generation t, we use the conventional notation as

$$\boldsymbol{x}_i^t = (x_{1,i}^t, x_{2,i}^t, ..., x_{d,i}^t), \tag{19.8}$$

which consists of d-components in the d-dimensional space. This vector can be considered as the chromosomes or genomes.

Differential evolution consists of three main steps: mutation, crossover and selection.

Mutation is carried out by the mutation scheme. For each vector $\boldsymbol{x}_i$ at any time or generation t, we first randomly choose three distinct vectors $\boldsymbol{x}_p$, $\boldsymbol{x}_q$ and $\boldsymbol{x}_r$ at t (see Fig. 19.6), and then generate a so-called donor vector by the mutation scheme

$$\boldsymbol{v}_i^{t+1} = \boldsymbol{x}_p^t + F(\boldsymbol{x}_q^t - \boldsymbol{x}_r^t), \tag{19.9}$$

where $F \in [0, 2]$ is a parameter, often referred to as the differential weight. This requires that the minimum number of population size is $n \geq 4$. In principle, $F \in [0, 2]$, but in practice, a scheme with $F \in [0, 1]$ is more efficient and stable. In fact, almost all the studies in the literature use $F \in (0, 1)$.

From Fig. 19.6, we can see that the perturbation $\boldsymbol{\delta} = F(\boldsymbol{x}_q - \boldsymbol{x}_r)$ to the vector $\boldsymbol{x}_p$ is used to generate a donor vector $\boldsymbol{v}_i$, and such perturbation is directed.

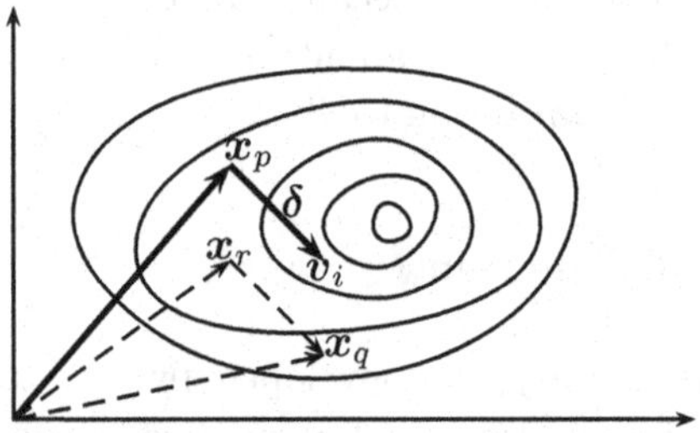

Fig. 19.6 Schematic representation of mutation vectors in differential evolution with movement $\boldsymbol{\delta} = F(\boldsymbol{x}_q - \boldsymbol{x}_r)$.

Crossover is controlled by a crossover parameter $C_r \in [0, 1]$, controlling the rate or probability for crossover. The actual crossover can be carried out in two ways: binomial and exponential. The binomial scheme performs crossover on each of the d components or variables/parameters. By generating a uniformly distributed random number $r_i \in [0, 1]$, the j-th component of $\boldsymbol{v}_i$ is manipulated as

$$\boldsymbol{u}_{j,i}^{t+1} = \begin{cases} \boldsymbol{v}_{j,i} & \text{if } r_i \leq C_r, \\ \\ \boldsymbol{x}_{j,i}^t & \text{otherwise,} \end{cases} \qquad j = 1, 2, ..., d. \tag{19.10}$$

This way, each component can be decided randomly whether or not to exchange with the counterpart of the donor vector.

In the exponential scheme, a segment of the donor vector is selected and this segment starts with a random integer k with a random length L which can include many components. Mathematically, this is to choose $k \in [0, d-1]$ and $L \in [1, d]$ randomly, and we have

$$\boldsymbol{u}_{j,i}^{t+1} = \begin{cases} \boldsymbol{v}_{j,i}^t & \text{for } j = k, ..., k - L + 1 \in [1, d], \\ \\ \boldsymbol{x}_{j,i}^t & \text{otherwise.} \end{cases} \tag{19.11}$$

As the binomial is simpler to implement, we will use the binomial crossover in our implementation.

Selection is essentially the same as that used in genetic algorithms. It is to select the most fittest; that is, the minimum objective value for a minimization problem. Therefore, we have

$$\boldsymbol{x}_i^{t+1} = \begin{cases} \boldsymbol{u}_i^{t+1} & \text{if } f(\boldsymbol{u}_i^{t+1}) \leq f(\boldsymbol{x}_i^t), \\ \\ \boldsymbol{x}_i^t & \text{otherwise.} \end{cases} \tag{19.12}$$

Differential Evolution

---

begin
Initialize the population $\boldsymbol{x}$ *with randomly generated solutions*
Set the weight $F \in [0, 2]$ *and crossover probability* $C_r \in [0, 1]$
while *(stopping criterion)*
  for $i = 1$ *to* n,
  For each $\boldsymbol{x}_i$, *randomly choose 3 distinct vectors* $\boldsymbol{x}_p$, $\boldsymbol{x}_r$ *and* $\boldsymbol{x}_r$
  Generate a new vector $\boldsymbol{v}$ *by DE scheme (19.9)*
  Generate a random index $J_r \in \{1, 2, ..., d\}$ *by permutation*
  Generate a randomly distributed number $r_i \in [0, 1]$
  for $j = 1$ *to* d,
  For each parameter $\boldsymbol{v}_{j,i}$ *(j-th component of* $\boldsymbol{v}_i$*), update*

$$\boldsymbol{u}_{j,i}^{t+1} = \begin{cases} \boldsymbol{v}_{j,i}^{t+1} & \text{if } r_i \le C_r \text{ or } j = J_r \\ \boldsymbol{x}_{j,i}^{t} & \text{if } r_i > C_r \text{ and } j \neq J_r \end{cases}$$

  end
  Select and update the solution by (19.12)
  end
end
Post-process and output the best solution found
end

Fig. 19.7 Pseudocode of differential evolution.

All the above three components can be seen in the pseudo code as shown in Fig. 19.7. It is worth pointing out here that the use of J is to ensure that $\boldsymbol{v}_i^{t+1} \neq \boldsymbol{x}_i^t$, which may increase the evolutionary or exploratory efficiency. The overall search efficiency is controlled by two parameters: the differential weight F and the crossover probability C_r.

Most studies have focused on the choice of F, C_r and n as well as the modifications of (19.9). In fact, when generating mutation vectors, we can use many different ways of formulating (19.9), and this leads to various schemes with the naming convention: DE/x/y/z where x is the mutation scheme (rand or best), y is the number of difference vectors, and z is the crossover scheme (binomial or exponential). So DE/Rand/1/* means the basic DE scheme using random mutation, one difference vector with either a binomial or exponential crossover scheme.

The basic DE/Rand/1/Bin scheme is given in (19.9). That is

$$\boldsymbol{v}_i^{t+1} = \boldsymbol{x}_p^t + F(\boldsymbol{x}_q^t - \boldsymbol{x}_r^t). \tag{19.13}$$

If we replace the $\boldsymbol{x}_p^t$ by the current best $\boldsymbol{x}_{\text{best}}$ found so far, we have the so-called DE/Best/1/Bin scheme

$$v_i^{t+1} = x_{\text{best}}^t + F(x_q^t - x_r^t). \tag{19.14}$$

There is no reason that why we should not use more than 3 distinct vectors. For example. if we use 4 different vectors plus the current best, we have the DE/Best/2/Bin scheme

$$v_i^{t+1} = x_{\text{best}}^t + F(x_{k_1}^t + x_{k_2}^t - x_{k_3}^t - x_{k_4}^t). \tag{19.15}$$

Furthermore, if we use five different vectors, we have the DE/Rand/2/Bin scheme

$$v_i^{t+1} = x_{k_1}^t + F_1(x_{k_2}^t - x_{k_3}^t) + F_2(x_{k_4}^t - x_{k_5}^t), \tag{19.16}$$

where F_1 and F_2 are differential weights in $[0, 1]$. Obviously, for simplicity, we can also take $F_1 = F_2 = F$. Following the similar strategy, we can design various schemes. For example, the above variants can be written in a generalized form

$$v_i^{t+1} = x_{k_1}^t + \sum_{s=1}^{m} F_s \cdot (x_{k_2(s)}^t - x_{k_3(s)}^t), \tag{19.17}$$

where $m = 1, 2, 3, ...$ and $F_s(s = 1, ..., m)$ are the scale factors. The number of vectors involved on the right-hand side is $2m + 1$. In the above variants, $m = 1$ and $m = 2$ are used.

On the other hand, there is another type of variants, which uses an additional influence parameter $\lambda \in (0, 1)$. For example, DE/rand-to-best/1/* variant can be written as

$$v_i^{t+1} = \lambda x_{\text{best}}^t + (1 - \lambda) x_{k_1}^t + F(x_{k_2}^t - x_{k_3}^t), \tag{19.18}$$

which introduces an extra parameter λ. Again, this type of variants can be written in a generalized form

$$v_i^{t+1} = \lambda x_{\text{best}}^t + (1 - \lambda) x_{k_1}^t + F \sum_{s=1}^{m} (x_{k_2(s)}^t - x_{k_3(s)}^t). \tag{19.19}$$

In fact, more than 10 different schemes have been formulated, and for details, readers can refer to the book by Price et al.

There are other good variants of DE, including self-adapting control parameter in differential evolution (jDE) by Brest et al, self-adaptive DE (SaDE) by Qin et al. and DE with the eagle strategy by Yang and Deb.

Chapter 20

Swarm Intelligence

Many algorithms such as ant colony algorithms and firefly algorithm use the behaviour of the so-called swarm intelligence (SI). All SI-based algorithms use the real-number randomness and the global communication among agents or particles. Therefore, they are also easier to implement as there is no encoding or decoding of the parameters into binary strings as those in genetic algorithms where real-number strings can also be used. In addition, SI-based algorithms are very flexible and yet efficient to deal with a wide range of problems.

20.1 Introduction to Swarm Intelligence

Many new algorithms that are based on swarm intelligence may have drawn inspiration from different sources, but they have some similarity to some of the components that are used in PSO and ant colony optimization (ACO). In this sense, PSO and ACO pioneered the basic ideas of swarm-intelligence-based computation.

From the algorithm point of view, the aim of a swarming system is to let the system evolve and converge into some stable optimality. In this case, it has strong similarity to a self-organizing system. Such an iterative, self-organizing system can evolve, according to a set of rules or mathematical equations. As a result, such a complex system can interact and self-organize into certain converged states, showing some emergent characteristics of self-organization. In this sense, the proper design of an efficient optimization algorithm is equivalent to finding efficient ways to mimic the evolution of a self-organizing system. In practice, all nature-inspired algorithms try to mimic some successful characteristics of biological, physical or chemical systems in nature.

Among all evolutionary algorithms, algorithms based on swarm intelligence dominate the landscape. There are many reasons for this dominance, though three obvious reasons are: 1) Swarm intelligence uses multiple agents as an evolving, interacting population, and thus provides good ways to mimic natural systems. 2) Population-based approaches allow parallelization and vectorization implementations in practice, and are thus straightforward to implement. 3) These SI-based algorithms are flexible and yet efficient enough to deal with a wide range of problems.

20.2 Ant and Bee Algorithms

Ant algorithms, especially the ant colony optimization developed by M. Dorigo, mimic the foraging behavior of social ants. Primarily, all ant algorithms use pheromone as a chemical messenger and the pheromone concentration as the indicator of quality solutions to a problem of interest. From implementation point of view, solutions are related to the pheromone concentration, leading to routes and paths marked by the higher pheromone concentrations as better solutions to problems such as discrete combinatorial problems.

Looking at ant colony optimization closely, random route generation is primarily mutation, while pheromone-based selection provides a mechanism for selecting shorter routes. There is no explicit crossover in ant algorithms. However, mutation is not a simple action as flipping digits as in genetic algorithms, the new solutions are essentially generated by fitness-proportional mutation. For example, the probability of ants in a network problem at a particular node i to choose the route from node i to node j is given by

$$p_{ij} = \frac{\phi_{ij}^{\alpha} d_{ij}^{\beta}}{\sum_{i,j=1}^{n} \phi_{ij}^{\alpha} d_{ij}^{\beta}}, \tag{20.1}$$

where $\alpha > 0$ and $\beta > 0$ are the influence parameters, and ϕ_{ij} is the pheromone concentration on the route between i and j, and d_{ij} the desirability of the same route. The selection is subtly related to some *a priori* knowledge about the route such as the distance s_{ij} is often used so that $d_{ij} \propto 1/s_{ij}$.

On the other hand, bee algorithms do not usually use pheromone. For example, in the artificial bee colony (ABC) optimization algorithm, the bees in a colony are divided into three groups: employed bees (forager bees), onlooker bees (observer bees), and scouts. Randomization are carried out by scout bees and employed bees, and both are mainly mutation. Selection is related honey or the objective. Again, there is no explicit crossover.

Both ACO and ABC only use mutation and fitness-related selection, they can have good global search ability. In general, they can explore the search space relatively effectively, but convergence may be slow because it lacks crossover, and thus the subspace exploitation ability is very limited. In fact, the lack of crossover is very common in many metaheuristic algorithms. In terms of exploration and exploitation, both ant and bee algorithms have strong exploration ability, but their exploitation ability is comparatively low. This may explain why they can perform reasonably well for some tough optimization, but the computational efforts such as the number of function evaluations can be very high.

As there is extensive literature about ant colony optimization and relevant variants, we will not discuss them in detail here. Interested readers can refer to more advanced literature.

20.3 Particle Swarm Optimization

Particle swarm optimization was developed by Kennedy and Eberhart in 1995 and has become one of the most widely used swarm-intelligence-based algorithms. The PSO algorithm searches the space of an objective function by adjusting the trajectories of individual agents, called particles, as the piecewise paths formed by positional vectors in a quasi-stochastic manner. The movement of a swarming particle consists of two major components: a stochastic component and a deterministic component. Each particle is attracted toward the position of the current global best $\boldsymbol{g}^*$ and its own best location $\boldsymbol{x}_i^*$ in history, while at the same time it has a tendency to move randomly.

When a particle finds a location that is better than any previously found locations, then it updates it as the new current best for particle i. There is a current best for all n particles at any time t during iterations. The aim is to find the global best among all the current best solutions until the objective no longer improves or after a certain number of iterations. The movement of particles is schematically represented in Fig. 20.1 where $\boldsymbol{x}_i^{*(t)}$ is the current best for particle i, and $\boldsymbol{g}^* \approx \min\{f(\boldsymbol{x}_i)\}$ for $(i = 1, 2, ..., n)$ is the current global best at t.

The essential steps of the particle swarm optimization can be summarized as the pseudocode shown in Fig. 20.2.

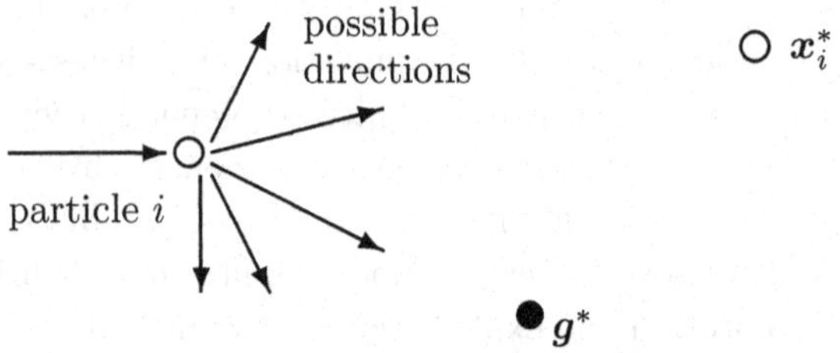

Fig. 20.1 Schematic representation of the motion of a particle in PSO, moving towards the global best g^* and the current best x_i^* for each particle i.

Particle Swarm Optimization

begin
Objective function $f(\boldsymbol{x})$, $\boldsymbol{x} = (x_1, ..., x_d)^T$
Initialize locations $\boldsymbol{x}_i$ *and velocity* $\boldsymbol{v}_i$ *of* n *particles.*
Find $\boldsymbol{g}^*$ *from* $\min\{f(\boldsymbol{x}_1), ..., f(\boldsymbol{x}_n)\}$ *(at* $t = 0$*)*
while *(criterion)*
  for *loop over all* n *particles and all* d *dimensions*
    Generate new velocity $\boldsymbol{v}_i^{t+1}$ *using equation (20.2)*
    Calculate new locations $\boldsymbol{x}_i^{t+1} = \boldsymbol{x}_i^t + \boldsymbol{v}_i^{t+1}$
    Evaluate objective functions at new locations $\boldsymbol{x}_i^{t+1}$
    Find the current best for each particle $\boldsymbol{x}_i^*$
  end for
  Find the current global best $\boldsymbol{g}^*$
  Update $t = t + 1$ *(pseudo time or iteration counter)*
end while
Output the final results $\boldsymbol{x}_i^*$ *and* $\boldsymbol{g}^*$
end

Fig. 20.2 Pseudocode of particle swarm optimization.

Let $\boldsymbol{x}_i$ and $\boldsymbol{v}_i$ be the position vector and velocity for particle i, respectively. The new velocity vector is determined by the following formula

$$\boldsymbol{v}_i^{t+1} = \boldsymbol{v}_i^t + \alpha \epsilon_1 [\boldsymbol{g}^* - \boldsymbol{x}_i^t] + \beta \epsilon_2 [\boldsymbol{x}_i^{*(t)} - \boldsymbol{x}_i^t], \tag{20.2}$$

where ϵ_1 and ϵ_2 are two random vectors, and each entry taking the values between 0 and 1. The parameters α and β are the learning parameters or acceleration constants, which can typically be taken as, say, $\alpha \approx \beta \approx 2$.

The initial locations of all particles should distribute relatively uniformly so that they can sample over most regions, which is especially important for multimodal problems. The initial velocity of a particle can be taken as zero, that is, $\boldsymbol{v}_i^{t=0} = 0$. The new position can then be updated by

$$\boldsymbol{x}_i^{t+1} = \boldsymbol{x}_i^t + \boldsymbol{v}_i^{t+1} \Delta t, \tag{20.3}$$

where Δt is the (pseudo)time increment. Since we are dealing with the discrete time increment or iteration in iterative algorithms, we can always set $\Delta t = 1$, and we have

$$\boldsymbol{x}_i^{t+1} = \boldsymbol{x}_i^t + \boldsymbol{v}_i^{t+1}. \tag{20.4}$$

It is worth pointing out that the above updating equations are based on the physical system of particles, but there is no need to consider the units of variables, from the mathematical or numerical point of view. This also applies to other algorithms that are based on swarm intelligence.

Although $\boldsymbol{v}_i$ can be any values, it is usually bounded in some range $[0, \boldsymbol{v}_{\max}]$.

There are many variants which extend the standard PSO algorithm, and the most noticeable improvement is probably to use inertia function $\theta(t)$ so that $\boldsymbol{v}_i^t$ is replaced by $\theta(t)\boldsymbol{v}_i^t$

$$\boldsymbol{v}_i^{t+1} = \theta \boldsymbol{v}_i^t + \alpha \epsilon_1 [\boldsymbol{g}^* - \boldsymbol{x}_i^t] + \beta \epsilon_2 [\boldsymbol{x}_i^{*(t)} - \boldsymbol{x}_i^t], \tag{20.5}$$

where θ takes the values between 0 and 1 in theory. In the simplest case, the inertia function can be taken as a constant, typically $\theta \approx 0.5 \sim 0.9$. This is equivalent to introducing a virtual mass to stabilize the motion of the particles, and thus the algorithm can usually be expected to converge more quickly.

20.4 Accelerated PSO

The standard particle swarm optimization uses both the current global best $\boldsymbol{g}^*$ and the individual best $\boldsymbol{x}_i^{*(t)}$. One of the reasons of using the individual best is probably to increase the diversity in the quality solutions; however, this diversity can be simulated using some randomness. Subsequently, there is no compelling reason for using the individual best, unless the optimization problem of interest is highly nonlinear and multimodal.

A simplified version which could accelerate the convergence of the algorithm is to use the global best only. The so-called accelerated particle swarm optimization (APSO) was developed by Xin-She Yang in 2008 and then has been developed further in recent years. Thus, in APSO, the velocity vector is generated by a simpler formula

$$\boldsymbol{v}_i^{t+1} = \boldsymbol{v}_i^t + \alpha(\epsilon - 1/2) + \beta(\boldsymbol{g}^* - \boldsymbol{x}_i^t), \tag{20.6}$$

where ϵ is a random variable with values from 0 to 1. Here the shift $1/2$ is purely out of convenience. We can also use a standard normal distribution $\alpha\epsilon_t$ where ϵ_t is drawn from $N(0,1)$ to replace the second term. Now we have

$$\boldsymbol{v}_i^{t+1} = \boldsymbol{v}_i^t + \beta(\boldsymbol{g}^* - \boldsymbol{x}_i^t) + \alpha\epsilon_t, \tag{20.7}$$

where ϵ_t can be drawn from a Gaussian distribution or any other suitable distributions. Here, α is a scaling factor that controls the step size or the strength of randomness, while β is a parameter controls the movement of particles.

The update of the position is simply

$$\boldsymbol{x}_i^{t+1} = \boldsymbol{x}_i^t + \boldsymbol{v}_i^{t+1}. \tag{20.8}$$

In order to simplify the formulation even further, we can also write the update of the location in a single step

$$\boldsymbol{x}_i^{t+1} = (1-\beta)\boldsymbol{x}_i^t + \beta\boldsymbol{g}^* + \alpha\epsilon_t. \tag{20.9}$$

The typical values for this accelerated PSO are $\alpha \approx 0.1 \sim 0.4$ and $\beta \approx 0.1 \sim 0.7$, though $\alpha \approx 0.2$ and $\beta \approx 0.5$ can be taken as the initial values for most unimodal objective functions. It is worth pointing out that the parameters α and β should in general be related to the scales of the independent variables $\boldsymbol{x}_i$ and the search domain. Surprisingly, this simplified APSO can have global convergence under appropriate conditions.

A further improvement to the accelerated PSO is to reduce the randomness as iterations proceed. This means that we can use a monotonically decreasing function such as

$$\alpha = \alpha_0 e^{-\gamma t}, \tag{20.10}$$

or

$$\alpha = \alpha_0 \gamma^t, \qquad (0 < \gamma < 1), \tag{20.11}$$

where $\alpha_0 \approx 0.5 \sim 1$ is the initial value of the randomness parameter. Here t is the number of iterations or time steps. $0 < \gamma < 1$ is a control parameter. For example, in most implementations, we can use $\gamma = 0.9$ to 0.99. Obviously, other non-increasing function forms $\alpha(t)$ can also be used. In addition, these parameters should be fine-tuned to suit your optimization problems of interest, and such parameter tuning is essential for all evolutionary algorithms.

A demo implementation can be found at the Mathworks website[1].

[1] http://www.mathworks.co.uk/matlabcentral/fileexchange/29725-accelerated-particle-swarm-optimization

20.5 Binary PSO

In the standard PSO, the positions and velocities take continuous values. However, many problems are combinatorial and their variables only take discrete values. In some cares, the variables can only be 0 and 1, and such binary problems require modifications of the PSO algorithm. Kennedy and Eberhart in 1997 presented a stochastic approach to discretize the standard PSO, and they interpreted the velocity in a stochastic sense.

First, the continuous velocity $\boldsymbol{v}_i = (v_{i1}, v_{i2}, ..., v_{ik}, ..., v_{id})$ is transformed using a sigmoid transformation

$$S(v_{ik}) = \frac{1}{1 + \exp(-v_{ik})}, \quad k = 1, 2, ..., d, \tag{20.12}$$

which applies to each component of the velocity vector $\boldsymbol{v}_i$ of particle i. Obvious, when $v_{ik} \to \infty$, we have $S(v_{ik}) \to 1$, while $S_{ik} \to 0$ when $v_{ik} \to -\infty$. However, as the variations at the two extremes are very slow, a stochastic approach is introduced.

Secondly, a uniformly-distributed random number $r \in (0, 1)$ is drawn, then the velocity is converted to a binary variable by the following stochastic rule:

$$x_{ik} = \begin{cases} 1 & \text{if } r < S(v_{ik}), \\ 0 & \text{otherwise.} \end{cases} \tag{20.13}$$

In this case, the value of each velocity component v_{ik} is interpreted as a probability for x_{ik} taking the value 1. Even for a fixed value of v_{ik}, the actual value of x_{ik} is not certain before a random number r is drawn. In this sense, the binary PSO (BPSO) differs significantly from the standard continuous PSO.

In fact, since each component of each variable takes only 0 and 1, this binary PSO can work for both discrete and continuous problems if the latter is coded in the binary system. As the probability of each bit/component taking one is $S(v_{ik})$, and the probability of taking zero is $1 - S(v_{ik})$, the joint probability p of a bit change can be computed by

$$p = S(v_{ik})[1 - S(v_{ik})]. \tag{20.14}$$

Based on the runtime analysis of the binary PSO by Sudholt and Witt, there are some interesting results on the convergence of BPSO. If the objective function has a unique global optimum in a d-dimensional space, and the BPSO has a population size n with $\alpha + \beta = O(1)$, then the expected number N of iterations/generations of BPSO is

$$N = O(\frac{d}{\log d}), \tag{20.15}$$

and the expected freezing time is $O(d)$ for single bits, and $O(d \log d)$ for nd bits.

One of the advantages of this binary coding and discritization is to enable binary representations of even continuous problems. For example, Kennedy and Eberhart provided an example of soving the 2nd De Jong function and found that 110111101110110111101001 in a 24-bit string corresponds to the optimal solution 3905.929932 from this representation. However, a disadvantage is that the Hamming distance from other local optima is large; therefore, it is unlikely that the search will jump from one local optimum to another. This means that binary PSO can get stuck in a local optimum with premature convergence. New remedies are still under active research.

Various studies show that PSO algorithms can outperform genetic algorithms and other conventional algorithms for solving many optimization problems. This is partially due to that fact that the broadcasting ability of the current best estimates gives a better and quicker convergence towards the optimality. However, PSO algorithms do have some disadvantages such as premature convergence. Further developments and improvements are still under active research.

It is worth pointing out that the discretization approaches outlined here to convert PSO into binary PSO can be used to convert all other algorithms into their binary counterpart. For example, we can use the above approaches to convert the cuckoo search algorithm into a binary cuckoo search algorithm.

Furthermore, as we can see in the next chapter in this book, other methods such as the cuckoo search (CS) and firefly algorithms (FA) can perform even better than PSO in many applications. In many cases, DE, PSO and SA can be considered as special cases of CS and FA. Therefore, CS and FA have become an even more active area for further research.

Chapter 21

Swarm Intelligence: New Algorithms

Swarm intelligence has attracted great attention in the last two decades and the literature has expanded significantly in the last few years. Many novel algorithms have appeared and new algorithms such as the firefly algorithm and cuckoo search have been found to be very efficient. In this chapter, we will briefly introduce some of the most recent algorithms.

21.1 Firefly Algorithm

Firefly Algorithm (FA) was developed by Xin-She Yang in 2008, which was based on the flashing patterns and behaviour of tropical fireflies. FA is simple, flexible and easy to implement.

The flashing light of fireflies is an amazing sight in the summer sky in the tropical and temperate regions. There are about 2000 firefly species, and most fireflies produce short and rhythmic flashes. The pattern of flashes is often unique for a particular species. The flashing light is produced by a process of bioluminescence, and the true functions of such signaling systems are still being debated. However, two fundamental functions of such flashes are to attract mating partners (communication), and to attract potential prey. In addition, flashing may also serve as a protective warning mechanism to remind potential predators of the bitter taste of fireflies.

The rhythmic flash, the rate of flashing and the amount of time form part of the signal system that brings both sexes together. Females respond to a male's unique pattern of flashing in the same species, while in some species such as *Photuris*, female fireflies can eavesdrop on the bioluminescent courtship signals and even mimic the mating flashing pattern of other species so as to lure and eat the male fireflies who may mistake the flashes as a potential suitable mate. Some tropical fireflies can even synchronize their flashes, thus forming emerging biological self-organized behavior.

We know that the light intensity at a particular distance r from the light source obeys the inverse square law. That is to say, the light intensity I decreases as the distance r increases in terms of $I \propto 1/r^2$. Furthermore, the air absorbs light which becomes weaker and weaker as the distance increases. These two combined factors make most fireflies visible to a limited distance, usually several hundred meters at night, which is good enough for fireflies to communicate. In addition, the attractiveness β of a firefly is in the eye of the other fireflies, which varies with their distance. To avoid singularity, we use the following form of attractiveness

$$\beta = \beta_0 \exp[-\gamma r^2], \tag{21.1}$$

where β_0 is the attractiveness at distance $r = 0$, and γ is the light absorption coefficient. Here, r can be defined as the Cartesian distance between the two fireflies of interest. However, for other optimization problems such as routing, the time delay along a route can also be used as the 'distance'. Therefore, r should be interpreted in the most wide and appropriate sense, depending on the type of problem.

The flashing light can be formulated in such a way that it is associated with the objective function to be optimized, which makes it possible to formulate new optimization algorithms.

Now we can idealize some of the flashing characteristics of fireflies so as to develop firefly-inspired algorithms. For simplicity in describing the standard Firefly Algorithm (FA), we now use the following three idealized rules:

- All fireflies are unisex so that one firefly will be attracted to other fireflies regardless of their sex;
- Attractiveness is proportional to their brightness, thus for any two flashing fireflies, the less brighter one will move towards the brighter one. The attractiveness is proportional to the brightness and they both decrease as their distance increases. If there is no brighter one than a particular firefly, it will move randomly;
- The brightness of a firefly is affected or determined by the landscape of the objective function.

For a maximization problem, the brightness can simply be proportional to the value of the objective function. Other forms of brightness can be defined in a similar way to the fitness function in genetic algorithms.

Based on these three rules, the basic steps of the firefly algorithm (FA) can be summarized as the pseudocode shown in Figure 21.1.

```
Firefly Algorithm
begin
Objective function f(x),     x = (x_1, ..., x_d)^T.
Generate an initial population of n fireflies x_i (i = 1, 2, ..., n).
Light intensity I_i at x_i is determined by f(x_i).
Define light absorption coefficient γ.
while (t <MaxGeneration),
for i = 1 : n (all n fireflies)
   for j = 1 : n (all n fireflies) (inner loop)
   if (I_i < I_j)
      Move firefly i towards j.
   end if
      Vary attractiveness with distance r via exp[-γr^2].
      Evaluate new solutions and update light intensity.
   end for j
end for i
Rank the fireflies and find the current global best g_*.
end while
Postprocess results and visualization.
end
```

Fig. 21.1 Pseudocode of the firefly algorithm (FA).

The movement of a firefly i is attracted to another more attractive (brighter) firefly j is determined by

$$\boldsymbol{x}_i^{t+1} = \boldsymbol{x}_i^t + \beta_0 e^{-\gamma r_{ij}^2} (\boldsymbol{x}_j^t - \boldsymbol{x}_i^t) + \alpha\, \boldsymbol{\epsilon}_i^t, \tag{21.2}$$

where the second term is due to the attraction, and β_0 is the attractiveness at zero distance $r = 0$. The third term is randomization with α being the randomization parameter, and $\boldsymbol{\epsilon}_i^t$ is a vector of random numbers drawn from a Gaussian distribution at time t. Other studies also use randomization in terms of $\boldsymbol{\epsilon}_i^t$ that can easily be extended to other distributions such as Lévy flights.

From the above equation, we can see that mutation is used for both local and global search. When $\boldsymbol{\epsilon}_i^t$ is drawn from a Gaussian distribution and Lévy flights, it produces mutation on a larger scale. On the other hand, if α is chosen to be a very small value, then mutation can be very small, and thus limited to a subspace. However, during the update in the two loops in FA, ranking as well as selection is used.

The author has provided a simple demonstration code of the firefly algorithm that can be found at the Mathworks website[1].

One novel feature of FA is that attraction is used, and this is the first of its kind in any SI-based algorithms. Since local attraction is stronger than long-distance attraction, the population in FA can automatically subdivide into multiple subgroups, and each group can potentially swarm around a local mode. Among all the local modes, there is always a global best solution which is the true optimality of the problem. FA can deal with multimodal problems naturally and efficiently.

From Eq. (21.2), we can see that FA degenerates into a variant of differential evolution when $\gamma = 0$ and $\alpha = 0$. In addition, when $\beta_0 = 0$, it degenerates into simulated annealing (SA). Further, when $\boldsymbol{x}_j^t$ is replaced by $\boldsymbol{g}^*$, FA also becomes the accelerated PSO. Therefore, DE, APSO and SA are special cases of the firefly algorithm, and thus FA can have the advantages of all these three algorithms. It is no surprise that FA can be versatile and efficient, and perform better than other algorithms such as GA and PSO.

21.2 Cuckoo Search

Cuckoo search (CS) is one of the latest nature-inspired metaheuristic algorithms, developed in 2009 by Xin-She Yang and Suash Deb. CS is based on the brood parasitism of some cuckoo species. In addition, this algorithm is enhanced by the so-called Lévy flights, rather than by simple isotropic random walks. Recent studies show that CS is potentially far more efficient than PSO and genetic algorithms.

Cuckoo are fascinating birds, not only because of the beautiful sounds they can make, but also because of their aggressive reproduction strategy. Some species such as the *ani* and *Guira* cuckoos lay their eggs in communal nests, though they may remove the others' eggs to increase the hatching probability of their own eggs. Quite a number of species engage the obligate brood parasitism by laying their eggs in the nests of other host birds (often other species).

There are three basic types of brood parasitism: intraspecific brood parasitism, cooperative breeding, and nest takeover. Some host birds can engage direct conflict with the intruding cuckoos. If a host bird discovers the eggs are not its own, it will either get rid of these alien eggs or sim-

[1] http://www.mathworks.co.uk/matlabcentral/fileexchange/29693-firefly-algorithm

ply abandon its nest and build a new nest elsewhere. Some cuckoo species such as the New World brood-parasitic *Tapera* have evolved in such a way that female parasitic cuckoos are often very specialized in the mimicry in colour and pattern of the eggs of a few chosen host species. This reduces the probability of their eggs being abandoned and thus increases their reproductivity.

In addition, the timing of egg-laying of some species is also amazing. Parasitic cuckoos often choose a nest where the host bird just laid its own eggs. In general, the cuckoo eggs hatch slightly earlier than their host eggs. Once the first cuckoo chick is hatched, the first instinct action it will take is to evict the host eggs by blindly propelling the eggs out of the nest, which increases the cuckoo chick's share of food provided by its host bird. Studies also show that a cuckoo chick can also mimic the call of host chicks to gain access to more feeding opportunities.

On the other hand, various studies have shown that flight behaviour of many animals and insects has demonstrated the typical characteristics of Lévy flights with power-law-like characteristics. A recent study by Reynolds and Frye shows that fruit flies or *Drosophila melanogaster*, explore their landscape using a series of straight flight paths punctuated by a sudden 90° turn, leading to a Lévy-flight-style intermittent scale free search pattern. Studies on human behaviour such as the Ju/'hoansi hunter-gatherer foraging patterns also show the typical feature of Lévy flights. Even light can be related to Lévy flights. Subsequently, such behaviour has been applied to optimization and optimal search, and results show its promising capability.

For simplicity in describing the standard Cuckoo Search, we now use the following three idealized rules:

- Each cuckoo lays one egg at a time, and dumps it in a randomly chosen nest;
- The best nests with high-quality eggs will be carried over to the next generations;
- The number of available host nests is fixed, and the egg laid by a cuckoo is discovered by the host bird with a probability $p_a \in (0, 1)$. In this case, the host bird can either get rid of the egg, or simply abandon the nest and build a completely new nest.

As a further approximation, this last assumption can be approximated by replacing a fraction p_a of the n host nests with new nests (with new random solutions). For a maximization problem, the quality or fitness of a solution can simply be proportional to the value of the objective function. Other forms of fitness can be defined in a similar way to the fitness function in genetic algorithms and other evolutionary algorithms.

From the implementation point of view, we can use the following simple representations that each egg in a nest represents a solution, and each cuckoo can lay only one egg (thus representing one solution), the aim is to use the new and potentially better solutions (cuckoos) to replace a not-so-good solution in the nests. Obviously, this algorithm can be extended to the more complicated case where each nest has multiple eggs representing a set of solutions. Here, we will use the simplest approach where each nest has only a single egg. In this case, there is no distinction between an egg, a nest or a cuckoo, as each nest corresponds to one egg which also represents one cuckoo.

Based on these three rules, the basic steps of the Cuckoo Search (CS) can be summarized as the pseudocode shown in Fig. 21.2.

Cuckoo Search via Lévy Flights

begin
Objective function $f(\boldsymbol{x})$, $\boldsymbol{x} = (x_1, ..., x_d)^T$
Generate initial population of n host nests $\boldsymbol{x}_i$
while *(t <MaxGeneration) or (stop criterion)*
    Get a cuckoo randomly
    Generate a solution by Lévy flights [e.g., Eq. (21.4)]
    Evaluate its solution quality or objective value f_i
    Choose a nest among n (say, j) randomly
    if $(f_i < f_j)$,
        Replace j by the new solution i
    end
    A fraction (p_a) *of worse nests are abandoned*
    New nests/solutions are built/generated by Eq. (21.3)
    Keep best solutions (or nests with quality solutions)
    Rank the solutions and find the current best
    Update $t \leftarrow t + 1$
end while
Postprocess results and visualization
end

Fig. 21.2 Pseudocode of the Cuckoo Search (CS) for a minimization problem.

CS uses a balanced combination of a local random walk and the global explorative random walk, controlled by a switching parameter p_a. The local random walk can be written as

$$x_i^{t+1} = x_i^t + \alpha s \otimes H(p_a - \epsilon) \otimes (x_j^t - x_k^t), \tag{21.3}$$

where x_j^t and x_k^t are two different solutions selected randomly by random permutation, $H(u)$ is a Heaviside function, ϵ is a random number drawn from a uniform distribution, and s is the step size. Here, the $\otimes$ means the entry-wise product of two vectors.

On the other hand, the global random walk is carried out by using Lévy flights

$$x_i^{t+1} = x_i^t + \alpha L(s, \lambda), \tag{21.4}$$

where

$$L(s, \lambda) \sim \frac{\lambda \Gamma(\lambda) \sin(\pi \lambda / 2)}{\pi} \frac{1}{s^{1+\lambda}}, \quad (s > 0). \tag{21.5}$$

Here $\alpha > 0$ is the step size scaling factor, which should be related to the scales of the problem of interest. It is worth pointing out that here $L(s, \lambda)$ is a random step size drawn from the Lévy distribution; it is a random number generator, not direct algebraic calculations, so we use $\sim$ to highlight this subtle difference.

A simple demonstration code of the cuckoo search is provided by the author, and the code can be found at the Mathworks website[2].

CS has two distinct advantages over other algorithms such as GA and SA, and these advantages are: efficient random walks and balanced mixing. Since Lévy flights are usually far more efficient than any other random-walk-based randomization techniques, CS can be very efficient in global search. In fact, recent studies show that CS can have guaranteed global convergence. In addition, the similarity between eggs can produce better new solutions, which is essentially fitness-proportional generation with a good mixing ability. In other words, CS has varying mutation realized by Lévy flights, and the fitness-proportional generation of new solutions based on similarity provides a subtle form of crossover. In addition, selection is carried out by using p_a where the good solutions are passed onto next generations, while not so good solutions are replaced by new solutions. Furthermore, simulations also show that CS can have autozooming ability in the sense that new solutions can automatically zoom into the region where the promising global optimality is located.

In addition, Eq. (21.4) is essentially the generalized simulated annealing in the framework of Markov chains. In Eq. (21.3), if $p_a = 1$ and $\alpha s \in [0, 1]$,

[2]http://www.mathworks.co.uk/matlabcentral/fileexchange/29809-cuckoo-search–cs–algorithm

CS can degenerate into a variant of differentia evolution. Furthermore, if we replace $\boldsymbol{x}_j^t$ by the current best solution $\boldsymbol{g}^*$, then (21.3) can further degenerate into accelerated particle swarm optimization (APSO). This means that SA, DE and APSO are special cases of CS, and that is one of the reasons why CS is so efficient.

In essence, CS has strong mutation at both local and global scales, while good mixing is carried out by using solution similarity, which also plays the role of equivalent crossover. Selection is done by elitism, that is, a good fraction of solutions will be passed onto the next generation. Without the explicit use of $\boldsymbol{g}_*$ may also overcome the premature convergence drawback as observed in particle swarm optimization.

21.3 Bat Algorithm

The metaheuristic bat algorithm (BA) was developed by Xin-She Yang in 2010. It was inspired by the echolocation behavior of microbats. It is the first algorithm of its kind to use frequency tuning.

Bats are fascinating animals. They are the only mammals with wings and they also have advanced capability of echolocation. It is estimated that there are about 1000 different species which account for up to 20% of all mammal species. Their size ranges from the tiny bumblebee bats (of about 1.5 to 2 g) to the giant bats with a wingspan of about 2 m and weight up to about 1 kg. Microbats typically have a forearm length of about 2.2 to 11 cm. Most bats uses echolocation to a certain degree; among all the species, microbats are a famous example as microbats use echolocation extensively, while megabats do not.

Most microbats are insectivores. Microbats use a type of sonar, called echolocation, to detect prey, avoid obstacles, and locate their roosting crevices in the dark. These bats emit a very loud sound pulse and listen for the echo that bounces back from the surrounding objects. Their pulses vary in properties and can be correlated with their hunting strategies, depending on the species. Most bats use short, frequency-modulated signals to sweep through about an octave, while others more often use constant-frequency signals for echolocation. Their signal bandwidth varies with species, and often increases by using more harmonics.

Studies show that microbats use the time delay from the emission and detection of the echo, the time difference between their two ears, and the loudness variations of the echoes to build up the three dimensional scenario of the surrounding. They can detect the distance and orientation of the target, the type of prey, and even the moving speed of the prey such as small insects. Indeed, studies suggested that bats seem to be able to discriminate targets by the variations of the Doppler effect induced by the wing-flutter rates of the target insects.

Though each pulse only lasts a few thousandths of a second (up to about 8 to 10 ms), however, it has a constant frequency which is usually in the region of 25 kHz to 150 kHz. The typical range of frequencies for most bat species are in the region between 25 kHz and 100 kHz, though some species can emit higher frequencies up to 150 kHz. Each ultrasonic burst may last typically 5 to 20 ms, and microbats emit about 10 to 20 such sound bursts every second. When hunting for prey, the rate of pulse emission can be sped up to about 200 pulses per second when they fly near their prey. Such short sound bursts imply the fantastic ability of the signal processing power of bats. In fact, studies show the equivalent integration time of the bat ear is typically about 300 to 400 μs.

If we idealize some of the echolocation characteristics of microbats, we can develop various bat-inspired algorithms or the bat algorithm. For simplicity, we now use the following approximate or idealized rules:

1. All bats use echolocation to sense distance, and they also 'know' the difference between food/prey and background barriers.
2. Bats fly randomly with velocity $\boldsymbol{v}_i$ at position $\boldsymbol{x}_i$. They can automatically adjust the frequency (or wavelength) of their emitted pulses and adjust the rate of pulse emission $r \in [0, 1]$, depending on the proximity of their target.
3. Although the loudness can vary in many ways, we can assume that the loudness varies from a large (positive) A_0 to a minimum value $A_{\min}$.

Another obvious simplification is that no ray tracing is used in estimating the time delay and three dimensional topography. Though this might be a good feature for the application in computational geometry, however, we will not use this, as it is more computationally extensive in multidimensional cases.

Each bat is associated with a velocity $\boldsymbol{v}_i^t$ and a location $\boldsymbol{x}_i^t$, at iteration t, in a d-dimensional search or solution space. Among all the bats, there exists a current best solution $\boldsymbol{x}_*$. Therefore, the above three rules can be translated into the updating equations for $\boldsymbol{x}_i^t$ and velocities $\boldsymbol{v}_i^t$:

$$f_i = f_{\min} + (f_{\max} - f_{\min})\beta, \tag{21.6}$$

$$v_i^t = v_i^{t-1} + (x_i^{t-1} - x_*)f_i, \tag{21.7}$$

$$x_i^t = x_i^{t-1} + v_i^t, \tag{21.8}$$

where $\beta \in [0, 1]$ is a random vector drawn from a uniform distribution.

The loudness and pulse emission rates are regulated by the following equations:

$$A_i^{t+1} = \alpha A_i^t, \tag{21.9}$$

and

$$r_i^{t+1} = r_i^0[1 - \exp(-\gamma t)], \tag{21.10}$$

where $0 < \alpha < 1$ and $\gamma > 0$ are constants. In essence, here α is similar to the cooling factor of a cooling schedule in simulated annealing.

Based on the above approximations and idealized rules, the basic steps of the Bat Algorithm (BA) can be summarized as the schematic pseudo code, as shown in Fig. 21.3.

```
Bat Algorithm
begin
Initialize the bat population x_i and v_i  (i = 1, 2, ..., n)
Initialize frequencies f_i, pulse rates r_i and the loudness A_i
while (t <Max number of iterations)
    Generate new solutions by adjusting frequency,
    Update velocities and locations/solutions [(21.6) to (21.8)]
    if  (rand > r_i)
        Select a solution among the best solutions
        Generate a local solution around the selected best solution
    end if
    Generate a new solution by flying randomly
    if  (rand < A_i & f(x_i) < f(x_*))
        Accept the new solutions
        Increase r_i and reduce A_i
    end if
    Rank the bats and find the current best x_*
end while
end
```

Fig. 21.3 Pseudocode of the bat algorithm (BA).

BA has been extended to multiobjective bat algorithm (MOBA) by Yang, and preliminary results suggested that it is very efficient. The author has provided a demo code of the bat algorithm that can be found at the Mathworks website[3].

In the BA, frequency tuning essentially acts as mutation, while selection pressure is relatively constant via the use of the current best solution $\boldsymbol{x}_*$ found so far. There is no explicit crossover; however, mutation varies due to the variations of loudness and pulse emission. In addition, the variations of loudness and pulse emission rates also provide an auto-zooming ability so that exploitation becomes intensive as the search moves are approaching the global optimality.

21.4 Flower Algorithm

The flower pollination algorithm (FPA), or simply flower algorithm, was developed by Xin-She Yang in 2012, inspired by the flower pollination process of flowering plants. It has been extended to multiobjective optimization problems and found to be very efficient.

It is estimated that there are over a quarter of a million types of flowering plants in Nature and that about 80% of all plant species are flowering species. It still remains a mystery how flowering plants came to dominate the landscape from the Cretaceous period. Flowering plants have been evolving for at least more than 125 million years and flowers have become so influential in evolution, it is unimaginable what the plant world would look like without flowers. The main purpose of a flower is ultimately reproduction via pollination. Flower pollination is typically associated with the transfer of pollen, and such transfer is often linked with pollinators such as insects, birds, bats and other animals. In fact, some flowers and insects have co-evolved into a very specialized flower-pollinator partnership. For example, some flowers can only attract and can only depend on a specific species of insects or birds for successful pollination.

Pollination can take two major forms: abiotic and biotic. About 90% of flowering plants belong to biotic pollination. That is, pollen is transferred by pollinators such as insects and animals. About 10% of pollination takes abiotic form which does not require any pollinators. Wind and diffusion help pollination of such flowering plants, and grass is a good example of abiotic pollination. Pollinators, or sometimes called pollen vectors, can be very diverse. It is estimated there are at least about 200,000 varieties of

[3]http://www.mathworks.co.uk/matlabcentral/fileexchange/37582-bat-algorithm-demo-

pollinators such as insects, bats and birds. Honeybees are a good example of pollinators, and they have also developed the so-called flower constancy. That is, these pollinators tend to visit exclusively certain flower species while bypassing other flower species. Such flower constancy may have evolutionary advantages because this will maximize the transfer of flower pollen to the same or conspecific plants, and thus maximizing the reproduction of the same flower species. Such flower constancy may be advantageous for pollinators as well, because they can be sure that nectar supply is available with their limited memory and minimum cost of learning, switching or exploring. Rather than focusing on some unpredictable but potentially more rewarding new flower species, flower constancy may require minimum investment costs and more likely guaranteed intake of nectar.

Pollination can be achieved by self-pollination or cross-pollination. Cross-pollination, or allogamy, means pollination can occur from pollen of a flower of a different plant, while self-pollination is the fertilization of one flower, such as peach flowers, from pollen of the same flower or different flowers of the same plant, which often occurs when there is no reliable pollinator available. Biotic, cross-pollination may occur at long distance, and the pollinators such as bees, bats, birds and flies can fly a long distance, thus they can be considered as the global pollination. In addition, bees and birds may behave as Lévy flight behaviour with jumps or fight distance steps obeying a Lévy distribution. Furthermore, flower constancy can be considered as an increment step using the similarity or difference of two flowers.

From the biological evolution point of view, the objective of the flower pollination is the survival of the fittest and the optimal reproduction of plants in terms of numbers as well as the most fittest. This can be considered as an optimization process of plant species. All the above factors and processes of flower pollination interact so as to achieve optimal reproduction of the flowering plants. Therefore, this may motivate us to design new optimization algorithms.

For simplicity in describing the flower pollination algorithm, we use the following four rules:

(1) Biotic and cross-pollination can be considered as a process of global pollination process, and pollen-carrying pollinators move in a way which obeys Lévy flights (Rule 1).
(2) For local pollination, abiotic pollination and self-pollination are used (Rule 2).

(3) Pollinators such as insects can develop flower constancy, which is equivalent to a reproduction probability that is proportional to the similarity of two flowers involved (Rule 3).
(4) The interaction or switching of local pollination and global pollination can be controlled by a switch probability $p \in [0, 1]$, with a slight bias towards local pollination (Rule 4).

These basic rules can be summarized as the basic steps as shown in Fig. 21.4.

In order to formulate proper updating formulae, we have to convert the above rules into mathematical equations. For example, in the global pollination step, flower pollen gametes are carried by pollinators such as insects, and pollen can travel over a long distance because insects can often fly and move in a much longer range. Therefore, Rule 1 and flower constancy can be represented mathematically as

$$\boldsymbol{x}_i^{t+1} = \boldsymbol{x}_i^t + \gamma L(\lambda)(\boldsymbol{g}_* - \boldsymbol{x}_i^t), \tag{21.11}$$

where $\boldsymbol{x}_i^t$ is the pollen i or solution vector $\boldsymbol{x}_i$ at iteration t, and $\boldsymbol{g}_*$ is the current best solution found among all solutions at the current generation/iteration. Here γ is a scaling factor to control the step size.

Here $L(\lambda)$ is the parameter that corresponds to the strength of the pollination, which essentially is also a step size. Since insects may move over a long distance with various distance steps, we can use a Lévy flight to mimic this characteristic efficiently. That is, we draw $L > 0$ from a Levy distribution

$$L \sim \frac{\lambda \Gamma(\lambda) \sin(\pi \lambda/2)}{\pi} \frac{1}{s^{1+\lambda}}, \quad (s \gg s_0 > 0). \tag{21.12}$$

Here $\Gamma(\lambda)$ is the standard gamma function, and this distribution is valid for large steps $s > 0$. This step is essentially a global mutation step, which enables to explore the search space more efficiently. In addition, s_0 is a step size constant. In theory, s_0 should be sufficiently large (base on mathematical approximations), but in practice $s_0 = 0.001$ to 0.1 can be used.

For the local pollination, both Rule 2 and Rule 3 can be represented as

$$\boldsymbol{x}_i^{t+1} = \boldsymbol{x}_i^t + \epsilon(\boldsymbol{x}_j^t - \boldsymbol{x}_k^t), \tag{21.13}$$

where $\boldsymbol{x}_j^t$ and $\boldsymbol{x}_k^t$ are pollen from different flowers of the same plant species. This essentially mimics the flower constancy in a limited neighborhood. Mathematically speaking, if $\boldsymbol{x}_j^t$ and $\boldsymbol{x}_k^t$ come from the same species or are selected from the same population, this equivalently becomes a local random walk if we draw ϵ from a uniform distribution in [0,1]. In essence, this is a local mutation and mixing step, which can help the algorithm to converge.

Flower Pollination Algorithm (or simply Flower Algorithm)

begin
Objective min *or* max $f(\boldsymbol{x})$, $\boldsymbol{x} = (x_1, x_2, ..., x_d)$
Initialize a population of n flowers/pollen gametes with random solutions
Find the best solution $\boldsymbol{g}_*$ *in the initial population*
Define a switch probability $p \in [0, 1]$
while *(t <MaxGeneration)*
  for $i = 1 : n$ *(all n flowers in the population)*
   if $rand < p$,
    Draw a (d-dimensional) step vector L from a Lévy distribution
    Global pollination via $\boldsymbol{x}_i^{t+1} = \boldsymbol{x}_i^t + \gamma L(\boldsymbol{g}_* - \boldsymbol{x}_i^t)$
   else
    Draw ϵ *from a uniform distribution in [0,1]*
    Do local pollination via $\boldsymbol{x}_i^{t+1} = \boldsymbol{x}_i^t + \epsilon(\boldsymbol{x}_j^t - \boldsymbol{x}_k^t)$
   end if
   Evaluate new solutions
   If new solutions are better, update them in the population
  end for
   Find the current best solution $\boldsymbol{g}_*$
end while
Output the best solution found
end

Fig. 21.4 Pseudocode of the proposed Flower Pollination Algorithm (FPA).

In principle, flower pollination activities can occur at all scales, both local and global. But in reality, adjacent flower patches or flowers in the not-so-far-away neighborhood are more likely to be pollinated by local flower pollen than those far away. In order to mimic this feature, we can effectively use a switch probability (Rule 4) or proximity probability p to switch between common global pollination to intensive local pollination. To start with, we can use a naive value of $p = 0.5$ as an initially value. A preliminary parametric showed that $p = 0.8$ may work better for most applications.

The author has provided a demo code of the basic flower pollination algorithm that can be downloaded from the Mathworks website[4].

Selection is achieved by selecting the best solutions and passing onto the

[4]http://www.mathworks.co.uk/matlabcentral/fileexchange/45112

next generation. It also explicitly uses $\boldsymbol{g}_*$ to find the best solution as both selection and elitism. There is no explicit crossover, which is also true for many other algorithms such as particle swarm optimization and harmony search.

21.5 Other Algorithms

Many other algorithms have appeared in the literature, including harmony search, eagle strategy, artificial immune systems and others. However, as this is not the main focus of this book, we will not delve into more details about these algorithms. Interested readers can refer to more advanced literature.

As nature-inspired algorithms have become hugely popular, the relevant literature is vast at all levels. For example, the book *Nature-Inspired Optimization Algorithms* by Xin-She Yang can be a good reference book to start with. Obviously, there are dozens of research level books and thousands of journal papers. Interested readers can refer to more advanced textbooks and edited books listed in the bibliography section of this book.

Bibliography

Abramowitz, M. and Stegun, I. A. (1965). *Handbook of Mathematical Functions* (Dover Publication).

Adamatzky, A. and Teuscher, C. (2006). *From Utopian to Genuine Unconventional Computers* (Luniver Press, UK).

Arfken, G. (1985). *Mathematical Methods for Physicists* (Academic Press).

Armstrong, M. (1998). *Basic Linear Geostatistics* (Springer, Berlin).

Atluri, S. N. (2005). *Methods of Computer Modeling in Engineering and the Sciences*, Vol. I (Tech Science Press, Forsyth, USA).

Basturk, B. and Karabogo, D. (2006). An artificial bee colony (ABC) algorithm for numerical function optimizaton, in: *IEEE Swarm Intelligence Symposium 2006*, May 12-14, Indianapolis, IN, USA.

Bathe, K. J. (1982). *Finite Element Procedures in Engineering Analysis* (Prentice-Hall, New Jersey, USA).

Bonabeau, E., Dorigo, M. and Theraulaz, G. (1999). *Swarm Intelligence: From Natural to Artificial Systems* (Oxford University Press, Oxford).

Bonabeau, E. and Theraulaz, G. (2000). Swarm smarts, *Scientific Americans.* March, pp.73-79.

Bridges, D. S. (1994). *Computatability* (Springer, New York).

Carrrier, G. F. and Pearson, C. E. (1988). *Partial Differential Equations: Theory and Technique*, 2nd edn. (Academic Press).

Carslaw, H. S. and Jaeger, J. C. (1986). *Conduction of Heat in Solids*, 2nd edn. (Oxford University Press).

Chatterjee, A. and Siarry, P. (2006). Nonlinear inertia variation for dynamic adapation in particle swarm optimisation, *Comp. Oper. Research*, **33**, pp.859-871.

Chong, C., Low, M.Y., Sivakumar, A. I. and Gay, K. L. (2006). A bee colony optimization algorithm to job shop scheduling, *Proc. of 2006 Winter Simulation Conference*, Eds Perrone L. F. et al, pp. 1954-1961.

Coello, C. A. (2000). Use of a self-adaptive penalty approach for engineering optimization problems, Computers in Industry, **41**, pp. 113-127.

Copeland, B. J., (2004). *The Essential Turing*, Oxford University Press, Oxford.

Courant, R. and Hilbert, D. (1962). *Methods of Mathematical Physics*, 2 volumes

(Wiley-Interscience, New York).
Crank, J. (1970). *Mathematics of Diffusion* (Clarendon Press, Oxford).
Cook, R. D. (1995). *Finite Element Modelling For Stress Analysis* (Wiley & Sons).
Das, B. M. (1983). *Advanced Soil Mechanics* (MicGraw-Hill, New York).
Davis, M. (1982). *Computability and Unsolvability* (Dover, New York).
De Jong, K. (1975). *Analysis of the Behaviour of a Class of Genetic Adaptive Systems*, PhD thesis, University of Michigan, Ann Arbor, USA.
Deb, K. (1995). *Optimisation for Engineering Design: Algorithms and Examples* (Prentice-Hall, New Delhi).
Drew, D. A. (1983). Mathematical modelling of two-phase flow, *A. Rev. Fluid Mech.*, **15**, pp. 261-291.
Dorigo, M. (1992). *Optimization, Learning and Natural Algorithms*, PhD thesis, Politecnico di Milano, Italy.
Dorigo, M. and Stützle, T. (2004). *Ant Colony Optimization* (MIT Press, Cambridge, USA).
El-Beltagy, M. A. and Keane, A. J. (1999). A comparison of various optimisation algorithms on a multilevel problem, *Engin. Appl. Art. Intell.*, **12**, pp. 639-654.
Engelbrecht, A. P. (2005). *Fundamentals of Computational Swarm Intelligence* (Wiley & Sons).
Flake, G. W. (1998). *The Computational Beauty of Nature: Computer Explorations of Fractals, Chaos, Complex Systems, and Adaptation* (MIT Press, Cambridge, USA).
Farlow, S. J. (1983). *Partial Differential Equations for Scientists and Engineers* (Dover Publications).
Fister, I., Fister Jr., I., Yang, X. S., Brest, J., (2013). A comprehensive review of firefly algorithms, *Swarm and Evolutionary Computation*, 13(1), 34-46.
Fletcher, C. A. J. and Fletcher, C. A. (1997). *Computational Techniques for Fluid Dynamics*, Vol. I, (Springer-Verlag, GmbH, Germany).
Forsyth, A. R. (1960). *Calculus of Variations* (Dover, New York).
Fowler, A. C. (1997). *Mathematical Models in the Applied Sciences* (Cambridge University Press, Cambridge).
Fowler, A. C. and Yang, X. S. (1998). Fast and slow compaction in sedimentary basins, *SIAM J. Appl. Math.*, **59**, pp. 365-385.
Gardiner, C. W. (2004). *Handbook of Stochastic Methods* (Springer, Berlin).
Geem, Z. W., Kim, J. H. and Loganathan, G. V. (2001). A new heuristic optimization algorithm: Harmony search, *Simulation*, **76**, pp. 60-68.
Gershenfeld, N. (1998). *The Nature of Mathematical Modeling* (Cambridge University Press, Cambridge).
Gill, P. E., Murray, W. and Wright, M. H. (1981). *Practical Optimisation* (Academic Press).
Gleick, J. (1988). *Chaos: Making a New Science* (Penguin).
Goldberg, D. E. (1989). *Genetic Algorithms in Search, Optimisation and Machine Learning* (Addison Wesley, Reading, Mass., USA).
Goodman, R. (1957). *Teach Yourself Statistics* (Teach Yourself Books, London).
Heitkotter, J. and Beasley, D. (2003). *Hitch Hiker's Guide to Evolutionary Com-*

putation http://www.cse.dmu.ac.uk/ rij/gafaq/top.htm.

Hinch, E. J. (1991). *Perturbation Methods* (Cambridge University Press, Cambridge).

Holland, J. (1975). *Adaptation in Natural and Artificial Systems* (University of Michigan Press, Ann Anbor).

Holland, J. (1995). *Hidden Order: How adaptation builds complexity* (Addison-Wesley, USA).

Jeffrey, A. (2002). *Advanced Engineering Mathematics* (Academic Press).

Jenkins, W. M. (1997). On the applications of natural algorithms to structural design optimisation, *Engineering Structures*, **19**, pp. 302-308.

Kant, T. (1985). *Finite Elements in Computational Mechanics*, Vols. I/II (Pergamon Press, Oxford).

Karaboga, D. and Basturk, B. (2008). On the performance of artificial bee colony (ABC) algorithm, *Applied Soft Computing*, **8**, pp. 687-697.

Keane, A. J. (1995). Genetic algorithm optimization of multi-peak problems: studies in convergence and robustness, *Artificial Intelligence in Engineering*, **9**, pp. 75-83.

Kennedy, J. and Eberhart, R. C. (1995). Particle swarm optimization. *Proc. of IEEE International Conference on Neural Networks*, Piscataway, NJ. pp. 1942-1948.

Kennedy, J., Eberhart, R. and Shi, Y. (2001). *Swarm intelligence* (Academic Press).

Karr, C. L., Yakushin, I. and Nicolosi, K. (2000). Solving inverse initial-value, boundary-valued problems via genetic algorithms, *Engineering Applications of Artificial Intelligence*, **13**, pp. 625-633.

Kardestruncer, H. and Norrie, D. H. (1987). *Finite Element Handbook* (McGraw-Hill).

Keane, A. J. (1995). Genetic algorithm optimisation of multi-peak problems: studies in convergence and robustness, *Artificial Intelligence in Engineering*, **9**, pp. 75-83.

Keener, J. and Sneyd, J. (2001). *A Mathematical Physiology* (Springer-Verlag, New York).

Kitanidis, P. K. (1997). *Introduction to Geostatistics* (Cambridge University Press, Cambridge).

Kirkpatrick, S., Gelatt, C. D. and Vecchi, M. P. (1983). Optimization by simulated annealing, *Science*, **220**, No. 4598, pp. 671-680.

Korn, G. A. and Korn, T. M. (1961). *Mathematical Handbook for Scientists and Engineers* (Dover, New York).

Korn, R. W. (1997). Pattern formation in the leaf of zebra grass, *J. Theor. Biol.*, **187**, pp. 449-451.

Koza, J. R. (1982). *Genetic Programming: On the Programming of Computers by Natural Selection* (MIT Press, Cambridge, USA).

Koziel, S. and Yang, X. S., (2011). *Computational Optimization, Methods and Algorithms*, Studies in Computational Intelligence, vol. 356, Springer, Heidelberg.

Kreyszig, E. (1988). *Advanced Engineering Mathematics*, 6th edn. (Wiley & Sons,

New York).

Kuhn, H. W. and Tucker, A. W. (1951). Nonlinear programming, *Proc. 2nd Berkeley Symposium*, University of California Press, pp. 481-492.

Langtangen, H. P. (1999). *Computational Partial Differential Equations: Numerical Methods and Diffpack Programming* (Springer, Germany).

Lee, K. S. and Geem, Z. W. (2005). A new meta-heuristic algorithm for continous engineering optimization: harmony search theory and practice, *Comput. Methods Appl. Mech. Engrg.*, **194**, pp. 3902-3933.

LeVeque, R. J. (2002). *Finite Volume Methods for Hyperbolic Problems* (Cambridge University Press, Cambridge).

Lewis, R. W., Morgan, K., Thomas, H. and Seetharamu, S. K. (1996). *The Finite Element Method in Heat Transfer Analysis* (Wiley & Sons, New York).

Matlab info, http://www.mathworks.com.

Meinhardt, H. (1995). *The Algorithmic Beauty of Sea Shells* (Springer-Verlag, New York).

Michaelewicz, Z. (1996). *Genetic Algorithm + Data Structure = Evolution Progamming* (Springer-Verlag, New York).

Mitchell, A. R. and Griffiths, D. F. (1980). *Finite Difference Method in Partial Differential Equations* (Wiley & Sons, New York).

Mitchell, M. (1996). *An Introduction to Genetic Algorithms* (MIT Press, Cambridge, USA).

Moler, C. B. (2004). *Numerical Computing with MATLAB* (SIAM, Philadelphia).

Murase, H. (2000). Finite element analysis using a photosynthetic algorithm, *Computers and Electronics in Agriculture*, **29**, pp. 115-123.

Moritz, R. F. and Southwick, E. E. (1992). *Bees as superorganisms* (Springer, New York).

Murch, B. W. and Skinner, B. J. (2001). *Geology Today - Understanding Our Planet* (John Wiley & Sons, New York).

Murray, J. D. ((1998). *Mathematical Biology* (Springer-Verlag, New York).

Nakrani, S. and Tovey, C. (2004). On honey bees and dynamic server allocation in Internet hosting centers, *Adaptive Behaviour*, **12**, pp. 223-240.

Ockendon, J., Howison, S., Lacey, A. and Movchan, A. (2003). *Applied Partial Differential Equations* (Oxford University Press, Oxford).

Octave info, http://www.octave.org.

Pallour ,J. D. and Meadows, D. S. (1990). *Complex Variables for Scientists and Engineers* (Macmillan Publishing Co., London).

Papoulis, A. (1990). *Probability and statistics* (Englewood Cliffs., UK).

Pearson, C. E. (1983). *Handbook of Applied Mathematics*, 2nd edn. (Van Nostrand Reinhold, New York).

Press, W. H., Teukolsky, S. A., Vetterling, W. T. and Flannery, B. P. (2002). *Numerical Recipes in C++: The Art of Scientific Computing*, 2nd edn. (Cambridge University Press, Cambridge).

Puckett, E. G. and Colella, P. (2005). *Finite Difference Methods for Computational Fluid Dynamics* (Cambridge University Press, Cambridge).

Rajaraman, A., Leskovec, J., Ullman, J. D., (2010). *Mining of Massive Datasets*, Cambridge University Press, Cambridge.

Riley, K. F., Hobson, M. P. and Bence, S. J. (2006). *Mathematical Methods for Physics and Engineering*, 3rd edn. (Cambridge University Press, Cambridge).

Ross, S. (1998). *A first Course in Probability*, 5th edn. (Prentice-Hall).

Sawaragi, Y., Nakayama, H. and Tanino, T. (1985). *Theory of Multiobjective Optimisation* (Academic Press, London).

Selby, S. M. (1974). *Standard Mathematical Tables* (CRC Press, USA).

Seeley, T. D. (1995). *The Wisdom of the Hive* (Harvard University Press, USA).

Seeley, T. D., Camazine, S. and Sneyd, J. (1991). Collective decision-making in honey bees: how colonies choose among nectar sources, *Behavioural Ecoloy and Sociobiology*, **28**, pp. 277-290.

Sirisalee, P., Ashby, M. F., Parks, G. T. and Clarkson, P. J. (2004). Multi-criteria material selection in engineering design, *Adv. Eng. Mater.*, **6**, pp. 84-92.

Smith, D. R. (1998). *Variation Methods in Optimization* (Dover, New York).

Smith, G. D. (1985). *Numerical Solutions of Partial Differential Equations: Finite Differerence Methods*, 3rd edn. (Clarendon Press, Oxford).

Spall, J. C. (2003). *Introduction to Stochastic Search and optimisation: Estimation, Simulation, and Control* (Wiley, Hoboken, NJ).

Storn, R. and Price, K., (1997). Differential evolution – a simple and efficient heuristic for global optimization over continuous spaces, *Journal of Global Optimization*, 11(4), 341–359.

Strang, G. and Fix, G. J. (1973). *An Analysis of the Finite Element Method* (Prentice-Hall, Englewood Cliffs, NJ).

Thomee, V. (1997). *Galerkin Finite Element Methods for Parabolic Problems* (Springer-Verlag, Berlin).

Swarm intelligence, http://www.swarmintelligence.org.

Weisstein, E. W., http://mathworld.wolfram.com.

Weinstock, R. (1974). *Calculus of Variations: with applications to Physics and Engineering* (Dover, New York).

Wikipedia, http://en.wikipedia.com.

Wolpert, D. H. and Macready, W. G. (1997). No free lunch theorems for optimization, *IEEE Transaction on Evolutionary Computation*, **1**, pp. 67-82.

Versteeg, H. K. and Malalasekra ,W. (1995). *An Introduction to Computational Fluid Dynamics: The Finite Volume Method* (Prentice Hall, New York).

Vapnik, V., (1995). The Nature of Statistical Learning Theory, Springer-Verlag, New York.

Wolpert, D. H. and Macready, W. G., (1997). No free lunch theorems for optimization, *IEEE Transaction on Evolutionary Computation*, 1(1), 67–82.

Wolpert, D. H. and Macready, W. G., (2005). Coevolutonary free lunches, *IEEE Trans. Evolutionary Computation*, 9(6), 721–735.

Yang, X. S. (2005). Engineering optimization via nature-inspired virtual bee algorithms, Lecture Notes in Computer Science, **3562**, pp. 317-323.

Yang, X. S. (2005). Biology-derived algorithms in engineering optimizaton (Chapter 32), in *Handbook of Bioinspired Algorithms*, edited by Olarius S. and Zomaya A. (Chapman & Hall/CRC), pp. 585-596.

Yang, X. S. (2005). New enzyme algorithm, Tikhonov regularization and inverse parabolic analysis, in: *Advances in Computational Methods in Science and Engineering*, Lecture Series on Computer and Computer Sciences, ICCMSE 2005, Eds T. Simos and G. Maroulis, **4**, pp. 1880-1883.

Yang, X. S. (2007). *A First Course in Finite Element Analysis* (Luniver Press, UK).

Yang, X. S., (2008). *Nature-Inspired Metaheuristic Algorithms*, Luniver Press, Frome.

Yang, X. S., (2009). Firefly algorithms for multimodal optimization, Proc. 5th Symposium on Stochastic Algorithms, Foundations and Applications, SAGA 2009, Eds. O. Watanabe and T. Zeugmann, Lecture Notes in Computer Science, vol. 5792, 169–178.

Yang, X. S., (2010). *Engineering Optimization: An Introduction with Metaheursitic Applications*, John Wiley and Sons, New Jersey.

Yang, X. S., (2011). Bat algorithm for multi-objective optimisation, *Int. J. Bio-Inspired Computation*, 3(5), 267–274.

Yang, X. S. and Deb, S., (2009). Cuckoo search via Lévy flights, in: *Proc. of World Congress on Nature & Biologically Inspired Computing* (NaBic 2009), IEEE Publications, USA, pp. 210–214.

Yang, X. S. and Deb, S., (2010). Engineering optimization by cuckoo search, *Int. J. Math. Modelling & Numerical Optimisation*, 1(4), 330-343.

Yang, X. S. and Deb, S., (2010). Eagle strategy using Lévy walk and firefly algorithms for stochastic optimization, in: Nature Inspired Cooperative Strategies for Optimization (NISCO 2010) (Eds. Cruz C, González, JR, Pelta DA, Terrazas G), Studies in Computational Intelligence 2010; vol. 284, Springer Berlin, pp. 101–111.

Yang, X. S. and Deb, S., (2013). Multiobjective cuckoo search for design optimization, *Computers and Operations Research*, 40(6), 1616–1624.

Yang, X. S., Karamanoglu, M., He, X. S., (2013). Multi-objective flower algorithm for optimization, *Procedia Computer Science*, vol. 18, 861–868.

Yang, X. S., (2010). A new metaheuristic bat-inspired algorithm, in: Nature Inspired Cooperative Strategies for Optimization (NICSO 2010), Springer, Studies in Computational Intelligence, vol. 284, pp. 65–74.

Yang, X. S. and Koziel, S., (2011). *Computational Optimization and Applications in Engineering and Industry*, Studies in Computational Intelligence, vol. 359, Springer, Heidelberg.

Yang, X. S., Cui, Z., Xiao, R., Gandomi, A. H., Karamanoglu, M., (2013). *Swarm Intelligence and Bio-inspired Computation: Theory and Applications*, Elsevier, London.

Yang, X. S., (2013). *Artificial Intelligence, Evolutionary Computing and Metaheuristics: In the Footsteps of Alan Turing*, Studies in Computational Intelligence, vol. 427, Springer, Heidelberg.

Yang, X. S., (2014). *Cuckoo Search and Firefly Algorithm: Theory and Applications*, Studies in Computational Intelligence, vol. 516, Springer.

Yang, X. S., (2014). *Nature-Inspired Optimization Algorithms*, Elsevier.

Zienkiewicz, O. C. and Taylor, R. L. (1991). *The Finite Element Method*, vol. I/II, 4th edn. (McGraw-Hill, London).

Index

Printed in the United States
By Bookmasters